ELECTRIC AND HYBRID VEHICLES

ELECTRIC AND HYBRID VEHICLES

Mohamed Mellal
Naira Hovakimyan

Kruger Brentt
Publishers
2025

Kruger Brentt Publishers UK. LTD.
Company Number 9728962

Regd. Office: 68 St Margarets Road, Edgware, Middlesex HA8 9UU

ISBN: 978-1-78715-410-0

For information on all our publications visit our website at http://krugerbrentt.com/

PREFACE

"Electric and Hybrid Vehicles" embarks on a journey into the transformative realm of electric vehicles (EVs), offering readers a comprehensive exploration of the principles, technologies, and applications driving the advancement of sustainable transportation. This book serves as an essential resource for students, researchers, engineers, policymakers, and industry professionals seeking to understand the fundamental concepts and emerging trends in electric vehicle engineering.

Electric vehicles represent a paradigm shift in the automotive industry, offering a cleaner, greener, and more sustainable alternative to traditional internal combustion engine vehicles. By harnessing electricity as a power source, EVs reduce greenhouse gas emissions, dependence on fossil fuels, and environmental impact, while also offering enhanced performance, efficiency, and driving experience.

The primary objective of this book is to provide readers with a comprehensive overview of electric vehicle engineering, covering key topics such as EV design and architecture, battery technology, electric drivetrains, charging infrastructure, vehicle-to-grid integration, and regulatory considerations. Each chapter offers insights, case studies, and practical examples to help readers understand the fundamental principles and applications of electric vehicle engineering.

Contributions from leading experts in the field explore various aspects of electric vehicle technology, including battery management systems, electric motors, power electronics, regenerative braking, and vehicle-to-grid communication protocols. Special emphasis is placed on emerging trends and innovations, such as autonomous electric vehicles, vehicle electrification in the aviation and maritime sectors, and the role of electric vehicles in smart grid ecosystems.

As electric vehicles continue to gain momentum and market adoption, it is essential for stakeholders to understand the opportunities and challenges inherent in this transformative technology. By fostering interdisciplinary collaboration, innovation, and sustainable practices, we can accelerate the transition to a cleaner and more efficient transportation system for the benefit of current and future generations.

We extend our sincere gratitude to the contributors whose expertise and dedication have enriched this book. It is our hope that "Electric and Hybrid Vehicles" will serve as a valuable resource and inspiration for readers committed to shaping the future of sustainable mobility through electric vehicle innovation and engineering excellence.

Mohamed Mellal

Naira Hovakimyan

CONTENTS

PREFACE V

1. A COMPREHENSIVE REVIEW OF VEHICLE TECHNOLOGY: ELECTRIC AND HYBRID 1

1. Introduction: An Overview 1

2. Classification of electric vehicles 4

3. Architecture of BEVs and HEVs 7

4. Electric motors for EVs 14

5. Power electronics driver topologies for EVs 15

6. Control strategies 17

7. Battery types 19

8. Concise Summary 20

References 20

2. ELECTRIC VEHICLES IN URBAN SETTINGS: ENVIRONMENTAL ADVANTAGES AND TECHNOLOGICAL-ECONOMIC OBSTACLES 23

1. Introduction: An Overview 23

2. Specifications of available electric vehicles 24

3. Electrical vehicles and the urban environment 26

4. Comparison of EV costs with alternative technologies 31

5. Obstacles in the implementation of electric car fleets 34

6. Policy alternatives and business framework for electric vehicle adoption 36

7. Sustainability of urban transportation 38

8. Concise Summary 39

References 40

3. PLUG-IN ELECTRIC VEHICLES A CENTURY LATER — HISTORICAL INSIGHTS ON WHAT HAS CHANGED AND WHAT REMAINS THE SAME? .. 45

1. Introduction: An Overview 45

2. Waves of History I: 1890s through the 1930s 46

3. Major Causes of Success or Failure 54

4. Waves of History II: Motivations for Re-introduction, 1965-2011 64

5. What is different this time, what is not? 72

6. References 76

4. WHAT IS THE SIGNIFICANCE OF ELECTRIC VEHICLES IN PROMOTING LOW CARBON TRANSPORTATION IN CHINA?.. 81

1. Introduction: An Overview 81

2. Methods 84

3. Results 88

4. Concise summary 92

References 93

5. PLUG-IN HYBRID AUTOMOBILES.. 95

1. Introduction: An Overview 95

2. History of EV and HEV 95

3. Hybrid vehicles 97

4. Why plug-in hybrid? 98

5. Methods for sizing PHEV components 100

6. Efficiency 111

7. Options 114

8. Concise summary 115

References 116

6. ELECTRIC CARS WITH FUEL CELLS 119

1. Introduction: An Overview 119

2. Fuel cell technique for use in transportation 124

3. The current market of FCVs 126

4. Vehicles configuration 134

5. CNR ITAE challenges and activities 136

References 149

7. SUPERCAPACITORS AS A POWER SOURCE FOR ELECTRIC VEHICLES 151

1. Introduction: An Overview 151

2. Supercapacitors vs. accumulator batteries and fuel cells 152

3. Characterization of supercapacitors 157

4. Applications of supercapacitors 159

5. Supercapacitors in regulated electric propulsion systems 162

6. Concise summary 166

References 167

8. INCORPORATION OF ELECTRIC VEHICLES INTO ELECTRIC UTILITY SYSTEMS.....171

1. Introduction: An Overview 171

2. Electric utility systems 173

3. Modeling of electric vehicles 187

4. Effects on primary energy use, fossil fuel utilization, greenhouse gas emissions, and electricity expenses 193

5. Concise summary 196

References 197

9. COMMUNICATION PERTAINING TO ELECTRIC VEHICLES.....199

1. Introduction: An Overview 199

2. Fixed and intersecting point: energy transfer 200

3. Essential dimensions for energy transmission applications 202

4. An information device for electric cars 205

5. E-mobility and roaming 209

6. Concise summary 214

References 214

10. UTILIZATION OF SR DRIVE SYSTEMS IN ELECTRIC VEHICLES 217

1. Introduction: An Overview 217

2. Design of SR motor on EV 218

3. SR drives power converter 225

4. Adaptive torque control system utilizing fuzzy logic for SRM 234

5. Result 238

6. Concise summary 244

References 245

11. CATHODE MATERIAL: LIFEPO4.....247

1. Introduction: An Overview 247

2. Feature of LiFePO4 249

3. Methods to improve performance 258

4. Application 264

5. Concise summary 266

References 266

12. A COMPREHENSIVE ELECTRIC VEHICLE CURRICULUM..........................271

1. Introduction: A.n Overview *271*

2. Curricula description *281*

3. Some implementation guidelines *290*

4. Concise summary *294*

References *294*

INDEX..299

Chapter 1 A COMPREHENSIVE REVIEW OF VEHICLE TECHNOLOGY: ELECTRIC AND HYBRID

1. INTRODUCTION: AN OVERVIEW

Internal combustion engine vehicles (ICEVs) have undergone persistent advancements in manufacturing technology, materials science, motor performance, vehicle control, driver comfort, and safety for over a century. The evolution of ICEVs was accompanied by the establishment of an extensive network of roadways, refueling stations, repair shops, and manufacturers, dealers, and sellers of replacement parts. Undoubtedly, these remarkable industrial activities and enterprises have played a pivotal part in molding the planet and, in numerous respects, society as well. Currently, the variety of ICEV types and applications is remarkable, encompassing small personal vehicles, large passenger buses, big freight trucks, and robust construction machinery. Contemporary internal combustion engine vehicles offer superior comfort, exceptional performance, and enhanced safety features at relatively affordable prices, making them highly appealing consumer items since their inception. Nonetheless, despite over a century of efforts by business and academia to enhance internal combustion engine efficiency, it remains, and will persist in being, exceedingly poor. As depicted in Fig. 1, only around 30% of the energy generated in the ICE combustion reaction is transformed into mechanical power. Approximately 70% of the energy released during burning is dissipated. Moreover, the energy inefficiency of thermal engines, commonly referred to as internal combustion engines (ICEs), is converted into heat within the motor and exhaust gases. The exhaust gases comprise mostly carbon dioxide ($CO2$) and, to a lesser degree, nitrogen oxides (NOx), hydrocarbons ($CxHy$), carbon monoxide (CO), and soot. Carbon dioxide is recognized for obstructing the Earth's radiation emissions into outer space, hence facilitating an increase in global temperatures—commonly referred to as the greenhouse effect. According to climate specialists, this is quietly inducing other global catastrophic alterations, such as sea level rise. Air pollution in metropolitan areas is a significant issue resulting from exhaust emissions, contributing to respiratory ailments, including lung cancer. The excessive noise level is another concern associated with large fleets of internal combustion engine vehicles in major urban areas. This presents an additional challenge for municipal administrators and authorities: everyday congestion, which may only be mitigated by mass transit systems (e.g., subways and trains). Regardless of the existence of the aforementioned issues, an immediate challenge had to be addressed: the limited supply of fossil fuels for a growing global fleet. As oil reserves deplete, the price of this commodity surges, further driven by geopolitical tensions in producing regions of the Middle East. Conversely, renewable energy sources, such as ethanol derived from sugarcane or maize, are being explored as alternate solutions in several countries. In Brazil, sugarcane biofuel is a well-established alternative, having been in use for over two decades, with internal combustion engine vehicles designed to operate alternately on gasoline or ethanol automatically. Drivers may select their preferred fuel type at the refueling station, largely influenced by pricing. This strategy faces criticism concerning the implications for food supply and pricing, as agricultural land is allocated for biofuel production rather than food cultivation. The emission of greenhouse gases and air pollution issues persist, but to a diminished degree.

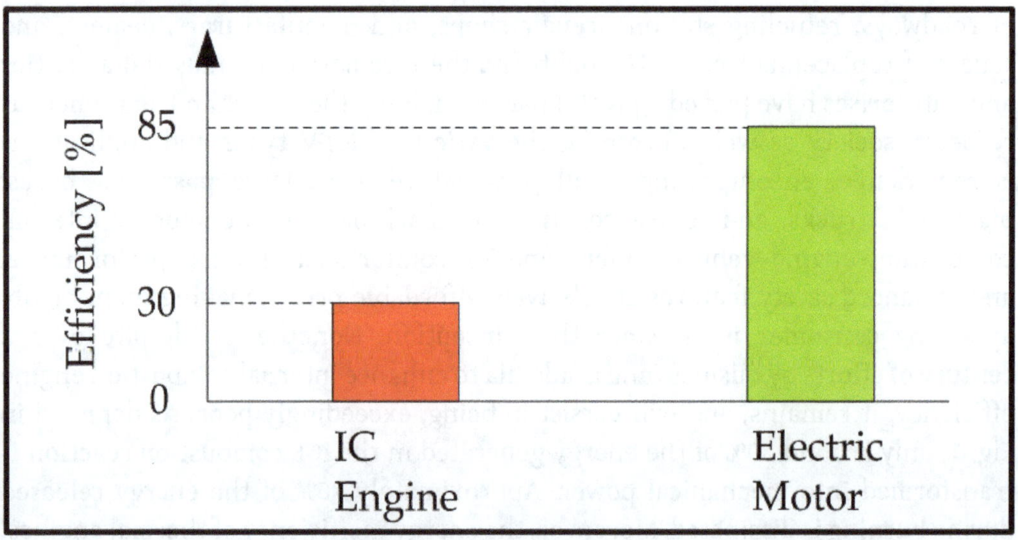

Figure 1. ICEs are very inefficient energy converters as compared to electric motors

An accurate look at Fig. 1 reveals that electric motors are far superior to ICE and could do an excellent job in propulsion of vehicles, helping to solve the serious climate, air pollution and noise problems created by ICEVs. As a matter of fact, electric vehicles (EVs) were invented in 1834, before ICE vehicles, being manufactured by several companies of the U.S.A, England, and France (Chan, 2007). Fig. 2(a) shows a picture of commercial EV in 1920. Poor performance of their batteries contrasting to fast development of ICE technology, extremely high energy density and power density of gasoline and petrol, and the abundance and low price offer of fossil fuel, all conspired against those days' EVs that rapidly became defunct. Interestingly, more than 150 years later, triggered by the world energy crisis in the 1970s, EVs entered the agendas of world's greatest carmakers, governments' energy and climate policy, and of worldwide non-governmental organizations worried about environmental pollution and greenhouse effect.

Today, although their sales are negligible in relation to that of ICEVs, pure EVs and hybrid EVs (HEVs), i.e., those that combine ICE with electrical machines fed by batteries or fuel cells (hydrogen derived electricity), are offered by world's greatest carmakers. The performance of HEVs, from the driver's standpoint, rivals or outdoes that of modern ICEVs. Their energy consumption ranges from circa 10% to 70% lower than that of an equivalent ICE car, depending on their power, battery size, control strategy, etc. For the sake of illustration, until 2008, Toyota Prius, the world's first commercially mass-produced and marketed HEV, sold over 500,000 units on the world's market (Xiang et al., 2008). Fig. 2(b) shows a photograph of a modern 2010 Toyota Prius HEV whose selling price begins at 23,000 USD.

The dramatic gain in energy efficiency, besides much lower or zero gas emission and noise-free operation, is due to the much higher efficiency of electric motors and control strategies such as regenerative braking and storage of excess energy from the ICE during coasting.

(a) (b)

Figure 2. a) 1920 Detroit Electric b) 2010 Toyota Prius (HEV) [Toyota Motor Co., 2011]

There are many reasons for EVs and HEVs to represent so low a share of today's car market. For EVs, the most important are their shorter range, the lack of recharging infrastructure, and higher initial cost. Though HEVs feature range, performance and comfort equivalent or better than ICEVs, their initial cost is higher and the lack of recharging infrastructure is a great barrier for their diffusion. Nevertheless, the energy efficiency of the latter, though far higher than that of ICEVs', seems not capable of solving the greenhouse gas emissions by world vehicle fleet. And this situation is expected to become worse and worse, given that world fleet is expected to triple by 2050, in relation to 2000, due to massive car use in countries such as China, India and Brazil. To limit the planet's average temperature to 2-2.4 ºC above the pre-industrial era level, scientists calculate a needed reduction of 50-85% in CO_2 emissions in all sectors by 2050. EVs may play a fundamental role in this struggle, given that the transportation sector is one of the largest emitters of CO_2 (Bento, 2010). To that end, industry, government, and academia must strive to overcome the huge barriers that block EVs widespread use: battery energy and power density, battery weight and price, and battery recharging infrastructure.

This chapter presents a synthetic review on the technology of modern EVs. This includes the types and classification of EVs, electric motor kinds employed by EV manufacturers, power electronics driver topologies, control strategies, battery types and performance, and infrastructure demands.

2. CLASSIFICATION OF ELECTRIC VEHICLES

A comprehensive taxonomy of the many sorts of electric vehicles is likely to emerge in the near future, driven by mass manufacturing initiatives from automotive associations and research teams around. A literature review clearly indicates that a convergence in nomenclature is readily observable. This nomenclature is stronger and more definitive when EVs classification is carried out based on either the energy converter type(s) used to propel the vehicles or the vehicles' power and function (Chan, 2007; Maggetto & van Mierlo, 2000). When referring to the energy converter types, by far the most used EV classification, two big classes are distinguished, as depicted in Fig. 3, namely: battery electric vehicles (BEVs), also named pure electric vehicle, and hybrid electric vehicles (HEVs). BEVs use batteries to store the energy that will be transformed into mechanical power by electric motor(s) only, i. e., ICE is not present. In hybrid electric vehicles(HEVs), propulsion is the result of the combined actions of electric motor and ICE. The various modes of hybridization result in distinct architectures: series hybrid, parallel hybrid, series-parallel hybrid, and complex hybrid, each elaborated upon in separate sections. As anticipated, no universal architecture can be deemed superior across all practical dimensions: energy efficiency, vehicle performance and range, driver comfort, manufacturing complexity, and production cost. Consequently, automotive manufacturers may select varying architectures to attain diverse objectives and satisfy specific transportation segment demands.

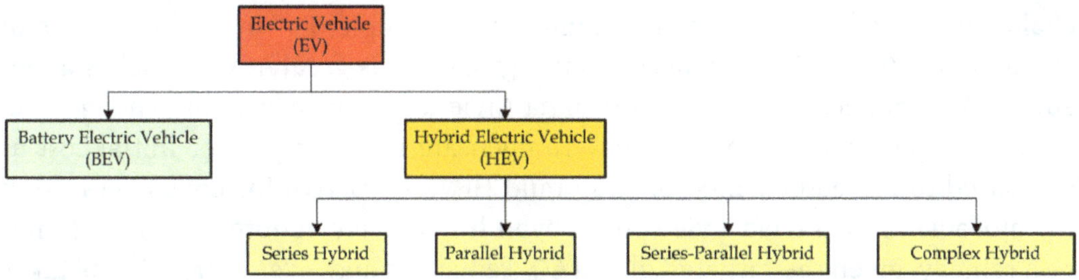

Figure 3. Classification of EVs according to the type(s) and combination (if any) of energy converters used (electric motor & ICE)

Under the large umbrella of HEVs, there is another category (not shown in Fig. 3) that utilizes a fuel cell instead of an ICE together with the electric motor, always in the series-hybrid architecture. This is the fuel cell vehicle (FCV). The following rationale is to justify FCVs absence in Fig. 3 and in the rest of this chapter. The hydrogen-based chemical reaction of FCVs generates the electricity either to be used by the EM or stored in battery or supercapacitor. The by-product of hydrogen and oxygen reaction is simply pure water, which renders FCVs emission-free and consequently an environmentally friendly technology (Gulhane et al., 2006). A recent research about the dynamic competition for market between plug-in HEVs (PHEVs) and FCVs showed that the early deployment PHEVs is almost certain to close the market for FCV in the future (Bento, 2010). Another study shows that from 2006 onwards auto makers decreased sharply the prototyping activities with FCVs and much of the public funding in the U.S.A. and other industrial countries shifted from FCVs to BEVs and PHEVs. Despite some controversy over the reasons, which range from the lack of a hydrogen infrastructure, absence of a technological breakthrough in hydrogen technology, to very high cost production of FCVs (Honda FCX Clarity, an FCV most close to market, costs circa 1 million USD, clearly not an attractive pricing), as a matter of fact these vehicles do not seem an option anymore (Bakker, 2010). Nevertheless, hopefully fuel cells may play a very important role in replacing ICE of stationary machines.

A second useful classification for HEVs (Fig. 4) places them into the following three categories, according to the electric motor power under the hood: *micro hybrid*, *mild hybrid*, and *full hybrid* (Chan, 2007). In effect, this classification is a measure of the hybridization degree of the HEV (Maggetto, 2000). In other words, it indicates how much important is the role played by the electric motor in the car propulsion. *Micro hybrids* use electric motor of about 2.5 kW at 12 V. The EM is only a helping hand to the ICE, in the start and stop operations, which dominate in city driving. Even in this driving mode, energy savings is of only about 5% to 10%. This is a very poor economy, obviously with a negligible impact on fossil fuel dependence, metropolitan area air pollution and greenhouse gas emissions, the challenging triad. C3 Citroen is a commercial example. EM in *mild hybrids* is of 10-20 kW at 100-200 V. As expected, energy savings is greater and reaches about 20%-30%. Commercial models are Honda Civic and Honda Insight. Though fuel (and thus operational) economy may compensate for their greater initial cost as compared to ICE equivalents, turning mild HEVs attractive for consumers, from the aforementioned triad's viewpoint, even if massively adopted, they could not be a remedy, given the targeted global CO_2 reduction and, even worse, if one takes into account that world fleet (vastly of ICE vehicles) is increasing more and more, as new consumers come into life in emerging countries. For the sake of illustration, only in Brazil, passenger car fleet doubled in the last decade. The last member of this category is the *full hybrid*, which embeds an EM of circa 50 kW at 200-300 V and, in city driving, yields energy saving of 30%-50%, thanks to complex control algorithms that manage to operate the ICE, when needed, always at maximum efficient region, directing the excess energy to batteries. Energy is collected and stored in the battery and/or supercapacitor during coasting and regenerative braking. The Toyota Prius is an authentic member of this lineage. While full hybrids can serve as a supplementary measure against the triad, their efficiency metrics are significantly inadequate to independently mitigate the triad, for the same reasons previously mentioned. In the author's view, they only postpone the climate catastrophe and provide psychological comfort to their possessors.

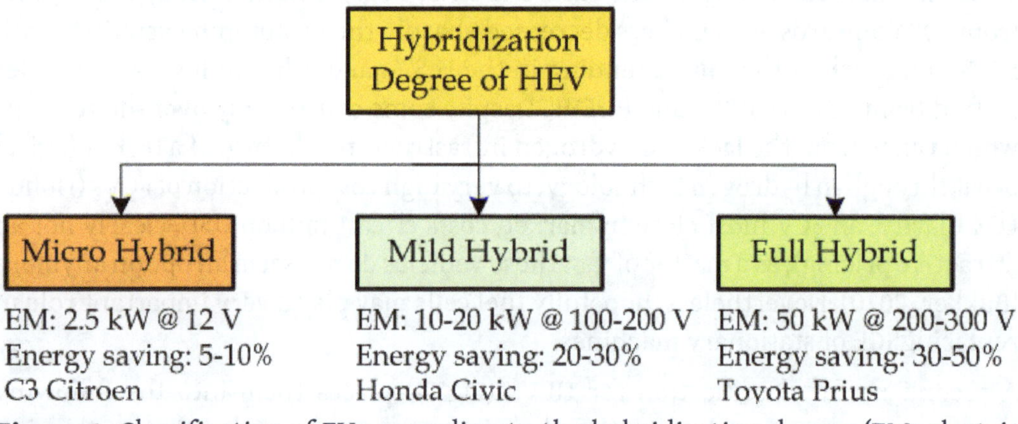

Figure 4. Classification of EVs according to the hybridization degree (EM: electric motor) (Chan, 2007)

A last classification for HEVs divides the automobile market into a number of categories (or segments) mostly based on their prices (Maggetto, 2000). Five segments are identified, as depicted in Fig. 5. HEVs of the *second family-car* segment are for frequent use in town and move a relatively low daily distance. If propelled mainly by ICE, in urban areas, the overall efficiency is very low. Conversely, if propulsion relies only on electric motor, high efficiency can be reached, and an effective combat to the triad (greenhouse gas emission, air pollution, and fossil fuel dependence) might be given. The ICE (with its fuel tank) might serve as a range extender, in practice, an efficient manner to null drivers' anxiety of being run out of charge. Even though this classification has been used for HEVs, it is interesting to extend it to BEVs since, as discussed just above, the latter can play a very important role in alleviating problems in metropolitan areas and world climate changes. The *intermediate car* segment is planned to be frequently the family's unique car. Therefore, it must be appropriate for use in town and present also a good road performance. Toyota Prius can be allocated in this segment. HEVs of the *high class car* segment are not convenient for city use. They feature extraordinary road performance, excellent technical performance and perfect comfort. Of course, the excellent technical performance does not take into account the environmental viewpoint. The *small delivery vehicle* segment is intended mainly for city use. However, unlike the second family car segment, vehicles of the former segment must be capable of moving a great number of relatively short-distance trips everyday. Therefore, high efficiency would be welcome, from the environmental, climate and fuel economy viewpoints. The *city bus* segment is dedicated to urban public transport, including tourist transport in urban areas. Vehicles of this segment feature low speed and circa 250-km driving range. Once more, the higher the hybridization degree the better for the environment and climate. Chinese manufacturer Dongfeng argues that its EQ6121HEV hybrid bus achieves 80 km/h and reduces in 30% greenhouse gas emission and in 20-30%, fuel consumption (Xiang et al., 2008).

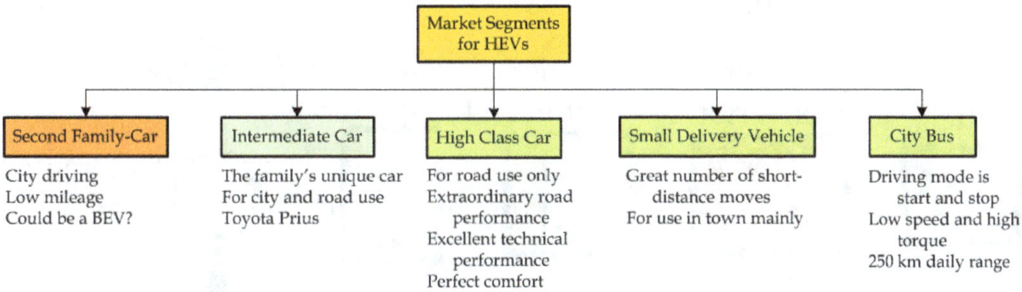

Figure 5. Classification of EVs according to the market segment

3. ARCHITECTURE OF BEVS AND HEVS

With the transition to electric vehicles, innovative design methodologies and powertrain configurations emerge to enhance performance based on criteria including energy efficiency, energy source types, energy storage devices, hybridization levels, driving range, power output, driver comfort, production costs, and ownership expenses (Chen et al., 2009). Given the varying demands across different global areas and the existence of diverse market sectors within each region, it is expected that the number of BEV and HEV models will continue to proliferate (Xiang et al., 2008; Gulhane et al., 2006). Automakers endeavor to develop automobile models that more effectively meet market demands while maximizing their revenue.

3.1. BEVs architectures

Fig. 6 illustrates one of the simplest topology for battery-electric vehicles. The energy stored in the battery (or in a battery pack) is used by the power converter to drive the electric motor. This, in turn, drives the two wheels by means of a fixed or changeable gear and a power splitting differential gear. The power converter unit may include a dc-dc converter and a motor driver. It all depends on the motor type and ratings and on the battery voltage, energy and power density. For maximum efficiency, the vehicle's kinetic energy must be converted to electrical energy by the motor/generator and stored in the battery pack via the power converter, whenever the break pedal is pressed and during coasting. Of course, the electronic detail of the power converter (e.g., topology, control strategy) is a function of the employed motor type, battery technology and ratings, etc. Anyway, in order to regenerate energy, the power converter must be able to control the power flow in both directions: from the battery to the motor as well as from the motor to the battery. If the battery type cannot be fast charged with the recovered kinetic energy, either a supercapacitor or a flywheel may be used for temporary energy storage. If possible, the changeable (or fixed) gear may be cut out, to diminish the mechanical parts counting. In this case, it is replaced by more complex variable speed controller for the motor.

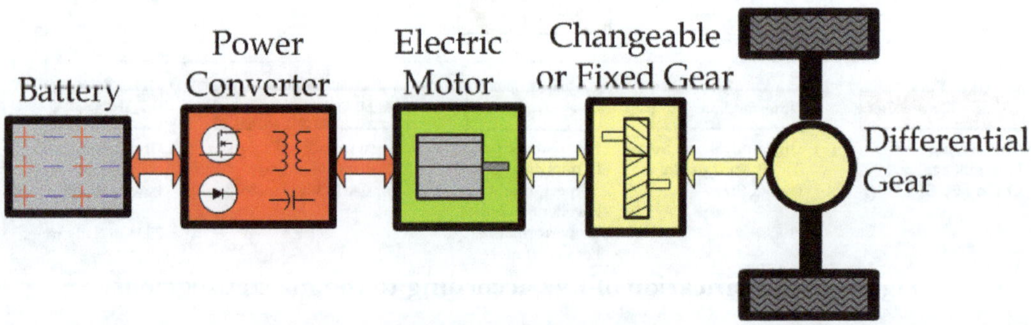

Figure 6. One-motor BEV

Pure electric vehicles may adopt two (or four) in-wheel motors in their powertrains, as sketched in Fig. 7. In this case, every motor is driven by a dedicated power converter that must control wheel's speed and torque. Moreover, a central electronic controller must coordinate speed differences (in steering wheels), whenever needed or as a result of wheel slippage, as long as a differential power splitting device is no more present. As expected, the simplification of the mechanical design is attained at the expense of increased complexity of the power electronics and controllers.

On the other hand, augmenting the motor number, for a desired vehicle power and performance, leads to significantly smaller motors and, what is less obvious, to lower rated power switches and passive electronic parts, which influence on drive cost and reliability. One interesting operating mode for multiple-motor BEVs is that the vehicle can continue to operate, though at a somewhat reduced power, if one of the motors (in case of two-motor BEVs) gets out of service. Comparing Figs. 7 and 6, one notices that in-wheel motor propulsion topology reduces radically EV's number of mechanical links.

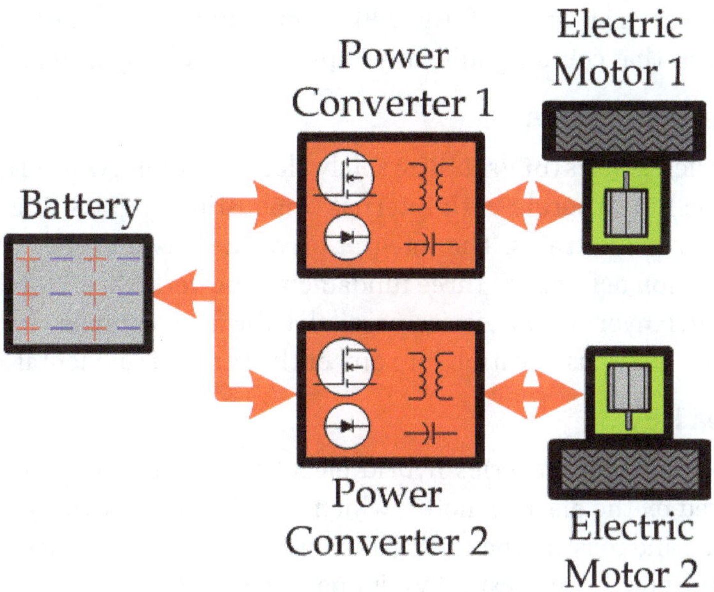

Figure 7. Two-motor BEV

To this date, owing to battery limitations such as high initial cost, relatively low energy and power density, and excessive weight, BEVs are not as attractive as HEVs, because of limited driving range, performance and comfort. Nevertheless, as BEVs are the only zero-emission cars, they must be viewed as an effective tool to combat greenhouse gas emissions, air pollution and petrol dependency. There are arguments to reinforce the idea that strong governmental incentive policies should be adopted in as many countries as possible to benefit owners of BEVs. Examples of such incentives are: government rebate to each BEV owner (say 10% of vehicle price), exemption of purchase tax, exemption of road maintenance fee, road passing fee and parking fee. In some countries some of these actions are under way (Xiang et al., 2008). Data of the U. S. Department of Transportation reveal that 50% of daily vehicle travel is less than 48 km and average daily vehicle trip is about 16 km (Kruger & Leaver, 2010). Today's batteries feature enough energy to easily enable *second-car family* BEVs (though this class was originally proposed to HEVs) to travel these distances without recharge. Consequently, there exists potential for substantial manufacturing and adoption of pure electric vehicles. Nonetheless, the extensive utilization of BEVs will be ineffective from a carbon emission perspective if fossil fuels (coal or petroleum) are employed to produce the power that finally charges the automobile batteries. For optimal efficacy, automotive batteries must be recharged using energy derived from carbon-neutral sources, including solar, wind, hydro, and nuclear power. Conversely, each nation must assess its grid capacity to accommodate a significant influx of new and specialized consumers. The extensive utilization of battery electric vehicles (BEVs) could significantly affect the power grid. In the future, BEVs may function as distributed energy storage systems that could significantly contribute to energy demand regulation.

3.2. HEVs architectures

Battery Electric Vehicles (BEVs) utilize solely electric motors, whilst Hybrid Electric Vehicles (HEVs) incorporate both internal combustion engines (ICE) and electric motors in their powertrains. The integration of these two energy converters for vehicle propulsion defines the three fundamental powertrain architectures: series hybrid, parallel hybrid, and series-parallel hybrid. A complex hybrid denotes structures that defy classification into any of the three fundamental kinds.

3.2.1. Series HEV

As illustrated in Fig. 8, in series hybrid electric vehicles (HEVs), the wheels are solely powered by the electric motor, which additionally functions as a generator during braking and coasting, hence enhancing total energy efficiency. This topology streamlines the powertrain design by eliminating the need for a clutch and reduction gear. Speed and torque regulation is achieved only through the management of the electric motor, which serves as a highly efficient power converter. The ICE is responsible for charging (or recharging) the battery and providing energy to the electric motor, consistently functioning at optimal efficiency. This is an additional method that enhances total energy efficiency.

Series HEVs are said to be ICE-assisted electric vehicles, for obvious reasons. An ICE, one generator and one motor are one of the main disadvantages of series HEV. Moreover, as the vehicles must be capable of cruising with maximum load against a graded road, all the machines, i.e., the ICE, the generator and, of course, the electric motor, must be powerful enough, which will result in relatively over-dimensioned machines. This leads to cost increase. As Eq. 1 indicates (Chen et al., 2009), given the constants for initial rolling force, F_0, the rolling coefficient, r, the drag coefficient, d, the total mass (vehicle's mass plus the passenger and luggage masses), m, and the gravity, g, the resistance force, F_{res}, the electric motor must be capable of surpassing increases with vehicle's speed, V, and the road's grade, θ.

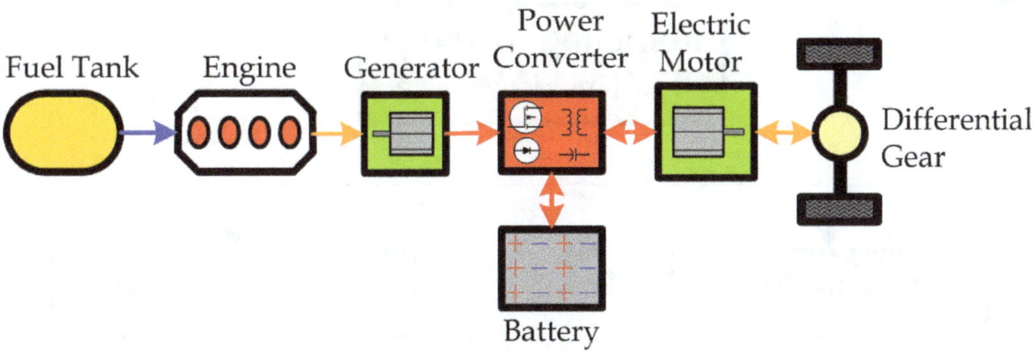

Figure 8. Architecture of series HEV

$$F_{res} = F_0 + rV + dV^2 + mg\ sin\theta \qquad (1)$$

On the other hand, as indicated by Eq. 2, the motor's torque is proportional to the inertia, J, and the first derivative of angular speed, ω, i.e., the angular acceleration. Eqs. 1 and 2 are interrelated to each other by the ratio of wheel to transmission radii. These two equations govern the vehicle's dynamic performance (acceleration power) and cruising speed. It is easy to note how stronger should be the powertrain if a desired series HEV had its maximum speed specification changed from, say, 80 km/h to 120 km/h. But, is such a performance always needed? As the ICE does not add its effort to aid in propelling the vehicle, this architecture is appropriate for small HEVs, as for instance, those of the *micro* category or *second-family car* segment already mentioned, for which cruising speed can be very modest.

$$T_m = J(d\omega/dt) \qquad (2)$$

Before proceeding to next section, it is worth making it clear that HEVs of all architectures can be recharged in two very distinct ways, as shown in Fig. 9: the so-called plug-in hybrids (PHEV) and the conventional HEVs. While PHEVs can have their batteries recharged directly from the power grid, which is an enormous advantage, the conventional HEVs have their batteries recharged by means of the ICE. In this case, the advantage is the omnipresence of gas stations. Studies indicate that conventional HEVs are potentially less eco-friendly than PHEVs. While the latter can take advantage of the ubiquitous power grid, the impact they can cause to the grid is far from being negligible and depends on the way charging and discharging (as PHEVs can return stored energy to the grid) are done: controlled or not by utilities companies (Clement-Nyns et al., 2011; Sioshansi et al., 2010; Kruger & Leaver, 2010). Moreover, if the electrical energy generated to the grid comes from fossil fuel plants, then to a great extent the environmental and climatic appeal of these vehicles is no more valid.

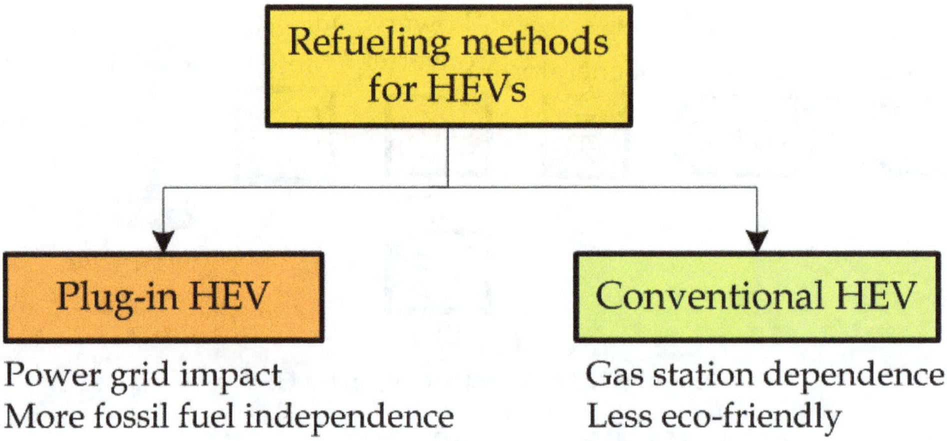

Figure 9. Recharging methods for HEVs

3.2.2. Parallel HEV

In parallel hybrid electric vehicles, propulsion can derive from torque produced concurrently by the internal combustion engine and the electric motor. As depicted in Fig. 10, this technology enables the autonomous operation of the internal combustion engine and electric motor, facilitated by the incorporation of two clutches. A principal characteristic of parallel hybrid electric vehicles (HEVs) is that, for equivalent vehicle performance, both the electric motor and internal combustion engine (ICE) can be considerably smaller than those utilized in series construction, resulting in a substantially more economical vehicle. Conversely, wheel propulsion by the internal combustion engine enhances the dynamic performance of this configuration. The complex powertrain controller may facilitate up to six distinct operational modes: electric motor activated and internal combustion engine (ICE) deactivated; ICE activated and electric motor deactivated; both electric motor and ICE activated, working in tandem to propel the vehicle; ICE activated to provide power for vehicle propulsion and to drive the electric machine, which functions as a generator to recharge the batteries using energy from the fuel tank (optimal energy savings are attained by operating the ICE at peak efficiency while transferring surplus energy to the batteries); ICE activated solely for battery recharging via the electric machine (i.e., the vehicle is stationary); and regenerative braking, wherein energy is captured and stored in the batteries (or in a supercapacitor) through the electric machine. The controller can efficiently manage this multitude of operational modes to enhance driving performance or fuel efficiency, for instance. Parallel HEVs are characterized as internal combustion engine cars augmented by electric motors, and their design is particularly suitable for high-end automobiles and full hybrids. Powertrain sizing is conducted according to the required dynamic performance of the vehicle, cruising speed, and several characteristics including maximum road grade, vehicle weight, and load. This activity relies significantly on computer simulation programs prior to the commencement of prototyping (Wu et al., 2011).

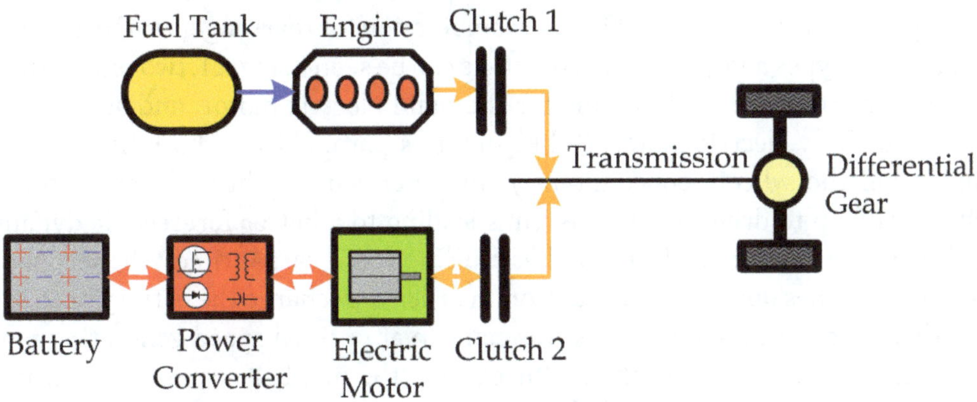

Figure 10. Architecture of parallel HEV

3.2.3. Series-parallel HEV

By incorporating an additional electric generator and a planetary gear, an intriguing powertrain architecture is achieved (Fig. 11) that integrates characteristics of both series and hybrid topologies, aptly termed series-parallel architecture. Despite being costlier than the parent architectures, series-parallel is a favored topology for HEVs, particularly when manufacturers aim for superior dynamic performance and elevated cruising speeds in their models. Similar to parallel HEVs, the degree of hybridization is modified as a compromise among performance, cruising speed, fuel efficiency, drivability, and comfort. A cursory examination of Fig. 11 reveals that multiple operational modes, exceeding six, are feasible for series-parallel HEVs, hence imposing challenges on the construction and testing of the controller. These are created and refined via computer simulations and expertise.

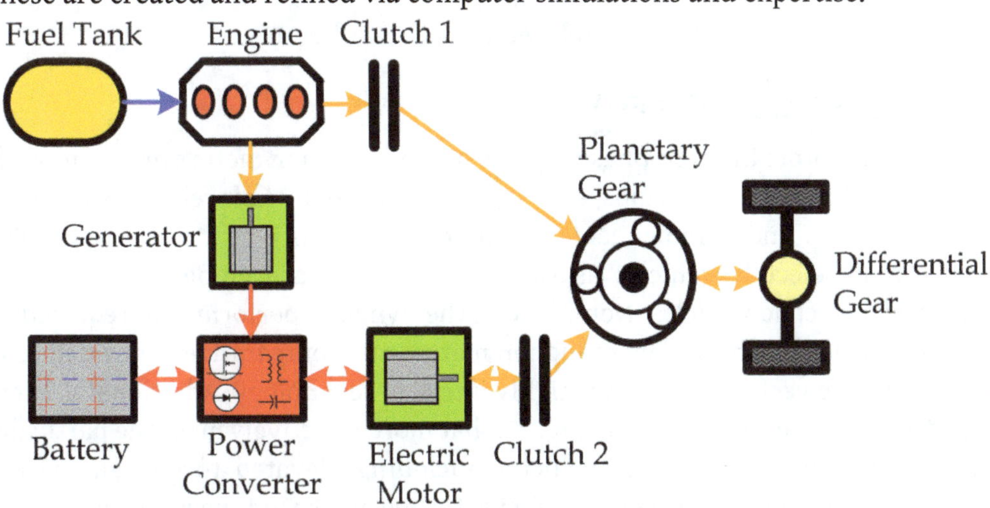

Figure 11. Architecture of series-parallel HEV

3.2.4. Complex HEV

Fig. 11 sketches an architecture named complex HEV. This name is reserved to the topologies that cannot be classified as a combination (or rearrangement) of the basic architecture types analysed to this point. As can be seen in Fig. 11, two bidirectional power converters are utilized, one for the main electric motor, and another one for the auxiliary electric motor. Unlike in series-parallel HEVs, both these motors can propel the wheels concomitantly. In other words, three different torque sources add up to drive the wheels, thus leading to a better foreseeable dynamic performance vehicle and clearly higher cruising speed car. At times, the secondary electric machines operates as generator, in order to recharge the battery or to save into this the excess ICE energy, as this can run at optimal speed generating more power than needed by the vehicle. Once more, the number of possible operation modes for the complex HEVs is half a dozen or greater. Component sizing (electric motors/generators, ICE, gears, battery, power converters, etc) is a very complex task. Control program development and test are highly challenging.

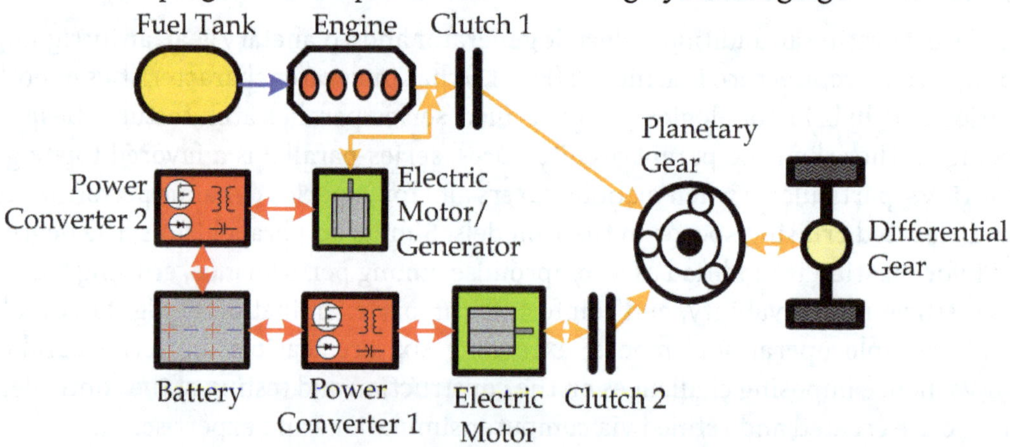

Figure 12. Architecture of complex HEV

4. ELECTRIC MOTORS FOR EVS

Squirrel cage rotors in three-phase asynchronous induction motors predominantly prevail in industrial applications, as is widely recognized. Their relatively low cost, great resilience, and commendable dynamic performance render them suitable for powering electric vehicles as well. Indeed, they are employed in various commercial electric vehicles. Nonetheless, the dynamic performance required by electric vehicles is achieved by induction motors at a comparatively elevated cost, as the requisite vector control is a highly intricate technology. Additionally, there exist drive alternatives, as seen in Fig. 13, that more effectively meet the particular requirements of some electric vehicles, including elevated torque and power density, high efficiency across an extensive torque and speed spectrum, and broad constant-power operational capability (Chau et al., 2008; Gulhane et al., 2006). The Permanent Magnet Brushless DC Motor (PMBL) is a highly promising technology that has seen extensive application in electric vehicles (EVs). This drive type is poised to become a significant market leader; nevertheless, automakers outside of China should exercise caution and explore alternate drive options, given that the global supplies of rare earths utilized in permanent magnets are predominantly located in China, where the government may impose export restrictions. Hybrid-field excited PMBL provides enhanced performance, as the field may be augmented and diminished. The consequence of this decision is elevated production costs and augmented control complexity.

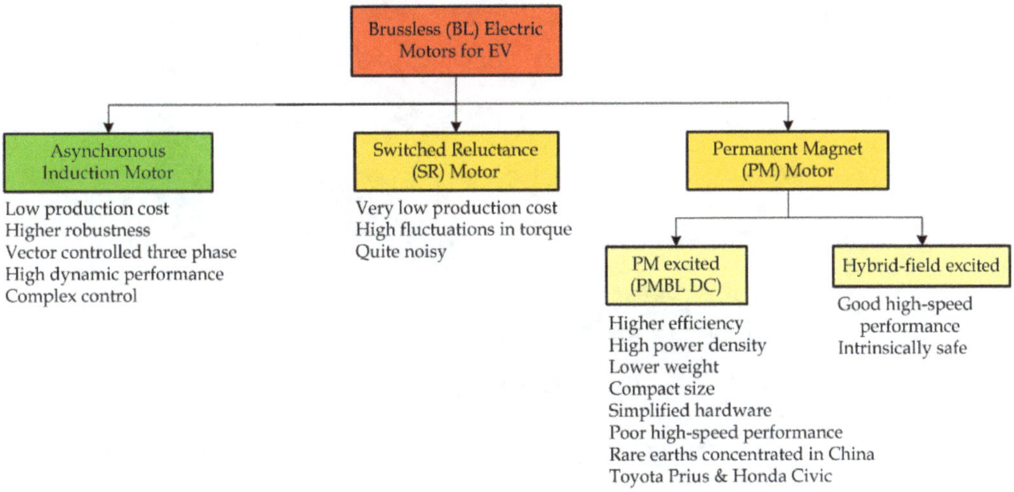

Figure 13. Electric motor for Evs

One final alternative for an electromagnetic torque generator in electric vehicles is the brushless switched-reluctance (BLSR) motor. The exceptionally cheap production cost of BLSR motors, which is even lower than that of induction motors, along with several other significant attributes, such as a broad speed range, positions them as a formidable contender for powering electric vehicles. Nonetheless, they are afflicted by (acoustic) noise and significant torque fluctuations, which may necessitate a more sophisticated (and costly) controller for compensation.

5. POWER ELECTRONICS DRIVER TOPOLOGIES FOR EVS

Power converters are specialized circuits designed with high-power electronic switches and control circuitry, both analog and digital, to transform one unregulated direct current (dc) voltage level into either a regulated and distinct dc voltage level or a regulated alternating current (ac) voltage level. The former are referred to as dc-dc converters, whereas the later are designated as dc-ac converters, commonly known as frequency inverters. Buck converters produce an output voltage that is lower than the input voltage, whereas boost converters provide an output voltage that exceeds the input value. Buck-boost converters can either decrease or increase the output voltage relative to the input, contingent upon the duty cycle of the control signal. Figure 14 depicts the utilization of power converters in a commercial hybrid electric vehicle (HEV). Converters are utilized to charge the battery pack from grid power (in PHEV), to replenish the battery pack from the fuel tank (involving ICE and generator), and to store energy in the battery pack (or ultracapacitor) during regenerative braking and coasting, as previously mentioned. They are utilized to power the electric motor(s) and supply energy to vehicle systems such as HVAC (heating, ventilation, and air conditioning).

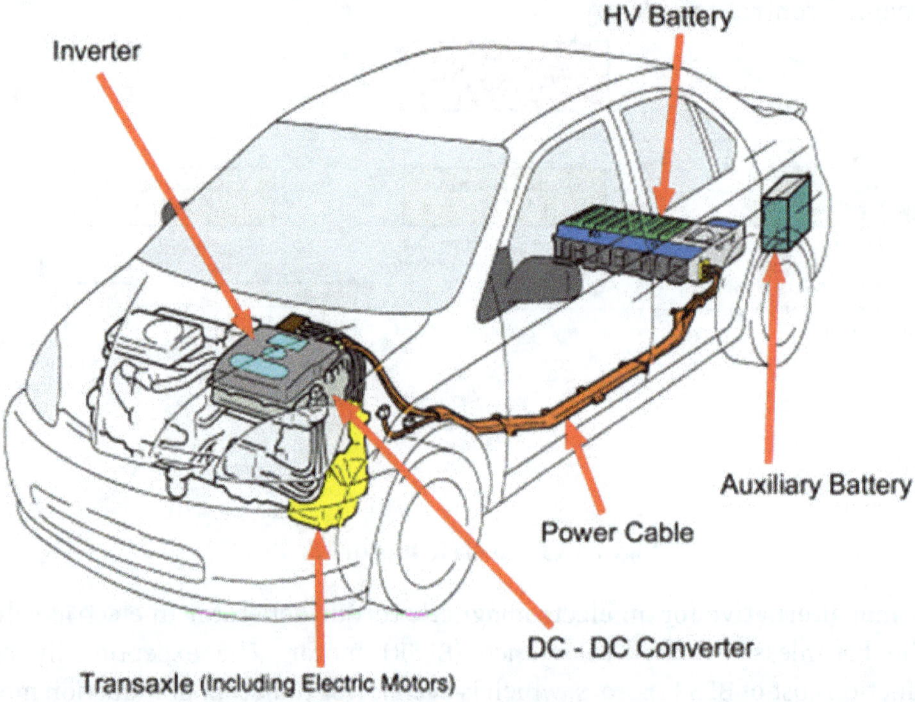

Figure 14. Power converters in a 2001 Toyota Prius HEV [Automobile Research Bolletin, 2008]

As illustrated in Fig. 15, classical power converter topologies, which are adequate to EVs, include the (transformer) isolated and non-isolated types and a family of bidirectional converters. Key characteristics of power converters for EVs are high efficiency (typically higher than 90%), high reliability, electromagnetic compatibility, and miniaturization (Bellur & Kazimierczuk, 2007). High-voltage, high-power, high temperature, fast switching and very low on-resistance semiconductor switches are of paramount importance in converters for EVs. These modern switches are metal-semiconductor oxide field-effect transistors (MOSFETs) and insulated-gate bipolar transistors (IGBT). Overall speaking, MOSFETs are faster than IGBTs, whereas these are capable of supporting high currents than MOSFETs.

A number of world-class semiconductor manufacturers (such as International Rectifier, Motorola, and ST Microelectronics) develop special power switches and auxiliary circuits (as gate drivers) appropriate for EV applications. Safety is a very critical issue in EVs, for the voltages of up to 600 V under the hood are lethal.

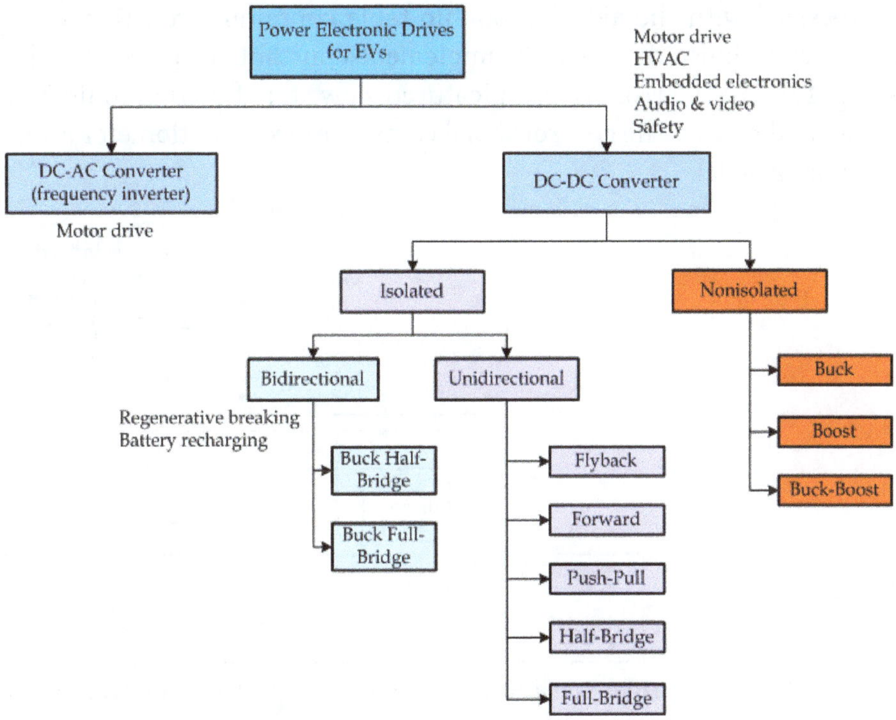

Figure 15. Power converters for EVs

6. CONTROL STRATEGIES

Control is a fundamental part of a successful EV design. Control engineering has matured for decades and nowadays counts on sophisticated microcontrollers and digital signal processors hardware and advanced integrated development environments. Despite all the available powerful tools and techniques, "efficient" control of EVs continues to challenge engineers and researchers, for the EVs embrace nonlinear processes (like battery behaviour), devices that are difficult to model (such as the ICE), and some conflicting goals, as for instance control for energy efficiency and control for better dynamic performance. It is not a coincidence that this area is one of the most prolific in the technical literature. Advanced digital control technique, such as optimal control and fuzzy control, are used by researchers and carmakers as they strive to improve EVs behaviour (Ambühl et al., 2010; Ohn et al., 2010).

Fig. 16 shows a power converter (frequency inverter) and a controller developed to equip a small off-road electric vehicle that is traditionally propelled by an ICE

(Lucena et al., 1997). The ICE was replaced by a 3-phase induction motor and a gearbox. High voltage, high power, fast switching MOSFETs were arranged to allow for the generation of 3-phase PWM voltage to feed the induction motor. Integrated bootstrap gate drivers facilitated MOSFET control by a microcontroller. The PWM was synthesized with the aid of a look-up table containing constant voltage-to-frequency ratio sinusoidal PWM, to implement constant torque at a wide speed range (Fig. 17). The control program featured slow start function to limit current in switches and motor. The control signal comes from a potentiometer attached to the accelerator pedal.

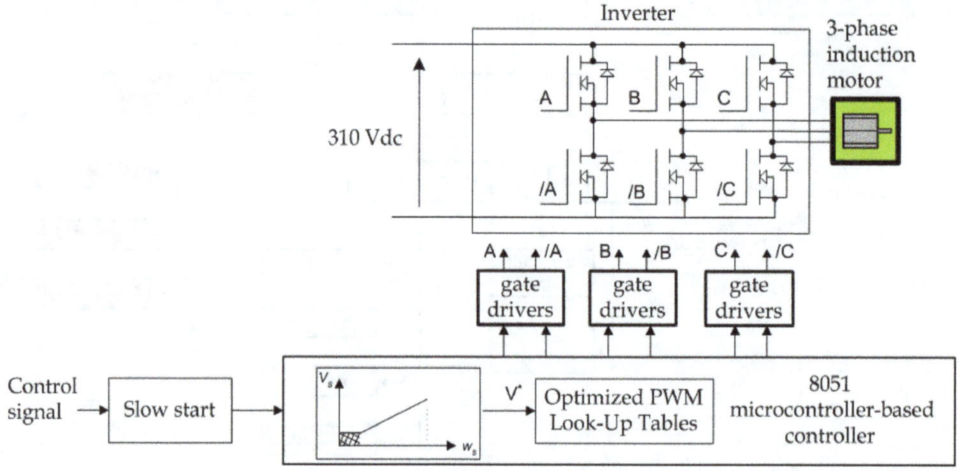

Figure 16. Power converter and controller for 3-phase induction motor

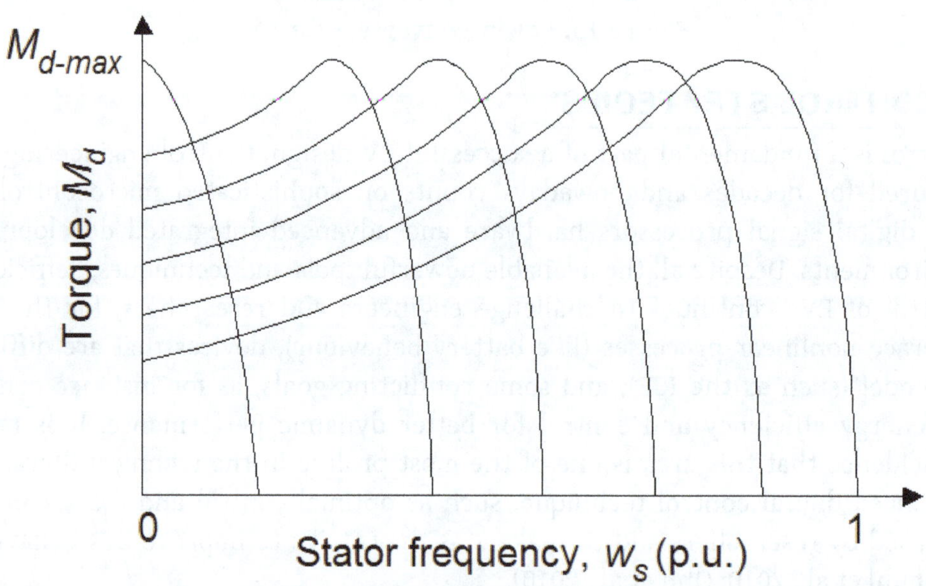

Figure 17. Induction motor torque versus stator frequency curves at different speeds

7. BATTERY TYPES

It is hoped that research on batteries would enhance their energy and power densities while substantially reducing production costs. In summary, these are the primary obstacles to the widespread adoption of BEVs, PHEVs, and traditional HEVs. Despite the suitability of contemporary technology for electric vehicles, in terms of driving range and performance, the cost remains prohibitively high for customers.As to the most promising technology for batteries, there seems to be no consensus among researchers. Some believe lithium-ion batteries will dominate the market for EVs (Burke, 2007), whereas others point out that nickel-metal hydride batteries are the best option (Wu et al., 2011). Meanwhile, commercial EVs are utilizing the following electrochemical technologies: Li-ion battery pack (388 V, 360 Ah), lead-acid batteries (12 V, 170 Ah), iron-lithium batteries (30 kWh) and sodium sulphate batteries (Xiang et al., 2008).

Carbon/carbon ultracapacitors have capacitance levels reaching 4000 F and a voltage rating of up to 3 V per cell (Gulhane et al, 2006). These high specific-power energy-storage devices may be fully charged in mere seconds and are optimal for regenerative braking and rapid vehicle acceleration, as they outperform batteries in speed. Regrettably, their insufficient energy density precludes them from serving as the primary storage devices. Commercially available electric vehicle charging stations are proliferating in nations such as the United States, offering simultaneous charging for several vehicles per station, along with authentication via RFID, IC cards, and synchronized mobile phones. Home charging systems and solar photovoltaic charging stations are also available. The latter may serve as a catalyst for a carbon-free future.

(a) (b) (c)

Figure 18. Commercial charging stations for EVs: a) Home battery charging station (240 V, 40 A), b) commercial battery charging station (240 V, 40 A), c) solar powered battery charging station [EV-Charge America, 2011]

8. CONCISE SUMMARY

Global apprehensions regarding climate change and the swift depletion of crude oil reserves, along with the deterioration of air quality due to vehicular emissions and noise pollution in metropolitan areas, ensure a persistent endeavor to transition from a noisy internal combustion engine fleet to a silent electric vehicle fleet in the forthcoming decades. Despite significant advancements in electric vehicle technology, the following obstacles must still be addressed prior to the mainstream adoption of EVs: Initially, the cost of electric vehicles (EVs), primarily attributed to battery expenses, must be reduced, potentially through ongoing and future research in battery technology. Secondly, the driving range of EVs needs substantial enhancement at competitive battery prices. Lastly, significant investments in EV infrastructure are imperative. The second is a very intricate issue that necessitates collaboration among governments, automotive manufacturers, technical organizations, researchers, and others to set standards, such as those for battery charging infrastructure and energy taxation within the power grid.

REFERENCES

1. Ambühl D. Sundström O. Sciarretta A. Guzzella L. 2010Explicit Optimal Control Policy and its Practical Application for Hybrid Electric Powertrains. Control Engineering Practice, 18 1429 1439
2. Automobile Research Bolletin 2008Toyota Prius Service Precautions. March 19, 2011, Available from: <http://www.tech-cor.net/AutoResBulletin/2000 8 /2000-8.htm>
3. Bakker S. 2010The Car Industry and the Blow-Out of the Hydrogen Hype. Energy Policy, 38 6540 6544
4. Bellur D. M. Kazimierczuk M. K. 2007DC-DC Converters for Electric Vehicle Applications, Proceedings of Electrical Insulation Conference and Electrical Manufacturing Expo, 286 293
5. Bento N. 2010Dynamic Competition between Plug-in Hybrid and Hydrogen Fuel Cell Vehicles for Personal Transportation. International Journal of Hydrogen Energy, 35 11271 11283
6. Burke A. F. 2007Batteries and Ultracapacitors for Electric, Hybrid, and Fuel Cell Vehicles. Proceedings of the IEEE, 95 4April 2007), 806 820
7. Chan C. C. 2007The State of the Art of Electric, Hybrid, and Fuel Cell Vehicles. Proceedings of the IEEE, 95 4April 2007), 704 718
8. Chen K. Bouscayrol A. Berthon A. Delarue P. Hissel D. Trigui R. 2009Global Modeling of Different Vehicles. IEEE Vehicular Technology Magazine, (June 2009), 80 89

9. Clement-Nyns K. Haesen E. Driesen J. 2011The Impact of Vehicle-to-Grid on the Distribution Grid. Electric Power System Research, 81 185 192

10. EV-Charge America. March 27, 2011Available from: <http://www.ev-chargeamerica.com>

11. Gulhane V. Tarambale M. R. Nerkar Y. P. 2006A Scope for the Research and Development Activities on Electric Vehicle Technology in Pune City, Proceedings of IEEE Conference on Electric and Hybrid Vehicles, 1 8

12. Kruger P. Leaver J. D. 2010Comparative Requirements for Electric Energy for Production of Hydrogen Fuel and/or Recharging of Battery Electric Automobile Fleets in New Zealand and the United States. International Journal of Hydrogen Energy, 35 11284 11290

13. Lucena S. E. de Marcelino M. A. Grandinetti F. J. 2007Low-Cost PWM Speed Controller for an Electric Mini-Baja Type Vehicle. Journal of the Brazilian Society of Mechanical Sciences and Engineering, 28 1 21 25

14. Ohn H. Yu S. Min K. 2010Spark Timing and Fuel Injection Strategy for Combustion Stability on HEV Powertrain. Control Engineering Practice, 18 1272 1284

15. Maggetto G. Van Mierlo J. 2000Electric and Electric Hybrid Vehicle Technology: a Survey, Proceedings of IEE Seminar on Electric, Hybrid and Fuel Cell Vehicles, 1

16. Sioshansi R. Fagiani R. Marano V. 2010Cost and Emissions Impacts of Plug-in Hybrid Vehicles on the Ohio Power System, Energy Policy, 38 6703 6712

17. Toyota Motor Corporation. March 19, 2011Available from: <http://www.toyota.com/prius/specs.html>

18. Xiang Z. Jia W. Jianzhong Y. Zhibiao C. Qinglin H. Yuanzhang H. 2008Prospects of New Energy Vehicles for China Market, Proceedings of Hybrid and Eco-Friendly Vehicle Conference, 1 8

19. Wu X. Cao B. Li X. Xu J. Ren X. 2011Component Sizing Optimization of Plug-in Hybrid Electric Vehicles, Applied Energy, 88 799 804

2 Chapter

ELECTRIC VEHICLES IN URBAN SETTINGS: ENVIRONMENTAL ADVANTAGES AND TECHNOLOGICAL-ECONOMIC OBSTACLES

1. INTRODUCTION: AN OVERVIEW

The mobility of individuals and things is a vital element of economic competitiveness; it is also a fundamental right of citizenship. Efficient transportation systems are crucial for societal prosperity, exerting considerable influence on economic growth, social development, and environmental sustainability. The objective of any sustainable transportation policy is to guarantee that our transport systems fulfill society's economic, social, and environmental requirements. In 2006, the transport industry accounted for 31% of total final energy consumption, with road transport contributing 82% of this figure, and was responsible for 25% of CO2 emissions in the EU-27. In 2007, road transport accounted for approximately 83% of overall passenger transport demand. Road travel is responsible for 71% of transport-related CO2 emissions, with passenger automobiles contributing 63% of these emissions. Presently, road transport is predominantly reliant on fuel oil (>90%), rendering it very susceptible to anticipated crude oil shortages, while also significantly contributing to air pollutants such as NOx, PM10, and volatile organic compounds.It is estimated that more than 80% of the developed world population lives in an urban environment and therefore it is in this environment where a larger concentration of vehicles are found. As example there were about 230 million passenger vehicles in the EU-27 in 2007 and the new vehicle sales were nearly 16 million vehicles in that year. Consequently the urban population is very much at risk by directly suffering the impact of conventional vehicles because their closeness to the pollutant source. Air pollution is one of the important external costs of transport as it impacts on the health of the population (it is estimated to be 0.75% of the EU GDP). On the other hand, the large concentration of vehicles causes traffic congestions in metropolitan urban areas that can be considered a threat to economic competitiveness (a recent study on the subject showed that the external costs of road traffic congestion alone amount to about 1.25% of the EU GDP) and it also increases the inefficiency of an overcrowded transport infrastructure.

Electric vehicles (EV) might offer a step change technology based on the much higher efficiency of electric motors compared to ICEs as well as the potential to de-carbonise the energy chain used in transportation and in particular in the well to tank pathway (JRC et al., 2008, Thiel et al., 2010). This will also open the possibility to use alternative energy paths to secure mobility and making the road transport more independent from crude oil.

This chapter examines the potential role of electric vehicles (including Battery Electric Vehicles - BEV and Plug-in Hybrid Vehicles - PHEV) in urban settings over short, medium, and long-term horizons, evaluates anticipated improvements in environmental performance, identifies key obstacles to their implementation, and considers the potential additional costs associated with the technology. This chapter analyzes potential business models and policy alternatives that could facilitate a more rapid market adoption of urban transport electrification. The ability of electric vehicles to mitigate transportation impacts varies by kind of impact and is also contingent upon the time scale. In other words, it does not serve as the definitive solution to the issues of environmental degradation and transportation inefficiencies (traffic congestion) in our metropolitan areas; therefore, it should be regarded as one option among many available to address this challenge. These possibilities encompass non-technical alternatives that, with the technological ones, must be evaluated within a holistic framework. The chapter concludes with a review and recommendations for positioning electric vehicles as a viable solution to mitigate the environmental impact of urban mobility.

2. SPECIFICATIONS OF AVAILABLE ELECTRIC VEHICLES

Customers are increasingly affected by the automobile industry's announcements of new electric car models, which appear to be making significant efforts to introduce electrified vehicles to the market. The examination of the technological characteristics of current and forthcoming electric vehicles is essential for comprehending their potential market penetration. The comprehension of their attributes (range, battery capacity, energy consumption, etc.) and restrictions will determine the client demographic drawn to this technology and the operations these cars will do. Recent literature sources have analyzed the automobile industry's plans for the deployment of electric vehicles (City of Westminster, 2009; Hacker et al., 2009). The realization of these ambitions in the short to medium term will depend on production capabilities and the quantity of car models offered to consumers. This final feature will undoubtedly influence the range of options available to consumers, hence affecting the likelihood of purchasing BEVs and PHEVs. A partial list of available car models is presented in Table 1. The facts in the table align with both the manufacturer's declarations and the available literature.

Table 1. Main features of the fully electric vehicles (cars and light duty vehicles) already present in the market or expected to be commercialised in the near-term (energy consumption is not well-to-wheel). Technical information has been retrieved from different official and non-official sources. Official sources have been reported in the references.

	Brand	Model	Capacity (kWh)	Range (km)	Consumption (kWh/100km)	Vehicle segment
Cars	Audi	e-Tron EV	42.40	248	17.10	Large
	BMW	MINI-E	35.00	180	19.44	Small
	BYD Auto	BYDe6	72.00	400	18.00	Large
	Chery Automobile	S18 EV	15.00	135	11.11	Small
	Chrysler	Dodge Circuit EV	26.00	175	14.86	Large
	CODA	Sedan-EV	33.80	180	18.78	Large
	Daimler	SmartED	14.00	125	11.20	Small
	Detroit	e63	25.00	180	13.89	Mid-Size
	Fiat	Panda	19.68	120	16.40	Small
	FIAT	500	22.00	113	19.53	Small
	Ford	Focus Ev	23.00	160	14.38	Mid-Size
	Ford	Transit Connect	24.00	160	15.00	Mid-Size
	Heuliez	WILL EV	18.00	300	6.00	Small
	Hyundai	i10 Ev	16.00	140	11.43	Small
	Lighting	GTS	35.00	175	20.00	Large
	Loremo EV	Loremo Ev	10.00	150	6.67	Mid-Size
	Lumeneo	Smera EV	10.00	150	6.67	Small
	Mercedes	SLS eDrive	48.00	160	30.00	Large
	MILES	ZX40S/ZX40ST	10.00	105	9.56	Small
	Mitsubishi	i-MIEV	20.00	160	12.50	Small
	NICE	Micro-Vett	10.50	80	13.05	Small
	Nissan	Leaf	24.00	160	15.00	Mid-Size
	Peugeot	iOn	20.00	140	14.29	Small
	Phoenix	SUV/SUT	35.00	209	16.73	Mid-Size
	Pininfarina	Bluecar	30.00	250	12.00	Small

	Brand	Model	Capacity (kWh)	Range (km)	Consumption (kWh/100km)	Vehicle segment
	Citroen	C-Zero	16.00	110	14.55	Small
	Renault	Kangoo	15.00	160	9.38	Small
	Renault	Zoe ZE	15.00	160	9.38	Small
	Renault	TwingoQuickshift E	21.45	129	16.60	Small
	Renault	Fluence	30.00	160	18.75	Mid-Size
	REVA	NXR	14.00	160	8.75	Small
	REVA	NXG	25.00	200	12.50	Small
	Rud. Perf. Roadstar	Spyder	16.00	125	12.80	Large
	SUBARU	R1e	9.00	80	11.25	Small
	SUBARU	Stella	9.00	80	11.25	Small
	Tata Motors	Indica EV	25.00	200	12.50	Small
	TESLA	Roadster/Model S	55.00	300	18.33	Large
	Think	City	28.50	180	15.83	Small
	Toyota	FT-Ev	11.00	150	7.33	Small
	Volkswagen	E-Up!	18.00	130	13.85	Small
	Volvo	C30 BEV	24.00	150	16.00	Mid-Size
	Zenn	CityZENN	52.00	400	13.00	Small
LDVs	Alke	ATX	8.40	70	12.00	LDV
	Piaggio	Porter	25.74	110	23.40	LDV
	Melex	XTR	4.32	60	7.20	LDV
	Modec	Delivery	50.00	100	50.00	LDV

3. ELECTRICAL VEHICLES AND THE URBAN ENVIRONMENT

The primary impetus for advocating the introduction of Electric Vehicles in the private car industry is their potential to diminish pollutant emissions in urban settings. This consideration is only partially applicable to greenhouse gases, specifically carbon dioxide (CO2). Given that a significant proportion of electric energy is generated by power plants utilizing fossil fuels, and recognizing the global implications of greenhouse gas emissions, it is prudent to assess the potential reduction, if any, in total CO2 emissions from the urban vehicle fleet. An assessment of electric car market penetration and its progression in an urban setting is essential to accomplish this.

3.1. Market penetration of electric vehicles

The deployment of electric vehicles will depend on a large variety of factors. This includes the performance and costs of batteries, the access to the distribution grid and its efficiency, the type of business model implemented to supply the consumer with reliable batteries and electricity, the acceptance by the consumer of new vehicle types and possible implied driving habits.

This diversity of, and interlinks between these factors make any market projection extremely difficult and impossible to define one single scenario about the penetration of electric vehicles. Several sets of assumptions can be made on the above-mentioned aspects, resulting in different expectations on the market penetration of electric cars.

In the open literature it is possible to find studies in which the market penetration estimation is very optimistic. In Clement et al. (2007-2008), PHEVs reach the 28% of the total Belgian vehicle fleet in 2030. In Hadley and Tsvetkova (2008), it has been estimated that by the year 2020, PHEVs will achieve a constant 25% market share, reaching the number of 50 million of vehicles in 2030 in the USA. Other studies also confirm these estimations although present fleet composition does not seem to support these penetration scenarios; however, as already stated above, the problem has too many degrees of freedom (as outlined also in Simpson, 2006).

More recently two studies addresses within the broader aim of the work the market penetration of electrical vehicles. In the first one (Perujo and Ciuffo, 2010) the approach was to make three scenarios and it was constraint to the case study of the city of Milan and its metropolitan area:

Scenario (1) assumed in 2010 that 0.5% of the vehicle fleet is made up of electric vehicles. Then the number of vehicles evolves in time assuming that the forecasted market share follows a logistic trend calibrated on the trend that methane (CNG) and Liquefied Petroleum Gas (LPG) powered vehicles have had in the period 2000-2009. This assumption is based on the idea that from the consumer perspective the electric technology has fairly the same appeal as the other "alternative" ones.

Scenario (2) assumed in 2010 that 1% of the vehicle fleet is made up of electric vehicles. Then the number of vehicles evolves in time assuming that the forecasted market share follows a logistic trend double than the one calibrated on the trend that CNG and LPG powered vehicles had in the period 2000-2009. This assumption is based on the idea that from the consumer perspective the electric technology has fairly the same appeal than the other "alternative" ones apart from the fact that electric vehicles do not suffer from the limited availability of service stations.

Scenario (3) did not considered a specific future trend, the impact of different percentages of electric vehicles on the whole fleet at a 2030 time horizon were evaluated (from 10 to 30%). This evaluation was carried out in order to show the impact on the electric supply system of a wider penetration of electric vehicles on the vehicle market, also according to the scenarios forecasted in Clement et al. (2007-2008) and in Hadley and Tsvetkova (2008).

With these assumptions the authors arrived to an EV-fleet share in the area of study in 2030 of 1.55 and 3.09% for scenarios (1) and (2) respectively.

The second study addresses the market share at European level. Having developed an enhanced version of the TREMOVE 3.1 model, Nemry and Brons, (2010) constructed and compared four market penetration projections taking into account two major drivers, i.e. technology progress of batteries and access to charging infrastructure. For each of them, two extremes scenarios (conservative and ambitious) were considered. The four projections are compared with a reference scenario in which the electric vehicle market doesn't develop. The energy efficiency of ICE cars gradually improves in accordance to the EU target on CO2 emissions. This means that by 2015 and 2020, new ICE cars average emissions in the EU would are respectively 135 g CO2/km and 115 g CO2/km. Then, from 2025 onwards, the emissions are limited to 95 g CO2/km.

In all four scenarios, the market deployment of pure electric cars and plug-in cars is endogenously determined by the cost efficiency (especially fuel costs and investment/maintenance costs) and by their effective range (determined by both battery capacity and access to charging).

Scenario assumptions on batteries cover two extreme future trends. In the conservative case, technical progress is slow and limited to a better durability while the usable SOC window remains unchanged. A continuous cost reduction is assumed, up to ~300 €/kWh. In the ambitious case progress is faster and more radical (200 €/kWh by 2030). Technology progress results in a much better durability and, also a higher useable SOC window.

With respect to infrastructure charging, given the already planned investments in various countries, the access to charging facilities is expected to increase in the future. At least, current charging possibilities - mainly at home, where garages exist - are already or will be extended in a relatively short term. These existing national plans are implicitly considered in the most conservative scenario but are not assumed to get much more ambitious in the future. In the second scenario (ambitious scenario), an even larger scale infrastructure charging deployment is assumed for all countries. It is to be noted that the potential role of fast charging is neglected in both scenarios.

Without surprise, the estimated market shares drawn by Nemry and Brons (2010) of electric cars (BEVs and PHEVs) are shown to increase when charging infrastructure deployment and battery progress are fast and significant. Charging infrastructure deployment, through a wide access to the grid at home and in other places (especially work places) contribute to offer to more car buyers a wide range of car options able to meet their need – not only conventional car but also electric cars. Battery progress seems to be the second-order driving factor and contributes to make the electric cars more performing and cost efficient so that it can better compete with its conventional counterparts.

The anticipated developments for these two elements indicate that, in both instances, the market share of BEVs remained constrained until 2020, ranging from 0.5% to 3%. PHEVs swiftly penetrate the market upon availability. This arises from the reality that battery and charging infrastructure impose more limitations on BEVs. The EV fleet share projected by both studies is consistent for the period 2020-2030, despite the differing research areas (the metropolitan region of Milan and the EU) and the distinct bases for scenario selection.

3.2. Potential EV impact on the overall CO2 emission in an urban environment

In 2009, both the European Union (EU) and G8 leaders agreed that CO2 emissions must be cut by 80% by 2050 if atmospheric CO2 is to stabilise at 450 parts per million (CO2 equivalent) keeping the global warming below what it is considered to be the safe level of 2ºC. But 80% decarbonisation overall by 2050 requires 95% decarbonisation of the road transport sector.

There are many options to achieved decarbonisation (through efficiency, biofuels and electric power-trains including hydrogen). However with a forecasted large increase of the number of passenger cars (rising up to 273 million only in Europe – and to 2.5 billion worldwide) by 2050, full decarbonisation may not be achievable through the expected improvements in the traditional internal combustion engine or alternative fuels alone. Furthermore if this scenario is combined with the increasing scarcity and cost of energy resources, it seems that electrification of road transport using low-carbon electric power-trains and hydrogen fuel cells is vital to ensure the long-term sustainability of mobility in Europe (European Commission, 2010a)

It is obvious that electric vehicles do not have tailpipe emissions of pollutants i.e. CO, NOx, THC, NMHC, particles or others (aldehyde and VOCs). However, the electricity needed to propel the vehicle needs to be produced somewhere and that energy production depending upon the type of power station used will contribute to the overall environmental impact of EV. Nevertheless, it can be argued that the pollutants mentioned above have a local impact and therefore the use of EV in the urban environment will contribute to a drastic reduction of those pollutants in the urban air. One major benefit of electric vehicles is the "displacement" of harmful air pollutants from urban to rural areas, where population exposure is lower. Noise levels are also lower, particularly in urban driving conditions. However, the GHG (here we are mainly referring to CO2) emissions have a more global effect and therefore the energy production needed to be used in EVs have a role in the overall global CO2 balance. This section addresses the levels of CO2 reduction that the introduction of electric vehicles could provide depending upon the different EV penetration levels in an urban vehicle fleet.

This duality of electrification of road transport and emissions from the power sector has been studied by Unger et al. (2009). They compared the overall impact on climate and air quality by using energy resources for electric power for vehicles with zero carbon intensity, such as wind and solar power, and those from standard power plants. Their study suggests that a 50 per cent reduction in road transport emissions as a result of using more electric vehicles will result in a cooling effect on the climate. Their conclusion is based on different combinations of the warming and cooling effects due to the contribution of road transport and power plants to climate change by emitting long-lived CO2 and short-lived pollutants. Non-CO2, short-lived pollutants also contribute to air pollution and include ground-level ozone and the fine aerosol particles: sulphates, organic carbon and black carbon. CO2, ozone and black carbon contribute to global warming, but sulphates and organic carbon reflect the sun's heat back into space, causing a cooling effect. They considered scenarios over 20-year and 100-year periods. For all scenarios, they estimated whether emissions from road transport and power generation would have a warming or a cooling effect on the climate. For both cases (no carbon base and standard power plants) a net overall cooling effect is achieved, albeit in the first case the level of cooling achieved is higher and in a shorter period.

The effect on CO2 reduction of different penetration level in the urban fleet has been studied recently for the case of Milan and its hinterland (Perujo & Ciuffo, 2010). They used for their calculation the Italian electricity mix that consist of 81% non-renewable sources, thus causing important emissions of CO2 to the atmosphere, and assuming that for 2030 the CO2 emissions due to electric energy production will not change as compared with the present values (worse case scenarios as it is expected that the mix will change to lower CO2 intensities). A similar approach was used for the evaluation of the CO2 emissions generated by a number of vehicles equal to the number of electric vehicles estimated for the year 2030 in the different scenarios (resulting from the estimated EV share in the passenger cars fleet as reported above). In this case, however, due to the constant technological improvements, it was not realistic to think that in 2030 the vehicles' CO2 emissions will have the same levels as today. For this reason they evaluated three cases: a)

2030 emission factors equal to 2005 ones (considering only EURO IV technology); b) 2030 emission factors reflecting European 2012 objective to have an average of 120 g CO_2/veh*km on the passenger cars fleet and a 50% emission reduction for LDVs, and c) 2030 emission factors reflecting European 2020 objective to have an average of 95 g CO_2/veh*km on the passenger cars fleet and a 50% emission reduction for LDVs. This three scenarios goes from a very pessimistic one (scenario a) to a very optimistic one (scenarios b) and c)), since the European objectives refer to a standard driving cycle whose emission factors are lower than those deriving considering an urban real driving cycle.

The findings of this exercise indicated that, even under the most favorable circumstances, emissions from internal combustion engine cars significantly exceed those from electrical power generation. The reduction of CO_2 emissions varies from 90% in scenario a) to 70% in the most optimistic scenario c). Additionally, the scientists assessed the average emissions threshold of vehicles below which the adoption of electric vehicles would not result in any reduction of emissions. An estimated CO_2 emission value of 40 g CO_2/km for internal combustion engine cars is significantly lower than that documented in prior research (e.g., Mackay, 2009). It is important to emphasize that these results reinforce the assertion that the prospective effects on emission reduction from the introduction of electric vehicles surpass those of merely advancing seemingly "clean" engines. The authors estimated that to achieve a 20% reduction in worldwide CO_2 emissions by 2030, electric vehicles should constitute around 25% of the total fleet of passenger cars and light-duty vehicles. While it may appear challenging to attain, this goal could serve as a feasible ambition for policymakers.

4. COMPARISON OF EV COSTS WITH ALTERNATIVE TECHNOLOGIES

Consumers acquire a new car for several reasons, including purchasing price (a primary concern for most buyers), depreciation rate, aesthetics, performance and handling, brand loyalty, and social perception. Nevertheless, vehicle proprietors sometimes underestimate the expenses associated with operating an automobile. While they are acutely cognizant of fuel expenses, road tax, and insurance, they frequently neglect to consider servicing, maintenance, and depreciation expenditures. Consequently, for those interested in comparing the costs of electric vehicles with other competing automotive technologies, the relevant metric should be the Total Cost of Ownership (TCO). The Total Cost of Ownership (TCO) encompasses not only the acquisition cost but also the operational expenses of the vehicle, including maintenance, replacement and repair costs, reliability, insurance premiums, taxes, and fuel or energy expenses, thereby delineating the financial implications throughout the vehicle's entire lifespan.At present the additional purchase costs of a plug-in hybrid electric vehicle as compared with a gasoline one is almost 11000 € and for the case of a pure battery electric vehicle the amount is more than 15000 €, this cost takes into consideration the underlying specific battery costs that is assumed to be 600 €/kWh for hybrid vehicles as well as the PHEV and BEV. This indicates that the high cost driver for both PHEV and BEV is the battery. In reality, the hybrid vehicles will likely use power batteries, while the batteries in the PHEV and BEV will likely be more biased towards higher energy capacity (JRC et al., 2008, Thiel et al., 2010). At the moment the additional cost born by BEV and PHEV is a challenge for the uptake of this class of vehicles.

Many studies have been published trying to look into the future (2020, 2030 horizon) cost of electric vehicles. Most of them include essentially three types of scenarios that can be described generally as a low, medium and high EV uptake (see for example Mc Kinsey, 2009 and Deutsche Bank, 2008).

A recent study (Thiel et al., 2010) makes forecasts of the cost of EV in the above indicated scenarios by taken into consideration the indicative improvement levels in vehicle technology for both EVs and ICEs (including a broad spectrum of vehicles technologies: gasoline, gasoline hybrid, diesel, diesel hybrid, PHEV and BEV). They considered that ICE powered vehicle would have 15% better energy efficiency in 2020 than in 2010, while for the BEV and PHEV no further efficiency improvement was anticipated for 2020 versus 2010 as these vehicles probably feature all near-term conceivable advanced efficiency measures.

In the 2030 time horizon no further energy efficiency improvements were assumed for any vehicle type as they considered that possible incremental improvements were equal for ICE powered vehicles, PHEVs and BEVs in this time frame. Hence, in the relative comparison this would not change the picture.

Learning effects and cost reduction by economies-of-scale are related to the volume production of vehicles. For 2010 it can be considered that all the compared vehicle types would have annual sales volumes above 100,000 units. This number needs to be understood as a proxy for wider market introduction as the 100,000 unit volumes might not be reached by every compared vehicle type exactly in 2010, but for some only in the following years. However, this would not change the comparison as the 2020 snapshot has to be understood as a proxy for the medium term and the 2030 snapshot should be seen as a longer term outlook. With realized production volumes for the years subsequent to 2010 the authors (Thiel et al., 2010) obtained learning effects that should reduce the costs of the newly introduced components. For the non-hybridized ICE vehicles, a learning rate of 5% was applied only on the newly introduced powertrain/vehicle components. The considered components were those contemplated in a previous study (JRC et al., 2008) and they are amongst others: (i) additional exhaust aftertreatment measures due to stricter emission limits, (ii) starter based stop-start systems, (iii) more sophisticated injection systems for gasoline direct injection but also downsized diesel engines and (iv) turbocharger for the downsized gasoline engine. The 5% learning rate was also applied on 50% of the costs of the ICE engine in the case of the PHEV as a dedicated range extender design of the ICE engine creates cost reduction possibilities. For PHEV and BEV, a learning rate of 10% was applied on the battery, electric motors and other vehicle upgrade costs that are directly linked to the electrification of the vehicle.

The possible cost reduction achievable by learning effect for the components necessary for vehicle electrification (i.e. cooling system upgrade, high voltage wiring, electric power steering, electric drive AC compressor, power electronics and modifications to enable regenerative braking) were based on the cumulative global sales volumes of the respective components. For the year 2020 only one volume scenario was used, while for 2030, two volume scenarios were used, a medium volume scenario and a high volume scenario for the number of BEVs and PHEVs.

These numbers are based on the assumption of 61 million new vehicle sales in 2010, 75 million new vehicle sales in 2020 and 90 million new vehicle sales in 2030, globally. The 2010 figures were used as a starting point for the subsequent calculation of the cumulated volumes (Mc Kinsey, 2009). The 2020 new sales volume of the BEV and PHEV were also derived from Mc Kinsey, 2009 using their mixed technology scenario. Advanced gasoline and diesel vehicles are already on the market today and it was assumed that they continue to penetrate the market reaching each 5 million global sales by 2020. For 2030 it was assumed that advanced diesel and gasoline new sales reach 15 million vehicles each. For these vehicle types, no distinction was made between the high and medium scenario.

The above assumptions, scenarios and learning rate leads to significant cost reductions for the BEVs and PHEVs. In the 2030 high scenario, their calculated purchase costs are already very close to the one of the diesel hybrid. However the additional purchase costs for EV versus the advanced gasoline vehicle in the 2030 high volume scenario is still over 2800 €. This value implies that the specific costs for the battery pack would reach a level below 200 € per kWh for the BEV and PHEV.

The above analysis only considered purchase costs, however concerning the TCO it must be recognized that apart from taxes and incentives, many of the above listed additional factors that influence the TCO most probably play further against the BEV and PHEV in the beginning. For example, the higher vehicle component costs in the BEV and PHEV lead to higher replacement costs and these again adversely influence insurance premiums. However, through continuous improvement and learning effects these disadvantages versus the conventional vehicles presumably reduce over time.

If one considers the long term energy prices (the cost of crude oil will always increase) the payback time for off-setting the higher initial investment for the car owner through the savings that will be achieved in the use phase as a result from the lower use of energy and lower energy prices for this technology can also be estimated. With a very much conservative calculation of 2030 oil price of 62.8 US $ per barrel crude oil (2010: 54.5 US $ per barrel; 2020: 61.1 US $ per barrel, all given in 2005 $) the estimated payback time for EV are about 20 years for 2010; however, for the time horizon 2020 the time is reduced to about 8 years while in 2030 (medium scenario) this become 6 years and for the high scenarios it reaches below 5 years. If the longer term oil price is significantly higher (as it can be expected) than the assumed 62.8 US $ per barrel, the payback period would further improve for the BEV and also the PHEV.

5.OBSTACLESINTHEIMPLEMENTATIONOFELECTRICCARFLEETS

Several reasons can obstruct or diminish the extensive adoption of electric vehicles. They can be categorized into factors that affect, on one hand, the appeal of electric vehicles (EVs) for prospective customers and the subsequent user experience of EV operators, and on the other hand, the commercial incentives for the industry to invest in EV development, production, sales, and in charging and maintenance infrastructures.The customer interest will be amongst others determined by:

- Purchase price or lease costs
- Total cost of ownership
- Market offers (brands, models, trim levels etc.)
- Driving experience
- Convenience of re-charging
- Safety perception
- Familiarity with EV technology

The commercial interest of the industry will be constrained by:

- Potential EV market size and its uncertainty
- Profit margin
- Investment needs
- Supply risks
- Risk averseness.

The majority of analysts concur that the costs associated with technology, particularly battery expenses, render the now available electric vehicles uncompetitive in the mainstream market relative to conventional automobiles, even when factoring in the total cost of ownership (TCO). Once this first barrier is surmounted, learning effects and subsequent technological advancements may yield acceptable payback periods for discerning customers in the long term (Thiel et al., 2010). A crucial element of the total cost of ownership (TCO) is the vehicle's residual value. The residual value of electric vehicles is significantly affected by the anticipated longevity and lifespan of the batteries. Suitable warranty programs can mitigate associated client apprehensions. Many private consumers often prioritize the purchase price above conducting a total cost of ownership (TCO) analysis, which will render the elevated purchase price a mitigating factor in the long term. The driving range constraints of completely electric vehicles are a significant consideration when juxtaposed with conventional automobiles. While this feature may not significantly influence urban and suburban car users today, it could deter potential consumers from selecting an electric vehicle if they are reluctant to compromise on the range offered by conventional vehicles. Rapid charging or battery exchange may serve as potential solutions to mitigate this disadvantage of contemporary electric vehicles. Additional driving factors, such as restricted maximum velocity and other conventional attributes of electric vehicle operation, are not anticipated to provide significant adoption challenges for electric vehicles, especially in urban and suburban environments. Electric vehicles (EVs) represent an innovative propulsion technology necessitating the establishment of a corresponding recharging infrastructure alongside the deployment of the vehicle technology. Research conducted by Flynn (2002) and Struben and Sterman (2008) has examined the interplay between infrastructure and vehicle deployment in greater depth. The primary conclusions drawn from these research indicate that robust synchronization is essential for ensuring sufficient coverage of recharging stations and the deployment of electric vehicles. Despite the prevalence of electricity distribution systems in urban and suburban regions, the primary challenges lie in the establishment of recharging stations, the implementation of standardized recharging interfaces, the development of vehicle-to-grid communication protocols, and the formulation of billing procedures and payment schemes. All these factors must be meticulously considered to facilitate convenient EV recharging for the user. In urban environments, suitable recharging methods must be identified for those who lack the capability to recharge their electric vehicles at home. A crucial consideration for prospective EV customers is that electric vehicles meet the same stringent safety standards as traditional vehicle alternatives. The recent introduction of electric vehicles that meet all relevant safety criteria and attained a high EURO-NCAP rating is likely to enhance the safety perception of electric vehicles. However, additional efforts are required to enhance or establish standards for EV safety, electromagnetic interference, and health. Prior to the extensive deployment of electric vehicles, public familiarity with this novel propulsion technology may provide a hurdle. Familiarity can be augmented through targeted marketing and media initiatives prior to achieving a critical mass of electric vehicles on the road, while word of mouth further amplifies public awareness. As previously stated in chapter 3.1, the future market size of electric vehicles is indeterminate, and forecasts are fraught with uncertainty. Historically, there have been instances of failed endeavors to introduce BEVs into the market. Several of these endeavors were accompanied by hopeful projections for the future implementation of electromobility; however, a comprehensive rollout of electric vehicles did not materialize (Frery, 2000). This ambiguity diminishes the industry's propensity to invest in electric vehicles and their associated infrastructure. Due to the capital-intensive nature of the automobile sector and the requisite infrastructure expenditure, industry participants exhibit a tendency towards risk aversion in this setting. The profit margin for the initial electric vehicles will be minimal. The initial generation of deployed electric vehicles (EVs) is anticipated to provide a detrimental business case for the industry, which can be rationalized as a preliminary investment in a prospective growth market. Numerous manufacturers are strategizing to enter the EV market; nevertheless, they will want to mitigate investment risk by first launching a restricted number of vehicles. This restricts the available options and may deter prospective buyers who have a particular preference for certain brands or models. Manufacturers can mitigate initial investment requirements by sharing common component sets across brands (e.g., Mitsubishi i-MIEV, Citroen C-Zero, Peugeot iOn) or by concentrating their deployment efforts on specific lead markets. The second choice will, on one hand, constrain the requisite investments in the dealer and maintenance network, but on the other hand, it will also diminish the pool of possible clients. Infrastructure providers for recharging will seek to secure a sufficient return on their investment, which may result in inadequate coverage initially. Supply chains must be established for the new electric vehicle-specific technologies and components. This may initially hinder the acceleration of EV deployment, but it should not result in a prolonged supply constraint. Material constraints are anticipated to provide challenges for permanent magnet motors (e.g., neodymium) and certain cathode materials for lithium-ion batteries (e.g., cobalt) (European Commission, 2010b).

6. POLICY ALTERNATIVES AND BUSINESS FRAMEWORK FOR ELECTRIC VEHICLE ADOPTION

It may be considered that the trend towards transport electrification is on its way and is irreversible. This is for instance suggested by the fact that every large automotive company has or is currently developing electric models and that a considerable number of countries have established plans to foster the development and deployment of EVs.

However, overcoming the challenges discussed in the previous section is essential to enabling a viable market for electric-drive vehicles. This requires strategic planning, public intervention and synergies with private initiatives.

Developing advanced common standards for safety, environmental performance and interoperability are seen as indispensable (European Commission, 2010a).

Both public and private initiatives are needed, and given that electric cars are expected to deploy faster in urban and sub-urban zones, such intervention would, at least in a first stage focus on such areas.

Public-private collaborative strategies at different levels (supra-national, national and local) are needed to address different types of barriers. For instance, within the Public Private Partnership (PPP) "European Green Car Initiative" (EGCI) which is part of the European Economic Recovery Plan these barriers are addressed through a mix of R&D funding and other instruments. A broad range of improvements of performance, reliability and durability of batteries need to be achieved to increase the attractiveness, range and affordability that will condition the consumer willingness to purchase electric-drive cars.

In parallel to those R&D funding initiatives, charging infrastructure needs to be deployed progressively, taking into account of travel patterns, achievable autonomy ranges, urban land use constraints and time availability for car charging at the different parking places, e.g. residential, workplaces, commercial centres, shopping, cinemas.

In Europe, several national or local governments have adopted charging infrastructure plans (e.g. Portugal, Denmark, Netherlands, Spain, Germany). As it is hard to predict how fast and to which extent the market will grow, achieving any "optimal" deployment is improbable. Continuous monitoring of the market, including on consumer attitudes should however guide public planning. Surveys often represent the available basis for establishing such plans. In a survey carried out on behalf the South and West London Transport Conference (Sweltrac), towns - followed by home, work and supermarkets - appeared to be the most popular location for charging points (SWELTRAC, 2007). In many cases, Governments plans are targeting specific areas and networks (first residential areas and urban zones) and niche markets. Several plans concentrate in cities (Berlin, Paris, London).

Besides charging spots in towns, incentives can also be created to broaden the access to the grid at home and at work place. For instance, the French Government plans to require, by 2012, new apartment's buildings with parking to include charging stations. It also plans to make the installation of charging sockets mandatory in office parking lots by 2015. Member States are introducing incentives to companies to install recharging spots (21.5% tax exemption is granted in Belgium).

The requirement of installing charging infrastructure could also be integrated into sustainability housing plans and renewable energy targets (see for instance Sheffield – UK).

Progress on battery performance, especially on energy density should help reducing the upfront costs of electric vehicles. In the meantime, innovative policy instruments and business models need to be envisaged and put into place for improving affordability and reducing risk perception associated with a non mature technology could be facilitated with different instruments.

Various business models are being explored and tested involving the automotive industry and new emerging business companies in order to spread the costs of batteries over several years. This includes Battery leasing, Mobile phone style subscription service. Vehicle leasing and Car-sharing also constitute solutions.

Subsidies targeted to niche markets (e.g. taxi fleet), and specific provisions for electric in public purchase procurement (Green Public Procurement) could be used as an instrument in favor of technology learning, experience acquiring of user attitudes, and consumer trust to the new technology.

In the short term, extending such subsidies to the mass market may be impractical due to the constraints of the public budget and perhaps harmful, particularly while technology has not yet reached full maturity. It is anticipated that internal combustion engine (ICE) vehicles would continue to constitute a significant portion of the future fleet, extending to 2030 and beyond. Consequently, their energy efficiency will predominantly influence the energy consumption and CO2 emissions of the transportation sector, particularly in road transport. In the long run, a stable fiscal and regulatory framework is essential to promote energy-efficient technologies and safeguard public finances, in alignment with the new fuel consumption revenues. A long-term perspective is essential concerning the stability and sustainability of the supply chain, particularly in relation to raw resources like lithium and rare minerals. These various regulations and programs must be formulated and executed based on ongoing insights from the emerging electric vehicle market, encompassing both production and supply perspectives as well as consumer behavior. Demonstration projects can enhance knowledge and comprehension of customer behavior.

7. SUSTAINABILITY OF URBAN TRANSPORTATION

Earlier sections have demonstrated that the electrification of vehicle transport, particularly in urban settings, has the capacity to diminish CO2 and other pollution emissions in our cities. This technical shift addresses only one of the three pillars of sustainability, namely the environmental component, while the other dimensions, economy and society, must also be considered to meet the sustainability problem. The notion of sustainable mobility originates from the broader concept of sustainable development. The sustainability of transportation can be assessed by analyzing the transport system's external effects—both good and negative—on the environment, public health, safety and security, land usage, congestion, economic growth, and social inclusion (OECD, 2000). The social aspect of transport sustainability is fundamental to the primary purpose of the transport system: to facilitate access to resources, services, and markets, which are essential for generating welfare. The concept of economically sustainable transportation is predicated on comprehensive cost accounting and pricing systems that incorporate economic factors stemming from transport activities that hinder sustainable development, specifically externalities, spillover effects, non-priced inter-sectoral linkages, public goods, uncompetitive markets, risk and uncertainty, irreversibility, and policy failures (Panaytou, 1992). Alternative definitions of economically sustainable transport assert that it must be "cost-effective and adaptable to continually evolving demands, allowing commercial and free market operations without substantial negative externalities and distributional impacts" (UN, 2001). To attain sustainable transportation, it is essential to address a diverse array of positive and negative impacts, including contributions to climate change, congestion, local air pollution, and noise. Research on public views on transportation (Goodwin and Lyons, 2010) indicates congestion as a significant concern and advocates for behavioral modification to mitigate environmental hazards. To mitigate these adverse consequences, three approaches can be delineated: (i) price strategies, predominantly road pricing; (ii) alternatives to automobile-based transportation, with a significant emphasis on investment in public transit; and (iii) innovative technologies and fuels. Implementing pricing measures will diminish transport demand and/or guarantee that the demand is "optimal," so positively influencing urban road congestion. However, to achieve widely accepted price, alternatives to automobile-based transportation must be evaluated. This may encompass, for instance, enhanced public transportation services that could facilitate a transition from automobile usage. This measure will enhance the public perception that non-coercive or "pull" strategies are fairer, more effective, and thus more acceptable than "push" strategies such as pricing (e.g., Eriksson et al., 2008). Moreover, strategies to minimize travel distance, such as telecommuting or spatial planning, are recognized as effective in decreasing the mileage of personal vehicles, thereby contributing to carbon reduction in the transportation sector and alleviating congestion in urban areas and roadways.

8. CONCISE SUMMARY

Given that over 80% of the European populace resides in urban areas, the necessity to ensure their mobility while also protecting their health and the environment presents a conundrum. Numerous comprehensive European programs in the energy and transportation sectors are attempting to resolve the tension between mobility and environmental concerns. The electrification of urban road transport can substantially diminish CO_2 emissions and other pollutants in city streets, as well as our near-total dependence on fossil fuels. This is predicated on the significantly greater efficiency of electric motors relative to internal combustion engines, with the ability to decarbonize the energy chain utilized in transportation, particularly in the well-to-tank pathway. Battery Electric Vehicles (BEVs) are significantly more advantageous regarding CO_2 Well-to-Wheel emissions, whereas Plug-in Hybrid Electric Vehicles (PHEVs) serve as a viable transitional alternative. The significant cost disadvantage associated with BEVs and PHEVs will persist until 2030, when learning effects may diminish this penalty to a threshold that ensures acceptable payback periods of less than six years for BEVs and aligns the cost penalties of PHEVs with those of other hybrids. If the replacement costs for components or insurance premiums remain elevated compared to traditional vehicles, it may prolong the duration required to achieve a competitive total cost of ownership (TCO). Consequently, a coherent fiscal and regulatory framework is essential to promote the most energy-efficient technological alternatives while safeguarding public finances in alignment with the new fuel consumption revenues. Furthermore, to achieve a more extensive deployment of electric vehicles, the general public's familiarity with this novel propulsion technology must be addressed. Familiarity can be augmented through targeted marketing and media initiatives prior to achieving a critical mass of electric vehicles on the road, while word of mouth further amplifies public awareness. Ultimately, it is important to note that the widespread adoption of electric vehicles alone will not inherently foster the establishment of a sustainable transportation system. It can indeed alleviate the environmental impact of road traffic; however, this constitutes but one facet of sustainable development. To effectively tackle the sustainability paradigm, it is essential to undertake steps that lessen the reliance on personal transportation (private vehicles) in favor of public transport. This entails shifting the decisional viewpoint from sustainable transport to sustainable mobility.

REFERENCES

1. Alke, 2009 http://www.alke.com/electric-vehicles.html.
2. Altairnano, 2009 http://www.altairnano.com.
3. Atea, 2009a http://www.atea.it/twingo-elettriche.htm.
4. Atea,2009b http://www.atea.it/panda-elettriche.htm.
5. City of Westminster, 2009Understanding electric vehicle recharging infrastructure, vehicles available on the market and user behaviour and

profiles.http://www.westminster.gov.uk/workspace/assets/publications/Electric-charging-and-EV-vehicles-1247227333.pdf

6. Clement K. Van Reusel K. Driesen K. 2007The consumption of Electrical Energy of Plug-in Hybrid Electric Vehicles in Belgium. European Ele-Drive Conference. Brussels, Belgium
7. Clement K. Heasen E. Driesen K. 2008The Impact of Charging Plug-in Hybrid Electric Vehicles on the Distribution Grid. Proceedings 2008 4th IEEE BeNeLux Young Researchers Symposium in Electrical Power Engineering. Eindhoven, The Netherlands.
8. Coda, 2009http://www.codaautomotive.com.
9. Deutsche Bank, 2008Electric Cars: Plugged In-batteries must be included.
10. Eriksson L. Garvill J. Nordlund A. M. 2008Acceptability of single and combined transport policy measures. The importance of environmental and policy specific beliefs. Transportation Research Part A 42; 2008. 1117 1128
11. European Commission, 2010aA European strategy on clean and energy efficient vehicles (COM(2010)186 final, April 2010.
12. European Commission, 2010bCritical raw materials for the EU- Report of the Ad-hoc Working Group on defining critical raw materials.
13. Flynn P. 2002Commercializing an alternate vehicle fuel: lessons learned from natural gas for vehicles. Energy Policy 30; 2002. 613 619
14. Fréry F. 2000Un cas d'amnésie stratégique : l'éternelle émergence de la voiture électrique, Actes de la 9ème Conférence Internationale de Management Stratégique, 2000 24 26mai, Montpellier.
15. Goodwin P. Lyons G. 2010Public attitudes to transport: interpreting the evidence. Journal of Transportation Planning and Technology: UTSG special 33 3 17
16. Hacker F. Harthan R. Matthes F. Zimmer W. 2009Environmental impacts and impact on the electricity market of a large scale introduction of electric cars in Europe- Critical Review of Literature, Report of the European Topic Centre on Air and Climate Change.
17. Hadley S. W. Tsvetkova A. 2008Potential Impacts of Plug-in Hybrid Electric Vehicles on Regional Power Generation. Oak Ridge National Laboratory. Oak Ridge, Tennessee, U.S.A.
18. Italiaspeed, 2009http://www.italiaspeed.com/.
19. J. R. C. E. U. C. A. R. C. O. N. C. A. W. E.2008Well-to-wheels analysis of future automotive fuels and power-trains in the European context, Available from http://ies.jrc.ec.europa.eu/about-jecLdv, 2009 http://www.ldv.com.

20. Lighting, 2009http://www.lightningcarcompany.co.uk/.
21. Logghe S. Van Herbruggen B. Van Zeebroeck B. Transport & Mobility Leuven, Emissions of Road traffic In Belgium. January 2006
22. Mackay D. J. C. 2009Sustainable Energy-Without the Hot Air. UIT Cambridge Ltd., Cambridge, England.
23. Mc Kinsey,2009Roads toward a low-carbon future: reducing CO2 emissions from passenger vehicles in the global road transportation system.
24. Miles,2009http://www.milesev.com/.
25. Mini,2009http://www.miniusa.com/minie-usa/.
26. Mitsubishi,2009http://www.mitsubishi-motors.com.
27. Nemry F., Brons M., Plug-in Hybrid and Battery Electric Vehicles.Market penetration scenarios of electric drive vehicles. European Commission, Joint Research Centre. Technical Note JRC 58748 (http://ipts.jrc.ec.europa.eu/) 2009http://www.nicecarcompany.co.uk/. Organization for Economic Co-operation and Development- OECD 2000Environmentally Sustainable Transport: futures, strategies and best practices. Synthesis Report of the OECD Project on Environmentally Sustainable Transport (EST)- International EST Conference 4th to 6th October 2000, Vienna, Austria.
28. Panaytou T. 1992Economics of Environmental Degradation. The Earthscan Reader in Environmental Economics. Markandya, A., and Richardson, J. Earthscan. London.
29. Perujo A. Ciuffo B. 2010The introduction of electric vehicles in the private fleet: Potential impact on the electric supply system and on the environment. A case study for the Province of Milan, Italy, Energy Policy 38 (8), 4549 4561
30. Piaggio C. H. 2009http://www.ch.vtl.piaggio.com/porter_el.htm.
31. Phoenix,2009http://www.phoenixmotorcars.com/.
32. Sinanet,2009http://www.sinanet.apat.it/it/sinanet/fetransp.
33. Smartgrids,2009http://www.smartgrids.eu/.
34. Struben J. Sterman J. 2008Transition challenges for alternative fuel vehicle and transportation systems, Environment and Planning B: Planning and Design 2008, 35 1070 1097
35. Tesla,2009http://www.teslamotors.com/.
36. Thiel C. Perujo A. Mercier A. 2010Cost and CO2 aspects of future vehicle options in Europe under new energy policy scenarios, Energy Policy 38 (11); 7142 7151

37. United Nations 2001Sustainable Transport Pricing and Charges- Principles and Issues. Economic and Social Commission for Asia and the Pacific/ Asian Institute of Transport Development.

38. Unger N. Shindell D. T. Wang J. S. 2009Climate forcing by the on-road transportation and power generation sectors. Atmospheric Environment. 43, 3077 3085

Chapter 3

PLUG-IN ELECTRIC VEHICLES A CENTURY LATER – HISTORICAL INSIGHTS ON WHAT HAS CHANGED AND WHAT REMAINS THE SAME?

1. INTRODUCTION: AN OVERVIEW

Fundamental trade-offs between gasoline and electric vehicles. Unlike internal or external combustion engines, the primary advantage of electric vehicles (EVs) powered by on-board battery-stored electricity is their quiet, efficient, and emissions-free functioning. While emissions arise from fossil fuel-based energy generating, these pollutants are spatially and temporally distanced from the operation of the electric vehicle. Another advantage is the high low revolution per minute (rpm) torque, which provides exceptional starting acceleration. The primary drawback is storage capacity. The fuel tank of a traditional automobile can store significantly more energy in a smaller volume than a battery pack, and at a considerably cheaper starting cost. The gasoline car can consequently refuel significantly faster and cover greater distances on a single tank. The limited top speed compared to gasoline cars constitutes a drawback. A significant characteristic of electric vehicles is their very high peak power capacity for brief intervals of a few seconds. Nonetheless, the peak power level often far exceeds the sustainable power level. The storage drawback of electric vehicles is significantly mitigated when used at low average speeds in urban environments. At these velocities, the battery pack may take an extended duration to drain. Moreover, as average speed decreases, the average energy demand per hour of operation diminishes far more swiftly than that of traditional gasoline and diesel engines, hence prolonging the duration of operation on a full charge. Regrettably, at these velocities, the fuel conserved per operational hour compared to gasoline and diesel is inferior to that at elevated average speeds (Vyas, Santini, and Johnson, 2009), necessitating significantly more hours of vehicle utilization to recoup battery pack expenses. Consequently, enhancing fuel savings per hour of operation can be achieved through intensive intra-urban utilization of electric vehicles outside the most densely populated metropolitan areas, rendering it more financially appealing (Santini et al, 2011).

2. WAVES OF HISTORY I: 1890S THROUGH THE 1930S

Personal use electric vehicles. In the United States in the 1890s and early 1900s, EVs competed successfully with gasoline and steam cars predominantly in the Northeastern U.S., the most densely developed part of the nation, but also in Chicago. The highest volume manufacturer of EVs at the turn of the century was the Pope Manufacturing Company of Hartford Connecticut (Sulzberger, 2004). New York was then and remains today the most densely developed metropolitan area in the United States. In 1900 the nationwide registration of 4192 vehicles in the U.S. was 1681 steam, 1575 electric and 936 gasoline (Mom, 2004, p. 31). According to Sulzberger, in 1899 electric vehicles outnumbered gasoline by two to one *in the major metro areas* – New York, Boston, and Chicago. A total of 2370 vehicles were in these three metro areas, so the start of the motor vehicle in the U.S. was clearly in relatively affluent, large major cities. The technological historian G. Mom (2004) indicated that the dollar value of production of electric cars in 1900 was more than half of the total, despite the share of unit volume being 38%. Half of all passenger cars were produced in New England. However, over the next two decades production of motor vehicles in the U.S. moved westward and significantly toward gasoline.

When EVs were in the market from the 1890s to 1920s, they consistently served urbanized areas, rather than rural households and businesses. A caveat, however, is that for the personal EV in the U.S. in about 1914 the share of "home kept" electrics rose as city population dropped, as did the market share of EVs (Mom, 2004, p. 254). A logical deduction is that home kept EV share increased as city density decreased and as the share of single family dwelling units rose. The availability of a parking spot within or beside the electrified dwelling unit was then, and can be expected to be in the future, a major determinant of market success for personal use EVs. Mom concluded that the electric car of 1914 "functioned as the affluent suburban family's second car" (p. 254) having been identified as "an environmentally friendly secondary car" (p. 250). At this time the EV in the U.S. held a share of the market similar to hybrids today (<3%, far below the turn of the century), but it was a shrinking rather than rising share. Mom also noted that in 1916 the EV was no longer successful in the Northeast — "the electric passenger car seemed to prefer the medium-sized town in the Midwest." (Mom, p. 261). Midwest EVs were supported by "active central stations" that were Electric Vehicle Association of America members. By 1920 the personal use EV was no longer sold in New York, the densest and largest of U.S. cities, with steep hills and long commutes from the suburbs to downtown being blamed (Mom, p 262).

Vyas, Santini and Johnson (2009) drew attention to the suburban target market for personal use EVs a century later, pointing out that the suburbs of U.S. metropolitan areas are where affluence is greatest, as are the numbers and proportions of garages and multi-vehicle households. Ironically, although the Western U.S. (California) and Northeast have adopted regulations designed to encourage EVs, and the U.S. West Coast is aggressively pursuing electric infrastructure, the largest shares of single family dwelling units are found elsewhere – in the South and Midwest regions (Vyas. Santini and Johnson, p. 60). Typically, regardless of region, about half of all garages and carports are found in suburbs, and half of the households with two or more vehicles are found there. Only about one fifth of either garages or multi-vehicle households are found in center cities (Vyas, Santini and Johnson, p. 61). For the EV market, Santini et al (2011) recently deduced that charging infrastructure costs can be significant, so electric commuting with both house and workplace charging is probably not the least cost market. They therefore examined vehicles not driven to work, on the assumption that the home charger could be used more than once a day. They estimated that only those EVs driven more intensively than average could be financially desirable. Greater affluence is associated with higher annual distance driven.

Mom concluded that 1910 infrastructure and maintenance costs were an important drawback for the individual household. Despite the fact that "most *newly built* (italics mine) houses came complete with a connection to the electricity grid" ... the battery and its charging equipment had an important indirect effect" ... pushing ... "purchase and maintenance costs of the electric passenger car far above an acceptable level for middle-class gasoline vehicles." (Mom, pp. 286-287). He also noted that at the time the middle class – the utilitarian user - could only afford one car and therefore could not make a "fleet" choice, purchasing and using both a gasoline and electric vehicle for their respective advantages.

Mom noted that the motorization of areas outside cities was far slower in Europe than in the U.S. The mild success of the personal passenger car EV in the U.S. from about 1905 to 1920 accompanied the wave of gasoline vehicle motorization and regional growth in the Midwestern U.S. Recent investigation of household charging infrastructure cost suggests that installation of suitable charge circuits is far less expensive when designed into new houses than when houses are retrofitted (Santini, 2010). Thus, the growing Midwest would have had the opportunity to install charging infrastructure as it grew and its expanding major cities electrified.

Nevertheless, by the early 1920s the personal electric vehicle in the U.S. was rapidly shrinking toward zero production. The counterintuitive computation that Santini et al (2011) made for the 4-5 passenger *personal use* EV of 2020 was that the rate of utilization (hours of driving) in dense center cities would not be adequate to pay off the added costs of the pure electric vehicle. This is a quantitative candidate explanation for the 1900-1920 failure of the personal-use EV in Europe while it succeeded mildly in the U.S. Congested stop and go driving has financial advantage for the EV only if it is driven many hours per day, such as by a commercial delivery vehicle.

When commercial applications of horses, EVs and gasoline trucks were studied in the U.S. in 1912, it was concluded that horse wagons remained the most cost effective option up to 19 km per day (Mom, p. 223). If the attainable average speed on the local roadway network in most European nations was then less than about 15 km/h, and if daily travel for personal activities was between one and one and a half hours, then the implication is that it would have been a financially correct decision at the time for European households to continue to use horses rather than either gasoline or electric vehicles. In reference to first generation electric taxi cab capabilities for Berlin and Cologne in 1907, Mom notes that "first generation taxicabs could go 15 km/h" which compared unfavorably to the 40-50 km/h for gasoline taxis that could only be achieved in practice at night (Mom, p. 142). It was also clear that the speed competition drove up the costs of operating EVs, since a 25 km/h top speed required pneumatic, rather than hard tires; a stronger heavier frame; and a larger battery pack to provide needed power. This increased consumption from 220-250 Wh/km to 350-425 Wh/km (Mom pp. 142-143). It was estimated that only larger fleets of taxicabs could afford to own and operate electric vehicles, a finding perfectly consistent with a conclusion that a less intensively used electric vehicle in a household fleet of one would not be economic, as Mom reiterated in his conclusions (p. 291).

The electric passenger car in 1913-14 was far more successful in the United States than in Europe. Mom reports a count of about 1600 electric passenger cars in Europe in 1914, compared to 20,000 in the U.S. Passenger cars were about 56% of European electric cars and trucks with more than three wheels, while the U.S. share was about 67% (Mom, p. 252).

Although the peak number of electric passenger cars produced in the U.S. in 1915, at 4,715 (Mom, p. 254), was well in excess of the production of 1900, at 1575 (Mom, p. 31), and can be called a success, the sales of 611,695 gasoline cars *in the first nine months* of 1915 (Mom, p. 283) was an overwhelmingly greater increase relative to the 936 gasoline cars produced in 1900 (Mom, p. 31). At this point, 43% of gasoline vehicles were sold with an electric starter, which had been introduced in 1912.

According to Mom, "it is not correct to claim that the electric starter motor meant the deathblow for the electric car. But it did mean the last nail in its coffin" (Mom, p. 283).

In the U.S., the Ford Model T (initiated in 1908), which did not adopt an electric starter, had accounted for more than half of total vehicle sales during most of its lifetime. However, during the early 1920s its market share began to erode as more powerful cars with electric starters gained market share at its expense. The "T" had also been designed for low speed operation on poor dirt roads, having large diameter wheels and a high ground clearance. The 1920s saw increasing adoption of gasoline taxes to support state road building (Majahan and Peterson). The needs of the electric vehicle for reliable tires with low rolling resistance had led to development of the bias ply tire, a technology that was then adapted for use by gasoline vehicles in 1915 (Mom, p. 260). As noted by Loeb (1995), the electric starter, which itself had been developed for hybrids (Mom, p. 282) allowed reliable starting of engines with power well in excess of that of the 15 kW Model T. The electric starter ultimately allowed higher cranking power than a human arm could provide, allowing reliable starting of an engine with a much higher compression ratio, which in turn enabled more efficient gasoline engines, once octane enhancers had been added to gasoline (Loeb, 1995). The higher average speeds attainable by gasoline vehicles with more powerful engines, good roads, better tires and reduced ground clearance (thus reducing aerodynamic drag) most likely played a role in the 1920s demise of the personal electric vehicle, whose sustainable top speed was inherently limited.

The widespread 1912-16 adaptation of a cost effective combination of electrical and mechanical features in the predominantly mechanical gasoline vehicle signaled the end of the electric passenger car about a decade later. While attempts to combine electric and mechanical drive in hybrids failed in the marketplace, about a century later the new question is whether an increase in electrification of the gasoline vehicle, in the form of hybrids and the plug-in hybrid will again keep the market potential of the pure electric vehicle to less than 3% of the U.S. and European markets in coming decades. Mom, quotes a vice president for research from Ford in his closing pages, saying that "the most cost- effective and efficient road to a greener world is through the gradual electrification of vehicles ... rather than switching to an all-electric powertrain."(Mom, p. 299). Mother commends the electric vehicle for compelling the gasoline automobile to evolve and improve, endorsing the investigation of alternative powertrains. Nevertheless, he fails to include an opposing viewpoint from another automotive executive endorsing the merits of a transition to an all-electric drivetrain.

Trucks and taxis. The decline of personal use electric vehicles does not signify the end of electric vehicles altogether. Mom illustrates that commercial trucks and industrial (non-road) trucks experienced sustained growth in use over the majority of the 1920s and, in certain applications, into the 1930s. The electric taxi was abandoned in the mid-1920s. Since the early 1900s, the growth rates of commercial trucks in New York and Chicago were significant and markedly exceeded those of passenger cars. The pattern resembled that of the personal use car. The motorization of services previously reliant on horse-drawn wagons and cabs occurred swiftly, resulting in a significant increase in the total number of both electric and gasoline-powered commercial vehicles. The proportion of motorized services attributed to electricity far exceeded that of personal cars, comprising "a quarter of the entire American truck fleet" in 1912. In 1914, one-fourth of all electric trucks in the United States were located in New York City, constituting 39 percent of the city's motorized vehicle fleet (Mom, p. 228).In general, this was accomplished by much greater utilization of the electric vehicles and the battery packs in the business vehicles than was the case for personal use vehicles. Many of the business fleets in both the U.S. and Europe used battery swapping, with more than one battery pack per vehicle (Mom, p. 94, 231). The importance of assuring intensive use via reliable operation, with low maintenance, led to expenditures on well trained battery maintenance staff and also more expensive, more reliable batteries in commercial applications. Mom said that "for the electric car owner ... the character of the lead battery formed a virtually insurmountable barrier ... looking after the battery ... really needed the constant attention of a physician and a trained nurse." (Mom, p. 287). As of about 1911, commercial fleets found the "tubular lead and Edison batteries to reduce maintenance, though at a much higher cost. The commercial vehicle fleets made intensive enough use of their batteries to make this trade-off pay off. A private owner could not (Mom. P. 288). This shows the potential problem of mistranslating good reliability in fleet applications to an argument that a technology will also be reliable in the hands of an individual consumer.

Commercial fleets frequently charged battery packs overnight, very often at much lower cost per kWh than for daytime charging. With battery swapping, a 24 hour operation could be implemented with overnight charging and maintenance and checking on one battery when out of the vehicle while the other battery was in use. With swaps made in a matter of minutes, it also allowed the vehicle to stay in service for many hours of operation per day, rather than slowly recharging during the day. This significantly increased fuel saving per vehicle per day, helping pay off the investment.

Even so, the key to economic viability was to find the appropriate field of application. A 1924 book on the merits of the electric truck was written by E.E. La Schum of the American Railway Express Company. In this book La Schum was effusive in his praise for the electric truck, "the speediest of trucks where stops or delivery are frequent and traffic congested." (Mom, p. 245). He predicted that electric trucks, which then were used in a normal range of 48 km (perhaps about 7 km/h for an 8 hour workday, less if the truck were used on two or more shifts with battery swapping) would increase their competitive daily range to 64 km and more. At the time the American Railway Express Company had 1225 electric trucks, 575 other electric vehicles, 8200 horse wagons, and 2,500 gasoline trucks. The superior average speed of the electric truck, when stops were involved, probably included an advantage of a quicker start once the driver returned to the truck, in part because the electric did not have to be put into gear with a manual transmission, nor shifted. If the competing gasoline truck were turned off at stops in order to save fuel, this would also slow the start-up process upon return to the truck.

Santini et al (2011) recently estimated that financial viability of hypothetical 2020 mass market pure electric passenger cars with from 120-160 km of range would require full depletion of the pack and some recharging during the day under recent U.S. average gasoline prices and electricity rates. Typical passenger cars would not be driven enough to cause full depletion of such an EV. Only those driven far more hours per day than average could fully deplete the pack on a normal day, enabling additional gasoline saving via a daily recharge. Santini et al note that such vehicles are far more likely to be driven in suburbs than center cities. Identical electrical rates of $0.10 per kWh were assumed for both nighttime and daytime charging. However, the goal of the U.S. Federal Energy Regulatory Commission is to enable and encourage implementation of time-of-day pricing in the U.S. This will increase summertime average daytime electric rates to above $0.20/kWh, but will lower overnight rates by only a few dollar cents. Should time of day rates become common, second charges during the day might actually lead to higher energy costs per mile for the second charge than if a full hybrid gasoline vehicle were used.

High daytime electric rates were a problem in the past. Mom noted that public garages that had to return fully charged vehicles to customers suffered when the customer demanded delivery "during expensive peak hours" (Mom, p. 217). In fact, utilities (central stations in Mom's terminology), under the guidance of Samuel Insull, chose to offer "multistep" rates to stimulate charging overnight charging between 10 pm and 7 am. Today's term is "off- peak" or "time-of-day" rates. The truck was preferred as a customer over the passenger car because the "truck used 400 to 933 times more energy than a light bulb" while the car "consumed only 107 times more energy" (Mom, p. 208). In part, this difference was due to the greater hours per day of operation of the truck, not only the greater vehicle mass.

Mom found that electric vehicles serving business were consistently found in fleets of much larger size than were gasoline vehicles (Mom, p. 246). These fleets found it necessary to hire "competent men to take care of the batteries" (Mom, 229; see also p. 216). Costly maintenance made it imperative to spread the maintenance expertise costs over a number of vehicles. The idea of implementing battery rental services that thereby spread the cost of maintenance over a large number of vehicles of several different owners was tried in the "Hartford system" set up in 1910 (Mom, p. 230). This system, which involved a fixed fee and a mileage charge, with battery pack exchange using packs charged overnight, was implemented in many cities by 1916, though only hundreds of vehicles were involved. Battery pack exchange was also common in taxi fleets in Europe, though this was implemented by the taxi fleet itself, not as a rental service such as the Hartford system.

Another cost that promoted larger fleets was the per vehicle infrastructure cost involved in setting up in-house charging facilities for relatively few vehicles. However, if the infrastructure costs could be paid, a large fleet could then get a discounted electricity price. At "a purchase of 50 kW or more and a garaged fleet of 75 to 150 electric cars", Commonwealth Edison of Chicago offered a rate of $0.02/ kWh (Mom, p 254).

In Detroit in 1914 a taxi company tried the innovation of "curb boosting" – recharging of taxis while waiting at taxi stands. This idea was also implemented in Chicago and St. Louis in 1917, but it involved considerably fewer vehicles than the Hartford battery rental system and did not become common practice. Daytime rates for electricity may have been a deterrent.

Britain came late to the electric delivery truck, but found the very successful market niche – low speed urban delivery with many stops. In particular, milk trucks, which made quiet, clean early morning deliveries, became electrified in large numbers. Growth was dramatic from 1934 through 1949, by which time nearly 20,000 electric trucks were in service (Mom, p. 268).

Today, it is recognized that a portfolio of powertrain technologies is likely to be necessary in coming decades, as nations of the world slowly switch transportation from oil to other fuels. In effect, this process took place from 1895-1945 as nations switched from the grain fed horse and from the coal fed iron horse (steam locomotive) to the automobile. The electric passenger car and truck competed against horse drawn vehicles to a much greater extent than it competed with the iron horse, which dominated intercity travel. For commercial trucking services, a careful, but optimistic assessment was made by researchers from the Massachusetts Institute of Technology, indicating that the electric truck was more costly than horse drawn wagons at short distances up to 19 km, but less costly than a gasoline truck up to its maximum range of 72 km (Mom, p. 223). The study optimism criticized by Mom involved an assumption of a cost of electricity only available to large fleets served by large central stations supportive of electric drive. The key point here is that the electric vehicle did not supplant the current technology – horse wagons – when delivery wagons were used in short daily distances.

Mom, discussing the 1915 time period, said that "only *after* the electric vehicle had broken the most ardent resistance of the horse economy could the gasoline rival invade the city" (p. 293) and "the gasoline car even stole the entire city car concept" (p. 298). The position here is that this is an overstatement at the least, and perhaps simply incorrect. The personal electric car did not work in dense urban environments with multi-family rental housing units, where short distances to needed services made the electric vehicle far more expensive than walking, the horse taxi, or other public transportation. The American Railway Express Company – a company that very carefully evaluated the most viable applications of electricity to its fleet and adopted electricity with enthusiasm – retained more horse wagons than gasoline and electric trucks combined in 1924. Thus, it seems dubious to assert that the horse economy had been broken in 1915, much less by 1924, when the personal electric passenger car – which apparently never succeeded in large dense center cities – was no longer available. It would appear that the horse economy was probably fully broken later by the gasoline vehicle, well after the horse economy had held its own against the electric vehicle.

This interpretation of the past is identical to the conclusion that Santini et al (2011) reached with regard to the probable status of the hypothetical electric passenger car of 2020. The less expensive existing technology - gasoline in the 2020 case - is financially superior in the event of a low daily utilization rate. Accordingly, the financially viable four-five passenger electric car of 2020 was found to be a heavily utilized suburban vehicle, not a "city car", contrary to recent graphics generated by several automakers (Berretta, 2009; Satyapal and Aceves, 2009; Suckow, 2009; Yokoyama, 2009). The fact that the personal use electric passenger car failed earlier in New York City than elsewhere in the U.S. was not a coincidence. That Mom concluded that the personal use electric vehicle was a second car for the urban affluent also was not a coincidence.

In Europe, electrification of taxi fleets was often promoted by municipalities via regulation, so financial viability was not the only criterion. The regulations were adopted in order to prevent the noisy and smelly gasoline vehicle from capturing the downtown urban market. Although Mom did not associate the city regulations with preferences of affluent customers, it seems likely that relatively affluent business leaders had a strong influence on these decisions. Taxis were more likely the for-fee transportation service chosen by the urban affluent, while the typical urban resident most likely used trolleys and wagons (Omnibuses). Generally speaking, the downtowns of the largest metro areas within a nation are habited on a business day by some of the nation's most affluent citizens. New York City certainly falls into this category within the United States. Thus, to the extent that the electric truck and electric taxi were chosen instead of horse taxis and wagons, and instead of gasoline taxis and trucks, there was most likely a relationship to the preferences of the affluent for better hygiene (vs. horse taxis and wagons) and quieter operation (vs. gasoline taxis and trucks). Reduction of Odor was probably a goal in both cases.

Mother referred to the commercial truck competition as the "decisive battle," highlighting that it occurred within the city. The advantages for certain niches included: "facilitated speed regulation (sweeping and sprinkling trucks), seamless stop-start functionality (door-to-door delivery, garbage trucks), and the lack of odor and noise (ambulances, transportation of food supplies)" (Mom, p. 285). Taxis may be included in the list of Odors and sounds. This pivotal battle was predominantly (though not entirely) lost by the 1930s. The gasoline vehicle advanced significantly during the 1925-1935 period (Naul, 1978; Naul, 1980), surpassing both the horse and the electric car, which will be examined subsequently. The hypothesis posits that the favorable environmental attributes of the electric vehicle contributed to its restricted success among the educated elite in developed industrial nations from 1895 to 1935; however, its cost and other deficiencies hindered it from becoming a mainstream vehicle for the broader populace. Mother observed that numerous proponents of electric vehicles believed that a segment of the issue was behavioral, suggesting that consumers who refrained from purchasing electric vehicles were imprudent, uninformed, or potentially uncultured. An alternate explanation posits that the market functioned effectively, supported by robust justifications rooted in fundamental financial and systems engineering principles, as well as rational consumer preferences, which elucidated the varying degrees of success and failure observed in electric driving.

3. MAJOR CAUSES OF SUCCESS OR FAILURE

This analysis of the initial waves of modest success for electric vehicles suggests that historical challenges associated with them are intrinsic, indicating a substantial risk that history may repeat itself, thereby preventing pure electric vehicles from becoming formidable competitors to gasoline or diesel passenger vehicles and trucks. Nonetheless, it is crucial to acknowledge that there are notable variances in the current situation, so the conclusion may differ. For interpretation of this discussion, it is useful to summarize the three vehicle failure vs. success factors identified broadly by A. Loeb in 1995, and in more detail by Mom in 2004 – (1) power, (2) energy storage, and (3) adequacy of infrastructure. In the case of infrastructure, multiple types of infrastructure-related constraints were discussed by Mom –

roads, electricity, water (with reference to the failure of steam cars), maintenance, refuelling methods (exchange charging systems for business vehicles), safety of interacting modes operating at different speeds, flexibility of destination options.

By the 1920s, the electric passenger car was relegated to being marketed as a "lady's car" for the affluent. "It was called a lady's car' it was said it wouldn't run up hills; it was said it couldn't go fast enough ... I would ride in an electric car if it were not for the fact that all my neighbors coming to the city pass me with their gasoline machines." (Mom, pp. 280-281).

Within the section on limited energy content, Mom paraphrased the 1928 statements of an executive of Accumalatorenfabrik AG, a German battery manufacturing company that had existed since 1887 —" the electric car fell into disuse for private purposes, because it presented problems when one wanted to use it as a touring car." (Mom, p. 288). For a 1914 400 km publicity run between Boston and New York, the "aerodynamic runabout" Bailey Roadster spent almost half of the 23 hours in five "boosts", implying recharging at about 80 km intervals. In the same paragraph it is stated that the manufacturer claimed that the Bailey had a range between 130 and 190 km (Mom, p. 256). This highlights the problem of electrics for intercity travel — such travel results in highest electricity consumption per km and lowest range, and a need to recharge frequently at a time when vehicle occupants place a high premium on average speed to destination (Santini, 2010).

Gasoline vs. Electric Vehicle Supporting Infrastructure. Aside from Mom's comprehensive explanations, historians predominantly see range and cost as the principal factors for the failure of EVs and the success of CVs. A.P. Loeb (1995) highlighted the lack of fuelling infrastructure for electric vehicles, in contrast to the established support for gasoline-powered vehicles, as an overlooked factor contributing to the swift proliferation of gasoline-fueled vehicles in the U.S. by 1904. In 1900, the United States had 4,192 registered automobiles (Sulzberger, 2004). In 1905, the population was 78,000 (Melaina, 2007). The overwhelming majority were powered by gasoline. Loeb (1995) asserted that the matter was resolved by 1904-1905. The author does not address the limitations of electrical availability until the reader is well into the book. He observes that in 1917, "7 million of the 22 million houses in the United States were connected to an electricity grid" (Mom, p. 233). Loeb observed that the swift growth from 1901 to 1904 was primarily attributable to the sales of the two-passenger, single-cylinder gasoline-powered Oldsmobile. This significant emergence of the gasoline automobile in the U.S., prior to the notably successful Model T (1908) and the electric starter (1912), is not referenced by Mom.As noted earlier, the year 1900 concentration of steamers within Boston, New York and Chicago was even greater than for electrics. Although also capable of using widely available petroleum products, steam cars proved to be limited in range and overall average speed by the availability of water. Winter temperatures below freezing were clearly problematic, as was high mineral content in the Midwest (Mom, p. 291). Sulzberger (2004) reports that the range of an early steam car, before requiring replenishment of water, was 25-30 miles, no more than an electric vehicle of the time. The development of condensers to allow reuse of water and a range of 150 miles were implemented too late – in the 1920s – and added cost. Loeb credits superb road infrastructure in France for early emergence of automobiles there. Mom noted that the improving roadway infrastructure in the U.S. appeared to be an enabling technology for expanding truck services (heavier vehicles than passenger cars) in the 1920s, while the previously existing roadway infrastructure had been adequate for well adapted light passenger cars. "A direct relation could be demonstrated between the number of trucks and the length of the paved roads in cities with more than 30,000 inhabitants." (Mom, p. 238). Although the improved U.S. roads were clearly not necessary for success of light gasoline passenger cars (the Model T in particular), they were probably sufficient (along with other developments) to help cause the demise of the electric passenger car. Majahan and Peterson (1985) examine the diffusion of the state gasoline tax in the United States, showing that it swept the nation in the 1920s. It seems doubtful that business truck interests alone could account for this sweeping support of improved roads. Thus, private vehicle users must have seen a benefit. I presume that the benefit was higher speed and greater reliability (reduced tire failures). The practical realization of the benefits of dramatic increase of power and top speed of gasoline vehicles from the mid-1920s to mid-1930s (Naul, 1978, 1980) – unmatched by electrics – was undoubtedly enabled as a practical matter by the improved roads that were built in the 1920s.

Over a decade after Loeb's examination of the role of gasoline infrastructure, Melaina (2007) provided more extensive details of early gasoline refueling infrastructure. He asserts that "a key issue during early phases of infrastructure development is the requirement to provide a fuel inexpensively and in small volumes from many locations dispersed across large geographic regions." Melaina shows clearly that gasoline was a relatively simple add-on to an extensive delivery infrastructure for kerosene. "100 refineries and vast networks of bulk storage facilities and tank wagons" existed. In 1906 Standard Oil operated 3573 bulk stations, receiving barrels and tank wagon loads, distributing refined products (mostly kerosene) locally. Gasoline, used as a solvent, was widely available to both urban and rural populations before the automobile. Gasoline was in excess supply, effectively a waste product, often dumped into nearby rivers.

That the electric vehicle market could not have grown as rapidly as gasoline vehicles in a largely rural nation should not be surprising. In 1907 less than 10% of households had electricity (U.S. Department of Commerce, 1975). Another problem of the time was that the competition between AC and DC electricity had not yet been resolved, making national standardization of charging infrastructure very unlikely. Perhaps the fear of electrocution, used by Edison to promote continued use of DC electricity instead of the ultimate switch to AC played a role in the reluctance of individuals to assume the risk of field repairs of a malfunctioning electric vehicle, while familiarity with powered steam farm equipment made the gasoline vehicle transition seem more manageable. The war of the currents was underway during the 1890s and was not settled until after the turn of the century (Wikipedia, 2011).

Melaina's infrastructure requirements for successful introduction are encouraging for implementation of electric drive over 100 years from the first attempt. Nearly 100% of households in Europe and the U.S. now have electricity. The distribution system cannot deliver energy to electric vehicles at anywhere near the rate of gasoline, but small amounts can be delivered over several hours, at the dwelling unit, in the same way that cans of gasoline were originally stored at the house to fuel early gasoline cars. Because of the advent of air conditioning, afternoon summertime cooling requirements in the U.S. have led to construction of many very efficient combined cycle natural gas power plants which sit idle overnight and in off seasons.

Technological developments in drilling technology have only recently led to significant increases in estimates of the proven reserves of U.S. natural gas, and great optimism about its potential elsewhere. Thus, as gasoline was in excess in 1900, U.S natural gas producers, along with utilities that own natural gas powerplants, are looking for new customers. Further, a movement toward the "smart" grid, with time of day rates encouraging use of electricity overnight via reduced price, can encourage electric vehicle use, although the required metering is not inexpensive. In any case, it is clear that for plug-in electric drive today, initial infrastructure is not the limiting factor it was in 1900.

Melaina observed that the emergence of the refueling station followed the emergence of the gasoline car by a couple of decades. He noted that "non-station refueling methods allowed vehicles to be mass-produced without sales being inhibited by consumer concerns over limited refueling availability" (Melaina, 2007, p. 4922). Cans, barrels, and home refueling pumps emerged concurrently with gasoline vehicles. Next came refueling at repair shops and curbside dispensers (both will be used to support EVs). The ability to "fast fuel" many vehicles at a location dedicated to refueling followed in 1915-24, long after the vehicles.

The working assumption at this time, however, is that electric vehicles must have a network of public charging stations in place before electric vehicles are to be sold, if electric vehicles are to be successfully introduced into the market. The Electrification Coalition (2009) contends that the minimum number of public chargers per electric vehicle to be 1.5 in 2010, 1.0 in 2020, and 0.5 in 2030. In Germany in 1914, there were 862 passenger EVs, 554 electric trucks, 270 commercial-and-mail three-wheeler EVs, and 3 private 3-wheeler EVs. There were 39 charging stations, 13 for taxis only, 13 for mail vans, and 11 for private car owners (Mom, p. 252). In other words, there were 11 passenger car charging stations for 862 passenger cars, which is 0.013 stations per car.

Powertrain vs. Vehicle Body Technology. The examination of the evolution of early gasoline automobiles demonstrates swift technological advancement, serving as an example for current and expected improvements in battery chemistry. Although the requirement for specific energy in batteries is widely recognized, it was not an issue for gasoline powertrains in the late 1800s; however, particular power posed a significant challenge. Presnell (1992) distinctly demonstrates the significance of swift enhancements in specific power for gasoline engines in Europe. The power of the original Benz automobile from 1885 increased from approximately 0.2 kW to about 2 kW in its mass-produced variant, including a single-cylinder engine and achieving a maximum speed of 19 km/h. The weight of the light vehicle was initially essential, as the engine was affixed in a "tricycle" configuration. By 1899, Daimler, the pioneer in manufacturing a dedicated automotive engine, had created an 18 kW four-cylinder racing vehicle capable of exceeding 80 km/h. The Daimler vehicle engine weighed one-sixth of a typical conventional stationary Otto four-stroke engine from 1886.De Dion-Bouton in France also started with a modest engine – ½ hp mounted in a tricycle, soon switching to four wheels with a single cylinder engine design rated at 2.5 kW by 1900, and 6 kW in 1905 (Presnell pp. 12, 14) . This engine type provided the foundation for the early French automotive industry, being used in "a hundred different makes of vehicle in the 1898-1908 period and launched many a respected marque", among them Renault (Pressnell, p. 12).

The single cylinder engine in a small vehicle was also the starting point for sales of thousands of vehicles from a single manufacturer in the U.S. The curved dash, single bench seat Oldsmobile was "a clever exercise in minimal motoring" with "long springs giving a comfortable ride on poor roads", with a "chug along" engine limited to 500 rpm (Presnell). The 1901 version of the Oldsmobile is reported by General Motors to have a 4 kW engine, increasing to 5 kW in the 1904 model (Generations of GM History: Heritage Center, 2011).

Loeb emphasized the importance of consumer reaction to increasing power and speed in the early phase of the development of the automobile in the U.S., followed by ascension of the automotive virtue of utility realized in the Model T Ford.

The desire for cost-effective mobility had been demonstrated by the Oldsmobile success. On both sides of the Atlantic, the power density of Otto cycle engines jumped in the 1890s, then leveled off in a mass produced engine design that was the foundation for production of tens of thousands of vehicles. Where France's start was via mass production of a De Dion-Bouton single cylinder engine used by many vehicle manufacturers, Ford took this a step further and mass produced the whole vehicle, providing affordable and reliable automotive transportation to the middle class (Loeb, p. 75). The Ford Model T engine produced 15 kW. With two bench seats, this mass market car seated four or more people. The Model T weighed 544 kg, the single bench seat Oldsmobile 318. The Model T engine had four cylinders and 2.9 liters of displacement, while the prior Oldsmobile had a single cylinder engine with 1.6 liters of displacement (Vivian, 1994). The Model T engine power rating was unchanged throughout its nearly two decade lifetime, according to Naul (1978).

Sulzberger stated that lead acid batteries of the time had to have 76 kg/kW. The 1897 Pope Columbia Electric Phaeton Mark III weighed 816.5 kg, 386 of which was battery. If the battery had achieved the best performance cited by Sulzberger, this would give 6.2 W per kilogram of vehicle weight. The Ford Model T had more than four times more kW/kg. Further, since the battery power number is likely a peak power rating, it is likely that the relationship of continuous power per kg was even more favorable toward the Model T. Thomas Edison spent millions of dollars and years to develop better batteries, but, despite considerable improvement, increased cost was a problem too great to overcome for personal automotive use.

The significant and rapid rise of specific power of Otto cycle engines in the 1890s was a key enabler of the initial success of the gasoline engine. More than two decades later, from 1926 to 1935 the power of a base model Ford went from 20 to 95. There were similar, though less pronounced increases across all models (Naul, 1978, 1980). This second jump in engine power was undeniably a key cause of the final demise of the personal electric vehicle and taxi, and later the commercial truck.

Today, the next wave of EVs is benefiting from a shift in battery chemistry to lithium-ion (li- ion), a chemistry that not only provides higher specific energy; it also can provide considerably higher specific power (Kalheimer et al, 2007) — nearly an order of magnitude more gravimetric specific power than the present NiMH battery chemistry in some circumstances. Dramatic improvement in performance capabilities of engines gave the gasoline vehicle its early foothold in small vehicles capable of carrying one or two passengers. A possible historical analogy is that an enthusiastic base of consumers finding the improved performance made possible by li-ion batteries has provided one anchor for the modern EV technology. The two-seat

Tesla Roadster and BMW Mini EVs have been received relatively enthusiastically by those who drive them. Based in large part on reactions by drivers in its Mini EV tests, BMW is proceeding with multiple EV designs in four passenger vehicles, one of which will use lightweight composites and aluminum - a significant redesign of the vehicle body comparable to what Tesla has done with the Roadster. Tesla is also putting into production a five-plus-two passenger sedan, the Model S, with a stamped aluminum body rather than composites. Neither manufacturer is seeking a middle class market, but both do intend to produce "family sized" vehicles for performance oriented higher income consumers.

The early mass market success of the gasoline engine occurred when the technology reached an acceptable plateau of capability that could be made available to middle income consumers via the cost reducing benefits of mass production. The De Dion-Bouton and Model T engines were produced in very large volumes, enabling cost reductions that in turn enabled vehicle pricing resulting in high volume sales. Nevertheless, these engines initially had to appeal to small markets, before mass production was achieved.

For its hybrid vehicle design, the long-term *possibility* of profits at high volume (realized after several years) with reasonable cost was seen by Toyota in the early 1990s. Electric drive has today obtained a foothold in the heart of the automotive market because of this long- term vision. The NiMH battery may not have been adequate for EV success, but it did allow the technology innovation of packaging of electric and conventional mechanical drive together in hybrids that has created the current general confidence in electric drive. In retrospect, no manufacturer in the 1990s was willing to gamble that the Nickel Metal Hydride battery chemistry would lead to levels of EV cost and performance that could result in mass market success.

Lesson: mass market success of an alternative powertrain requires a technological leap in capability, initially supporting low volume sales to innovators and early adopters (most of them reasonably affluent), leading to mass production and cost reductions for its most critical components, making the technology affordable.

Judgments of participants interviewed for the IEA HEV&EV Implementing Agreement's *Lessons Learned in Market Deployments of Hybrid and Electric Vehicles* study was that production in many tens of thousands – perhaps triple digits – is necessary for batteries to be cheap enough to allow EVs to be mass marketed. Predictions of cost reductions as a function of production volume by Kromer and Heywood (2007) and by Santini, Gallagher, and Nelson (2010) for lithium ion based battery chemistries are quantitatively consistent with these opinions. Nissan is taking the gamble that high volume production of lithium ion based battery packs will reduce costs adequately. In contrast to the choices of Tesla and BMW regarding lightweight body materials, Nissan is relying on a conventional steel body to keep costs low. The gamble is "Ford" like in the sense that Nissan is the only manufacturer committed to production and sale of one model of electric vehicle within the leading sales size class in the world. Nissan follows Toyota's example, adopting significant corporate ownership of production of battery packs for the vehicle. Both the powertrain/storage system and vehicle body are to be produced in much higher volume than any other manufacturer has presently committed to.

One question for the gasoline vehicle of the early 1900s is what might have happened had the specific power of engines not increased to the levels that made the four-plus-passenger Model T and numerous French body alternatives on standardized De Dion-Bouton engines possible? Would steam or electric powertrains have succeeded, or would the horse have held its large share of the market? It can be argued that attributes such as specific power and specific energy of the powertrain and fuel storage device(s) dictate the approach that must be taken with the rest of the vehicle body. Looking back *at the start* of the gasoline vehicle from 1885-1900, the EV-related efforts of Swiss manufacturers in the 1990s look similar, but success did not follow. Swiss EVs were small, lightweight, low speed, low acceleration vehicles to enable adequate range and performance. An emphasis was placed on light weight. Many experiments, with multiple battery chemistries, were tried. No successful standard model emerged. The Mendrisio experiment found that the majority of consumers chose a single vehicle type (Peugeot 106) from the established vehicle manufacturer that provided the best service. Among vehicle manufacturers interviewed in the IEA HEV & EV Implementing Agreement's "Lessons Learned" study, Peugeot had made the biggest commitment to volume production in the 1990s, building a factory capable of producing 20,000 vehicles. Yet this commitment, seen to lead to the most success among Mendrisio participants, was not adequate, given consumer response to the vehicle capabilities and costs. Only thousands of Peugeot electrics were sold in the best year.

Multiple participants in the "Lessons Learned" study have said that changes in consumer behavior are necessary if EVs are to succeed. Mom shows that such thinking by electric vehicle proponents was also common in the early 1900s – the EV advocates attempting to convince potential customers that lower performance was a desirable attribute. Toyota participants expressed the opinion that consumer preferences would have to change, and a Swiss consultant argued that a change to a "future oriented attitude" was necessary. It is clear that over a century's time households (and entire economies) did adapt their behavior to the features of the gasoline vehicle. However, the question is one of cause. Did the vehicle entice a change in behavior, or did consumer behavior shifts enable the vehicle to succeed? The former direction of cause seems more plausible. Mom saw the gasoline automobile "culture" as one that was imposed on all other modes of travel, pushing them aside in favor of the needs of the gasoline automobile. "The building of an automobile only highway network was *forced on* the users ... the highly *functional* flexibility ... led to the collapse of one of the densest regional tramway systems in the world." (Mom, p. 296).

Lessons: Batteries — even lithium ion — are inadequate to allow consumers to purchase EVs without adapting their behavior. Since large changes in behavior are unlikely in a short period of time, EV designs must provide a large fraction of the mobility provided by the competing means of travel. If an EV design competes with a small volume gasoline vehicle type (such as a two-seat passenger car), it will not gain a large share of the national market even if it is successful against its competition.

A fundamental question is whether the powertrain/storage system dictates the vehicle body, or does the vehicle body dictate the powertrain/storage system? For the Nickel Metal Hydride (NiMH) chemisty, the Toyota Prius, quite conventional in many respects, adapted the vehicle body a bit, and the powertrain/storage system a lot, and captured half of the market for hybrid powertrains in the U.S. The Prius designers did choose to avoid too much weight and cost in the sense that the battery was made as small as possible and no plug-in feature was attempted. In this case, the relatively advanced battery design was adapted to rigid short term consumer expectations and behavior, assuming that only slight changes in the vehicle body would be accepted. Mom saw the success of the gasoline powertrain as one of successful adaptation first, using existing coachwork, roadways and fueling infrastructure. This led to the lowest cost among the competing powertrains during the take- off phase, only later leading to establishment of redesigned coachwork, roads and fueling infrastructure.

To allow batteries available to succeed in an electric vehicle in the 1990s, GM chose to develop an entirely new 2 seat body design that would hopefully provide an enticing combination of performance attributes — even with lead acid batteries. Here the attempt was to spend money on the body in order to make the initially inexpensive, proven battery chemistry workable. This body and battery package was very popular with very few consumers. The limitations of the batteries required plastic and aluminum rather than the steel used in the Prius body, and a body shape achieving a coefficient of drag of 0.19, well below the 0.26 value for a Prius. With far less financial resources, multiple Swiss innovators also attempted to develop an entirely new vehicle body configuration, emphasizing very light weight to enable lead acid, nickel cadmium, and sodium sulfur batteries to provide adequate performance to meet relatively inflexible consumer expectations. Neither GM nor the Swiss were able to achieve sales rates that promised reaching production volumes necessary to succeed. They had in common an initial attempt to succeed in a very small market niche – a smaller than average utilitarian vehicle with conventional (or worse) range and top speed in comparison competing gasoline vehicles. Recognizable power fade as the battery depleted was an issue.

Attempting to take advantage of the greater specific energy and power of li-ion more than a decade later, the $100,000+ Tesla roadster EV designers sought a different two-seat vehicle niche market – the high cost, high performance segment. Tesla engineers recognized there was still a need for a very lightweight body. Aluminum and carbon fiber are used for light weight, with a few parts common to the Lotus Elise, a lightweight sports car that uses a similar aluminum frame, but does not use carbon fiber body panels (Siry, 2008). Performance and range well beyond that of an EV1 were demonstrated in a two seat vehicle. The much less expensive (than the Roadster) base version of the coming Tesla Model S sedan will use a less expensive aluminum body. Though slower and with less range than the Roadster, it will accelerate faster and achieve greater range than an EV1, and be capable of seating 5 adults and two children, but its price is to be more than $20,000 more than the EV1's nominal $33,000 price (EV1's were leased, not sold).

A study of the design of the EV1 and the Chevrolet Volt demonstrates a lot of commonality in component placement and configuration. Acceleration capability is no longer the primary selling point, and the vehicle has four seats, more suitable for the middle class market. Nevertheless, a gasoline engine is included because of concerns over charging infrastructure. While prices of Prius HEVs are within the reach of the U.S. middle class, the Volt, produced in tens of thousands, at a base list price double that of the Prius, is not. The Nissan Leaf, to be produced in hundreds of thousands uses a conventional steel body. It is priced below the level of the Volt, but still expensive relative to a conventional gasoline vehicle of the same size, and compared to the Prius. U.S. subsidies of $7500 per vehicle will help, but battery costs must come down (or oil and gasoline prices rise) for high volume cost competitiveness in the U.S. market.

In 2020, the projected expenses for an electric vehicle akin to the Leaf, manufactured at an annual volume of 100,000 units, include a "generic" advanced lithium-ion battery pack cost of $9,340, which constitutes 27% of the estimated total cost of $34,845 for the vehicle and its associated infrastructure (based on simulations supporting Santini et al., 2011 [vehicle] and Santini, Gallagher, and Nelson, 2010 [li-ion battery pack]). In his 1924 book, La Schum cited the cost of an electric truck at $3,030 without the battery, with an average battery price of $970 (Mom, p. 245). The proportion of battery cost at that time was 24%, which is lower than the aforementioned projection for 2020. Consequently, the pervasive challenge of elevated capital expenditure for batteries persists nearly a century later. The restricted range and maximum velocity compared to comparable gasoline vehicles continue to pose a potential issue, although the top speed currently seems significantly closer to that of gasoline vehicles than it was in the early 1900s.

4. WAVES OF HISTORY II: MOTIVATIONS FOR RE-INTRODUCTION, 1965-2011

In a recent presentation, Mitsubishi dates three "waves" of modern interest in EVs (Wing, 2010). The first wave was in the 1970s, in response to the U.S. "Muskie Act of 1970", which dealt with tailpipe emission reductions. The second started in 1990 as a response to emerging concerns over global warming, and to California's Zero Emissions Vehicle (ZEV) regulation. The third was dated as starting in 2002 as a response to oil dependency. This section discusses these motivations for re-introduction of electric vehicles, along with the evolution of the technologies attempted.

In the 1890s the electric passenger car did not prove to be a viable competitor for personal transportation – even in urban areas. However, it is not true that electrified transportation failed. Quite the contrary - electrified subways and street railways were built in significant numbers in major urban areas in the 1890s and early 1900s (Middleton, 1974), sharply reducing the urban waste problem from both the horse and the iron horse. Horse manure and decaying horse carcasses were both significantly reduced. Smoke from steam powered street railways was not eliminated, but it was removed to more distant central generation stations. In later decades, gasoline buses and cars first helped in finally eliminating the horse, and eventually eliminated electrified urban transportation, in many cases contributing positively to reducing particulate emissions from power plants. Particulate emissions reduction efforts came first because this was an obvious pollutant, dirtying clothes and building facades. Only after the automobile became dominant and scientific discoveries about the deleterious effects of tetraethyl lead and ozone accumulated, did this less obvious pollution from the tailpipe of the passenger car become evident.

A new reason for considering EVs is their absence of dependence on oil, now an import concern in both the U.S. and Europe. Fuel imports were not a concern in the U.S. in the early 1900s. In fact, oil discoveries in the U.S. clearly played a positive role in the adoption of the gasoline vehicle at that time. Soon after WWII, automakers in two nations without domestic oil resources produced EVs for a short while. Nissan mentions that its founding company offered an EV for a few years after WWII (Nissan, 2011). PSA also mentioned to this Annex that it had developed an EV in 1945. Otherwise, it appears that no post-WWII EVs were *commercialized* by major automakers until the 1990s, after several noteworthy developments in the late 1980s. However, EV research began earlier. Due to air pollution concerns that first became evident in California, EVs began to be investigated by automakers again in the 1960s.

In fact, the very success of the gasoline-fueled internal combustion ICE in the U.S., in one of the leading oil producing states at the time, contributed to the emergence of the second most populous city, Los Angeles, on the West Coast, facing Asia. That location later played strongly into interactions with Japan. Where New York — a state without oil resources — had been developed at high density with considerable use of electricity for transport via a sophisticated subway and electrified commuter rail network, the early electric commuter system in Los Angeles was abandoned for the bus. Los Angeles thrived and grew rapidly, but the emissions of gasoline vehicles, trapped within a basin surrounded by mountains, led to unacceptable air pollution, in the form of ozone.

In the 1960s, California began studying the effect of gasoline vehicle related emissions of hydrocarbons and nitrogen oxides on ozone, finding that both were important contributors. Regulatory institutions were put into place and regulations were adopted, first to reduce hydrocarbon emissions from tanks that stored gasoline and other hydrocarbons, then from gasoline vehicles themselves. The emerging research and success in developing emissions reducing technology in California led to recognition nationwide that gasoline vehicle emissions would have to be reduced sharply if the nation was to continue to rely on the automobile as the foundation for its transportation. In 1970, the "Muskie Act", the Clean Air Act Amendments of 1970, was passed. Amending an original 1963 law, this law has recently been cited by both Toyota and Mitsubishi as a watershed event affecting their work on future powertrain technology for the automobile. Both Takehisa Yaegashi (revered within Toyota as 'the father of the hybrid') and Masatami Takimoto (Fairley, 2009) said that this Act was instrumental in causing Toyota's engineering department to begin reevaluating the powertrain for automobiles. Electric vehicles and hybrids were among the powertrains evaluated at the time. Takimoto dates Toyota's evaluation of "all kinds of hybrid systems" - series, parallel, mild, full - from 1969. Since 1969 precedes the passage of the Muskie act, we presume that Toyota was tracking the events in California and Los Angeles and regarded these as potentially important for its long-term market development.

Mitsubishi also cited the Muskie Act of 1970, and mentioned their Delica EV (a passenger van) and Minica EV (a two door sedan) at that time (Wing, 2010). General Motors' recent placement of its "first" EV, in a historical timeline, was the 1966 Electrovan (Mathe, 2010). This date also precedes the Muskie act, suggesting that emerging air pollution concerns in CA were having an effect on GM as well. The date also opens the possibility that Toyota and Mitsubishi were partially responding to GM and California initiatives and the Muskie Act only reinforced the desire to investigate alternative methods for tailpipe emissions reduction. A recent timeline on the history of the electric car by America's Public Broadcasting system says that in 1966 Congress introduces the earliest bills recommending use of electric vehicles as a means of reducing air pollution. A Gallup poll indicates that 33 million Americans are interested in electric vehicles.

The 1966 co-dating of GM's Electrovan and introduction of bills in Congress (introduction does not mean that the bill became law) and the Gallup poll suggests that tailpipe emissions concerns were already a significant U.S. national issue before 1970.

GM's timeline also shows one of the most successful low volume EVs ever, the 1972 Lunar Rover (Matthe, 2010). PSA reported to the IEA Annex that it had prototype electrified versions of the 17 and 104 models in 1972. These vehicles used lead acid batteries. BMW (Schamer, Lamp and Hockinger, 2010) dates its first EV at 1972, using lead acid and attaining a range of 30 miles. Their next BMW citation was 1987, based on the sodium sulfur battery chemistry, which had taken 20 years of development before being put into an automobile. These actions clearly predate the 1973-74 world oil price shock, subsequent 1974-75 collapse in automotive sales, and recession.

In May, preceding the October 1973 attacks on Israel by Egypt and Syria, and the subsequent Arab Oil Embargo that precipitated the oil price shock, Lee Iaccoca of Ford had solicited a long-term assessment of automotive powerplant options. The study solicitation award was made in Dec. of 1973. This study – "Should we have a new engine?" – was completed in August of 1975 by the solicitation winner, the Jet Propulsion Laboratory (Stephenson, 1975). The study concluded that it was clear that Brayton and Stirling engines should receive research funding as improvements were made to the internal combustion engine until these technologies could succeed. Electric vehicles and hybrids were regarded as undesirable (Lindsley, 2006). As the study progressed, Electric Vehicle Symposium Number 3 was held in 1974 in Washington DC. The Electric Auto Association (2005) considers the introduction of the two seat Sebring-Vanguard CityCar in Feb. (five months after the initiation of the Arab Oil Embargo) at the Symposium as a noteworthy event. The CitiCar had a top speed of 64 kph.

Despite the Jet Propulsion Laboratory's recommendation that research not be pursued on electric vehicles, the U.S. Public Braodcast System (PBS) (2009) indicated that, a year later, in 1976 Congress passes the Electric and Hybrid Vehicle Research, Development, and Demonstration Act. The law is intended to spur the development of new technologies including improved batteries, motors and other hybrid electric components.

Several electric vehicles - generally small and low volume - were produced worldwide in the 1970s (Anderson and Anderson, 2010), though none by large OEMs. These appear to have been supported and perhaps inspired by the high oil prices of the period. None are mentioned after 1983 (About.com, 2011, Public Broadcast System, 2009, Anderson and Anderson, 2010, Electric Auto Association, 2005). Variants of the CitiCar were produced until 1982, with total production about 4000 vehicles. Oil prices peaked in 1981, declined steadily until 1985, then dropped precipitously.

In 1985 the Swiss initiated the "Tour de Sol", a Swiss solar car race that was held every year until 1993, promoting development of solar technology. This was the first solar car race. Mercedes Benz sponsored the winning entry (Muntwyler, 2011). In 1987 the first World Solar Challenge race in Australia was run, over a distance of 1877 miles. General Motors sponsored the winning car in this race, the Sunraycer. Also in 1987, after a period of 15 years, BMW developed its second EV conversion vehicle, a 325 model with a sodium sulfur battery (Schamer, Lamp and Hockinger, 2010).

In the United States, an unusually hot summer in 1988 was accompanied by a jump in average national ozone levels, after several consecutive years of decline. In 1988 Roger Smith of General Motors "agrees to fund research efforts to build a practical consumer electric car" (Public Broadcasting System, 2010). From 1988-1990, oil and gasoline prices once again rise significantly, though not as severely as in 1973-74 or 1978-81. Nevertheless, a U.S. recession follows. The Aerovironment company prototype arising from the 1988 agreement, the two seat lightweight aerodynamic electric sports car, the "Impact" is introduced at the Los Angeles auto show in 1990 and the California Air Resources Board passes its Zero Emissions Mandate, requiring 2% of the state's sales of vehicles to consist of vehicles with zero tailpipe emissions in 1998, rising to 10% by 2003.

Another influence was also emerging. In 1988 the United Nations established the Intergovernmental Panel on Climate Change (IPCC), and in 1990 the first Assessment Report of the Panel was released (IPCC.org). This report served as the basis for negotiating the 1992 United Nations Framework Convention on Climate Change (UNFCCC) in Brazil. In Europe and Japan, the significant emerging concern over global warming being codified by the United Nations also promised significant change to come in the those markets.

BMW developed an "E1" EV in 1991 using the sodium sulfur battery, and another EV based on the 325 model in 1992, now using the high temperature sodium nickel chloride (Zebra) battery. By 1993, Germany had set up field tests of 60 electric vehicles on Reugen Island. In 1992, Ford placed into service a fleet of 80 small vans – the ECOSTAR – in Europe and the U.S., using the Zebra battery.

Based on interviews conducted by the IEA HEV&EV Implementing Agreement's "Lessons Learned" study, Japan and its automakers – who held a significant share of the California market – reacted strongly to the announcement of the GM Impact and the ZEV mandate. Throughout the 1990s, worldwide concern over GHGs began to emerge, along with agreements to develop greenhouse gas reduction strategies. In mid-decade, domestic pressure on automakers in Japan to meet previously agreed national fuel efficiency goals and to show significant progress before the 1997 Kyoto Japan meeting on climate change led to acceleration of Toyota efforts to implement electric drive technology to improve fuel efficiency.

Early in his administration, Al Gore, U.S. Vice President, began promoting research on very high efficiency vehicles. Congress funded this multi-agency, multi-manufacturer "Partnership for a New Generation of Vehicles (PNGV)" research in 1993, but would not support U.S. participation in international agreements to reduce GHGs. Hybrid powertrains were among the technologies chosen to enable very significant improvements in fuel efficiency, but significant research on electric vehicles was not a part of the program due to probable functional limitations including range, speed of "re-fueling", package space and infrastructure concerns. Battery research was supported, but not vehicle research.

Toyota responded to pressures from its government, California's government, and the U.S. research program supporting three of its competitors with an aggressive effort to develop a much more fuel efficient mass market vehicle that would allow Japanese consumers to move up to a larger, but considerably more fuel efficient vehicle than their leading world seller, the Toyota Corolla. This vehicle was named the Prius, which in Latin means "to go before". The first generation of the Prius was only sold in Japan. After a degree of reliability was assured, the Prius was sold in the U.S. In each generation it became larger, faster, and more fuel efficient. It moved from an initial U.S. size classification of compact car up into the midsize category in 2004.

In the late 1990s electric vehicles were produced and evaluated by Toyota in both Japan and California, but were abandoned – for reasons similar to those given by PNGV for not focusing on electric vehicles.

After the oil price shock of 1988-90, oil prices had been relatively stable for nearly a decade. Concerns over availability of oil had subsided. Test fleets of EVs were placed in service in California in the late 1990s. Volumes produced by each manufacturer were generally less than 1000. Automakers decided to oppose the introduction of EVs in other states and legally opposed any expansion of California's ZEV mandate to other states in the U.S. At the turn of the century, production was halted and cars were reclaimed by some manufacturers. All but Nissan used Nickel Metal Hydride or lead acid batteries. Nissan produced a few EVs using lithium ion batteries, now regarded as very promising. Among the participating manufacturers, Nissan was under the greatest financial stress. It too halted EV production. Gasoline vehicles had become far more efficient and far cleaner. Once again, improvements in the gasoline vehicle held the electric vehicle back.

Even an oil price increase from 1998 to 2000 did not revive interest in EVs by automakers and Congress. The PNGV project had led to production of a fuel cell hybrid by GM. This was also a zero tailpipe emissions vehicle which did not have the range limitations of an EV. California, major world automakers and oil companies agreed to a Fuel Cell Partnership to develop hydrogen fuel cell vehicles. The ZEV regulations were restructured. For several years, attention turned to fuel cell vehicles.

During this period, the lithium ion battery chemistry was completely supplanting NiMH in consumer electronic applications. The price and performance of cells with this chemistry were rapidly improving. Theory said that this chemistry could realize greater energy and power density than NiMH (Kalheimer et al, 2007). This was being proven in practice.

In 2003, four years after California and major automakers had shifted attention to fuel cells, Tesla motors was formed with the intention of producing a high performance two seat electric sports car. In a 2006 presentation to the California Air Resources Board, Tesla compared its coming roadster to high performance sports cars selling for prices of $100,000 to half a million and more (Eberhard, 2005). A key point of the presentation focused on delivery of miles of service per unit of original feedstock by the roadster in comparison to conventional high performance vehicles powered by internal combustion engines. Perhaps the key slide was the one comparing miles of service from wind, solar, hydroelectric and geothermal power via electricity vs. hydrogen. Although it must be remembered that the vehicles compared differed significantly in terms of range, the Tesla comparison dramatically favored the electric pathway.

Another slide highlighted the side effect of greater efficiency – less use of land resources. It may not be obvious, but this point is one that goes beyond computation of oil saving, tailpipe emissions reduction, and GHG reduction. For both an electricity-to-electric-vehicle pathway and an electricity-to-hydrogen fuel cell vehicle pathway, the oil savings, tailpipe emissions reductions, and GHG reductions will be about equal per mile. For tailpipe emissions and GHG reductions, since the pathways have nearly zero emissions, comparisons of pathways with different numbers of miles of operation will look the same if expressed on a percentage basis – about 100% reduction. However, the pathways may differ considerably in another respect, even when compared on the basis of percentage change. The fourth respect is "sustainability" – the more miles of service from a given feedstock, the more sustainable the resource base.

Distilled, what Tesla was arguing is that if the vision is a sustainable future for transportation based on use of renewable fuels, Tesla had identified a market niche where that transportation can be provided at lower cost with greater levels of service than hydrogen (or gasoline or biomass). Looked at another way, the argument was that, for any single renewable fuel examined, more miles of service can be provided by electric drive in modest range electric vehicles than if fuel cell hydrogen vehicles were used for the same purpose. Although it may be true that an electric vehicle cannot be anticipated to be a universal replacement for gasoline, due to its range and refueling time limitations, it does now have a widespread refueling infrastructure available and it can be started in market niches at much lower cost than fuel cell vehicles.

In the following year at the 23rd Electric Vehicle Symposium in Anaheim CA, a paper was presented that showed that this general argument also holds true for fossil fuels competing with oil, most notably for natural gas used to generate electricity in combined cycle power plants (Gaines et al, 2008). It appears that a proper generalization is that once a fossil or biomass fuel feedstock is gasified, it is more efficient to use the gas to generate electricity and provide electric drive than to turn that gas into a liquid for use in internal combustion engines, or to convert it into clean hydrogen for use in a fuel cell vehicle. For wind, solar, hydro and geothermal, the lesson is to never produce a gas (hydrogen), use the electricity directly. The critical caveat remains that this applies to probable niche vehicles with modest amounts of electric range compared to typical gasoline and fuel cell vehicles.

Lesson: for any single fuel/feedstock pathway, with technologies plausible in the near term, more miles of service can be provided via that feedstock by electric drive in modest range electric vehicles than if fuel cell hydrogen vehicles (or liquid fuels from that feedstock in ICEs) were used for the same purpose.

Tesla, selling its high performance roadster at a cost that undercuts many exotic sports cars, has since sold over 1200 of the roadsters. GM produced less EV1s and did not sell any. It only leased them. Though Tesla has yet to sell as many roadsters as 1970s CitiCars sold, it undoubtedly sold far more kWh of battery pack capacity and far more dollars of value of EVs by early 2011, then being the U.S. leader in these terms. Tesla production took advantage of a degree of pre-existing volume production for components. The roadster was produced by re-engineering a Lotus Elise body and frame. As of 2003, the Elise had been in production since 1995. 17000 had been produced (Conceptcars.com., 2003). Lithium ion cells used in the battery pack of the Tesla were standard commercial cells that had been perfected via several years of production for consumer electronic applications.

Price remains an issue. The Lotus Elise sells at a price of about $50,000, while the Tesla Roadster sells at a price of over $100,000. The Tesla Roadster accelerates to 60 mph faster than the Elise [3.7 (Tesla, 2011) vs. 4.4 (Zero to Sixty Times, 2011)], but has a lower top speed (125 mph vs. 150 mph). Its "acceleration feel" to an owner may be superior to the Elise because of the high initial torque available for start-up acceleration.

As Tesla developed their roadster for the performance market niche, oil prices continued a steady rise, reaching levels in 2008 that caused a widespread international collapse in automobile sales. Due to the encouraging long-run environmental and sustainability arguments on behalf of electric drive relative to hydrogen fuel cells, governments began to shift funds and commitments toward electric drive in plug-in hybrids and electric vehicles. Thanks in part to subsidies and to oil prices through 2008, the world market share of hybrids rose steadily through 2009. First the U.S. adopted subsidies, then Japan.

The technical possibility to convert a 2004 generation Prius to a plug in hybrid was demonstrated by the organization CalCars, using lead acid batteries. Multiple companies then produced prototypes making use of lithium ion battery packs. In 2008 the battery manufacturer A123 purchased the company Hymotion, then safety certified and produced a 5 kWh lithium iron phosphate battery pack Prius plug-in conversion for $10,000. U.S. Government testing of a fleet of plug-in Prius vehicles is demonstrating some of their strengths and weaknesses.

For European high performance vehicle manufacturers, electric drive offers the opportunity to meet ever tightening carbon dioxide emissions regulations while still selling vehicles with the historical level of performance customers expect. Several European OEM's that focus on high performance are now developing extended range electric vehicles conceptually similar to the Chevrolet Volt, but with considerably higher power.

To overcome the battery pack cost problems of the Tesla, assuming middle class customers, the Leaf uses a battery pack whose much larger, next generation "prismatic" cells are designed for automotive use. The new battery cell and pack redesign requires very high volume production to allow moderately competitive costs. Battery research is progressing steadily, with promise of favorable lifetime cost reductions for selected customers of plug-in vehicles using coming generations of lithium-ion-based automotive batteries. Few OEMs expect plug-in vehicles to become dominant technologies in the next decade or two. However, many now expect them to succeed in large enough numbers, at low enough costs, that the risks of not producing them are greater than the risks of producing them. Many are choosing to pursue a portfolio of electric drive technologies, including hybrids, plug-in hybrids, and electric vehicles.

The ambition of both established and emerging automakers to create and manufacture automobiles that significantly decrease oil consumption has intensified. In light of rising concerns regarding greenhouse gasses, the aspiration for minimal emissions—approaching zero—has expanded beyond solely the exhaust to encompass the entire fuel delivery process. This shift in mindset has adversely affected the hydrogen fuel cell option, altering the approach to use both renewable and fossil feedstocks for delivering vehicle kilometers of service. Electricity, when effectively utilized, seems to be the most viable immediate choice for improving the sustainability of personal light-duty cars. Regrettably, the expense associated with electric driving will ensure that less sustainable alternatives maintain market dominance for the foreseeable future.

5. WHAT IS DIFFERENT THIS TIME, WHAT IS NOT?

The vehicle sector is currently well-established, featuring substantial manufacturers. Nissan intends to achieve substantial electric vehicle production within a brief timeframe. According to the sole comparison presented, the incremental initial expense of electric vehicles, in percentage terms, seems unlikely to differ significantly from that of the 1920s. Consequently, the necessity to extensively employ the vehicle to recoup the additional acquisition expenses is paramount (Kley, Dallinger and Weitschel, 2010; Santini et al, 2011). The financially lucrative market niches for electric vehicles today are likely analogous to those in the early 1900s. Nonetheless, the magnitude of these markets has significantly increased. The current rivalry is solely with the internal combustion engine utilizing refined petroleum products, rather than with horses. The proportion of the people residing in U.S. suburbs is significantly higher, as is the overall wealth of this demographic. The performance of the Nissan Leaf electric remains to be evaluated by auto magazines and the U.S. Department of Energy, but initial information indicates that it will be competitive or better than gasoline vehicles of the same size with base engines (My Nissan Leaf, 2011; Autoblog, 2011). It is clear that this generation of electric vehicles using lithium ion battery packs (Nissan Leaf, BMW Mini-E, Tesla Roadster) has *significantly* better acceleration performance than comparably sized vehicles using nickel metal hydride battery packs in the 1990s (Idaho National Laboratory, 1996a&b, 1999a&b, 2009, My Nissan Leaf, 2011), and higher top speed. A Nissan auto show presentation indicates that the Leaf has the fastest 0- 48 km/h time of any Nissan vehicle sold (Nissan, 2011). Thus, the response of consumers in everyday urban and suburban driving, on neighborhood, feeder, and arterial roads with stop signs and stop lights, and speed limits of 88 kph and less may be very favorable.

Based on interviews of those who tested the BMW Mini-E, the range of today's electric vehicle using lithium ion batteries is adequate for most needs, but consumers want a charging infrastructure, apparently to be able to use the electric on days when driving distance exceeds the range (Presse Box, 2011). Unless consumers have a strong preference for the EV for its rapid initial acceleration capabilities, financial calculations imply that driving less kilometers per day than the range of the electric vehicle will not be financially desirable in the United States at current and somewhat higher gasoline prices (Santini et al, 2011). Recent evaluations for Europe indicate that fuel taxes (much higher than in the U.S.) will cause EVs and PHEVs to be financially attractive there. However, with "untaxed numbers no PHEV or EV was selected for any battery price." (Kley, Dallinger and Weitschel, 2010). As has been discussed, Europeans drive less kilometers per day on average than in the U.S., and at lower average speed, which tends to offset the EV favoring effects of higher fuel prices there. Further, expectations for top speed in some nations with limited access highways allowing much higher speed than in the U.S. may work against these EVs, which continue to have somewhat limited top speed relative to competing gasoline vehicles. For metropolitan area driving on limited access highways, it appears that coming EVs will have adequate top speed (135-145 kph). In most U.S. urban areas speed limits on such highways are 88 kph, though actual speed often significantly exceeds the limit. For inter-city travel on U.S. Interstates, speed limits vary, but consistently range from 104 to 120 kph, with higher speeds not unusual. Modern full function EVs using lithium ion battery packs will be capable of going fast enough on U.S. Interstates, but the effects on range will be a significant issue.

Many households now own a fleet of vehicles, so it is now possible for many middle income households to mix a gasoline and electric vehicle in a two car fleet, optimizing the use of the pair of vehicles. Electric service is available in almost every dwelling, though garage and carport space is not. The proportion of households living in urban and suburban areas is far greater than it was in the early 1900s. While the capability of driving off-road and on dirt roads remains a selling point for some consumers today, it is no longer a need of the majority of customers for motor vehicles, as it was in the U.S. in the early 1900s.

Culturally, the car is less an "adventure machine" than in the early 1900s. Aircraft are often used for trips out of town, rather than the highway vehicle. Those who are very affluent are likely to use air travel to a significant degree. From a financial viability perspective this will actually hurt the EV for this customer base, because the EV will be used less days per year than by less affluent consumers who do not fly as often.

An adequate road network exists today, with very great functional flexibility in choice of destinations. As in the 1900s, there remains a need for reliable low rolling resistance tires particularly for EVs. EVs and "extended range electric vehicles" (EREVs) are consistently using lower rolling resistance tires than are gasoline vehicles.

For the U.S., the establishment of a petroleum products delivery infrastructure before the advent of the gasoline car was an advantage, which was reinforced by the discovery of abundant oil supplies. Today, the U.S. has built a considerable number of efficient combined cycle natural gas powerplants to serve air conditioning demands, creating a high summertime peak and a deep and wide summertime overnight trough. In their recent assessment of the use of electric power by plausible, but optimistic numbers of plug-in hybrids, Argonne National Laboratory scientists (Elgowainy et al, 2010) estimated that the vast majority of power would be provided by *already existing* combined cycle natural gas powerplants. In the meantime, significant new resources of shale gas have become available in the U.S. (and probably elsewhere in the world) as a result of developments in drilling technology (Energy Information Administration, 2011). Thus, today the plug-in electric vehicle also has the benefit of a widely available existing electric delivery infrastructure whose electricity can be generated by an abundant resource, natural gas. The petroleum delivery infrastructure today appears to be at risk of dependence on expensive oil resources whose production may be reaching a worldwide plateau, while worldwide demand continues to rise.

Environmental motivations by the affluent today are far different than in the early 1900s. Due to dramatic improvements in the gasoline vehicle, reduction of local noise and smell are much less a concern today, though they remain a factor. Nitrogen oxides and particulate emissions of the diesel have become a concern in Europe, where diesel emissions regulation had been more lax than for gasoline. However, the leading new environmental concern for many affluent vehicle consumers and many national governments is global warming. The perception of the environment has changed. Escape from this environmental problem by moving to a different location (such as suburbanization in the U.S. in part to escape dirty industrial core cities) is no longer a possibility. Thus, changing the choice of technology to one with less global warming effect – rather than moving away from pollution – is a higher priority for those affluent consumers who wish to contribute to mitigating this problem. Plug-in electric vehicles are seen as enabling technology that can enhance the technical and economic feasibility of electrical generation with wind and solar power, two ultimate clean sources of such power. Combined cycle natural gas powerplants, relatively clean among fossil fueled power plants, have technical flexibility to vary load rapidly, creating the possibility of synergism with fluctuating wind and solar.

Thus, as in the early 1900s, the perception of the electric vehicle as a clean environmentally friendly vehicle remains important, though with a significant change in perspective.

Neither the U.S., nor Europe is growing as rapidly as in the early 1900s. New single family dwelling units, which can most inexpensively be designed to allow for plug-in vehicle charging – retrofit costs for existing units being much higher – are certainly not being built at a rate proportional to the growth in the early 1900s, so neighborhood and dwelling unit charging infrastructure costs will be relatively higher.

Since solar and wind resources are consistently exploited locally, these ultimately clean resources also have the benefit of reducing oil imports for the U.S. and Europe, which is a much greater concern than it was in the early 1900s. Similarly, shale gas also appears to offer many nations an enhanced opportunity to substitute another domestically produced transportation energy source for imported oil (Energy Information Agency, 2011).

The final key difference is that the hybrid electric vehicle has established a relatively steadily increasing market niche since the 1990s, while this technology was unsuccessful relative to the electric vehicle in the early 1900s. For the Kreiger hybrid of a few years after 1900, the battery pack accounted for 25% of the vehicle mass (Mom, p. 126); for the 2004 Prius, the pack accounted for 3.5% of vehicle mass. Undoubtedly, numerous other significant advancements have facilitated the success of hybrids; but, reducing the required pack size is undeniably a crucial factor. A plug-in version of a hybrid is being tested, showing that "electric vehicles" can be converted to incorporate an engine and generator, utilizing gasoline to expand their range. Engineering and cost assessments of various configurations of plug-in hybrid and range extender electric vehicles have been performed (Kromer and Heywood, 2007; Passier et al., 2007; Moawad et al., 2009; Shiau et al., 2009; Axsen, Kurani, and Burke, 2010; Kley, Dallinger, and Weitschel, 2010; Propfe and de Tena, 2010; Santini et al., 2011). The findings of the studies analyzing costs indicate that the plug-in hybrid with a 4-8 kWh battery pack is more cost-effective than the extended-range electric vehicle with a 12-16 kWh battery pack, which is, in turn, more cost-effective than the electric vehicle with a range of 160-320 km and a battery pack of 24 kWh or more. As battery costs decline, the economic feasibility of vehicles with greater battery pack capacity rises (Shiau et al., 2009; Kley, Dallinger, and Weitschel, 2010; Propfe and de Tena, 2010). The reduction in battery pack costs does neither diminish the appeal of plug-in hybrids nor guarantee the supremacy of electric vehicles; rather, it enhances the attractiveness of a more varied assortment of plug-in vehicles. Based on projected 2020 battery pack costs and historical oil prices, unsubsidized personal light-duty electric vehicles with 45 passengers are not expected to be financially appealing to most buyers. Consequently, the engineering cost assessments indicate that the initial phase in the forthcoming wave of motor vehicle electrification is the adaption of hybrids, involving a progressive electrification of the conventional powertrain rather than an immediate shift towards a focus on pure electric propulsion. Should electrics be adopted, the selection of optimal market niches would be paramount, much to the early 1900s, with first market shares anticipated to be minimal.

REFERENCES

About.com. *The History of Electric Vehicles.* http://inventors.about.com/library/weekly/aacarselectric1a.htm. Accessed April 7, 2011.

Anderson, C.D. and J. Anderson (2010) Electric and Hybrid Cars: A History 2nd Ed. MacFarland and Company. NC

Autoblog (2011). http://www.autoblog.com/2010/10/22/unofficial-nissan-leaf-does-0-60- mph-in-7-seconds-tops-out-at/. Accessed April 7.

Axsen, J. K. Kurani and A. Burke. (2011). *Are batteries ready for plug-in hybrid buyers?* Paper 10-0732, presented at the 89th Annual Meeting of the Transportation Research Board, Washington, DC. (Jan. 9-13)

Berretta, R. (2009). *Daimler AG Fuel Cell Vehicle Technology for Passenger Vehicles.* California Air Resources Board ZEV Symposium, Sacramento CA. Sept. 21.

Conceptcars.com. (2003). *Lotus Announce Elise for USA.* http://www.conceptcarz.com/vehicle/z6643/Lotus-Elise.aspx accessed Oct. 10, 2010.

Eberhard, M. (2006). *Tesla Motors.* Presented at the California Air Resources Board ZEV Symposium, Sacramento, CA. Sept. 26.

Electric Auto Association (2005). *Electric Vehicle History* (May). http://www.eaaev.org/Flyers/eaaflyer-evhistory.pdf

Electrification Coalition (2009). *Electrification Roadmap*http://www.electrificationcoalition.org/electrification-roadmap.php. Nov.

Elgowainy et al, (2010). *Well-to-Wheels Analysis of Energy Use and Greenhouse Gas Emissions of Plug-in Hybrid Electric Vehicles.* Argonne National Laboratory Report ANL/ESD/10-1. Argonne, IL. June.

Energy Information Agency (2011). World Shale Gas Resources: An Initial Assessment of 14 Regions Outside the United States. http://www.eia.doe.gov/analysis/studies/worldshalegas/. April 5.

Fairley, P. (2009). *Toyota's Secret: The Clean Air Act of 1970* http://spectrum.ieee.org/energywise/energy/renewables/toyotas_secret_the_cle an_air_a.html. accessed March 16.

Gaines, L., et al. (2008). *Sorting Through the Many Total-Energy-Cycle Pathways Possible with Early Plug-In Hybrids.* World Electric Vehicle Journal, 2 (1): 1 74-96.

GenerationsofGMHistory:HeritageCenter(2011). http://history.gmheritagecenter.com/ wiki/ index.php/ 1904_Curved_Dash_Owne r%E2%80%99s_Manual. accessed April 7, 2011.

Idaho National Laboratory (1996a). *Toyota RAV4EV.* http://avt.inel.gov/pdf/fsev/eva/toyrav96.pdf. accessed April 7.

Idaho National Laboratory (1996b). *Toyota RAV4EV w/NiMH.* http://avt.inel.gov/pdf/fsev/eva/toyrav98.pdf. accessed April 7.

Idaho National Laboratory (1999a). *1999 General Motors EV1 w/NiMH.* http://www1.eere.energy.gov/vehiclesandfuels/avta/pdfs/fsev/eva_results/ev1_eva.pdf. accessed April 7.

Idaho National Laboratory (1999b). 1999 Ford Th!nk Urban Electric Vehicle. http://avt.inel.gov/pdf/uev/thinkcityueva.pdf. accessed April 7.

Idaho National Laboratory (2009). *BMW Motors 2009 Mini E.* http://avt.inel.gov/pdf/uev/fact2009bmwmini.pdf. accessed April 7.

Kalheimer F. et al (2007). *Status and Prospects for Zero Emissions Vehicle Technology: Report of the ARB Independent Expert Panel 2007.* Prepared for the State of California Air Resources Board.

Kley, F., D. Dallinger, and M. Weitschel (2010). *Optimizing the charge profile – considering user's driving profiles.* Working Paper Sustainability and Innovation No. S. 6/2010. Fraunhofer ISI, Germany.

Kromer, M. A. and J. B. Heywood (2007). *Electric Powertrains: Opportunities and Challenges in the U.S. Light-Duty Vehicle Fleet,* Laboratory for energy and the environment publication No. LFEE 2007-03 RP. Massachusetts Institute of Technology, Cambridge MA. May.

Lindsley, E.F. (2006) *Lifting the Hood on Future Car Engines.* Popular Science, 208 (4) pp. 82-86; 153. April.

Loeb, A.P. (1995). *The Birth of the Kettering Doctrine: Fordism, Sloanism and the Discovery of Tetraethyl Lead.* Business and Economic History 24 (1) pp. 72-87.

Mahajan, V. and R.A. Peterson (1985). *Models for innovation diffusion.* Sage Publications. Beverly Hills CA.

Matthe, R. (2010). *Vehicle integration of the Chevrolet Volt Battery.* Advanced Automotive Battery Conference, Orlando, FL. May 17-21.

Melaina, M.W. (2007). *Turn of the century refueling: A review of innovations in early gasoline refueling methods and analogies for hydrogen.* Energy Policy 35 pp. 4919-4934

Middleton, W.D. (1974). When the Steam Railroads Electrified, Kalmbach Publishing Co. Milwaukee WI.

Mom, G. (2004). The Electric Vehicle: Technology and Expectations in the Automobile Age. The Johns Hopkins University Press, Baltimore.

Moawad et al (2009). *Impact of Real World Drive Cycles on PHEV Fuel Efficiency and Cost for Different Powertrain and Battery Characteristics.* The 24th World Battery,

Hybrid and Fuel Cell Electric Vehicle Symposium and Exhibition (EVS-25). Stavenger, Norway May 13-16.

Muntwyler, U. (2011). *EV and PV: a winning combination has its 25 years anniversary.*

International Advanced Mobility Forum, Geneva. March 8-9.

Naul, G.M., ed. (1978). The Specification Book for U.S. Cars, 1920-29. Motorbooks International, Osceola, Wis.

Naul, G.M., ed. (1980). The Specification Book for U.S. Cars, 1930-69, Motorbooks International, Osceola, Wis.

Nissan Official U.S. Media Newsroom (2011). *Nissan's pursuit for zero emissions mobility: the unplugged story.* http://www.nissannews.com/pressrelease/790/1/nissan's-pursuit-zero-emission- mobility-unplugged-story. accessed April 7.

Passier, G. et al. (2007). *Status Overview of Hybrid and Electric Vehicle Technology (2007): Final Report Phase III, Annex VII, IA-HEV,* International Energy Agency, Paris, France Dec.

Presse Box (2011). *Mini E UK Field Trial Switches Off.* http://www.pressebox.com/pressreleases/bmw-ag-0/boxid/412690. accessed April 7.

Pressnell, J. (1992). Great Cars of the World. PRION – Multimedia Books Ltd.

Propfe, B. and D. L. de Tena (2010). *Perspectives of electric vehicles: customer suitability and renewable energy integration.* Proceedings of the 25th World Battery, Hybrid and Fuel Cell Electric Vehicle Symposium and Exhibition (EVS-25)., Shenzhen, China. Nov. 5-9.

Public Broadcast System (PBS) (2009). *Timeline: History of the Electric Car*. Oct. 30. http://www.pbs.org/now/shows/223/electric-car-timeline.html

Santini, D.J., K.G. Gallagher, and A.P. Nelson (2010). *Modeling of Manufacturing Costs of Lithium-Ion Batteries for HEVs, PHEVs, and EVs.* The 25th World Battery, Hybrid and Fuel Cell Electric Vehicle Symposium and Exhibition (EVS-25). Shenzhen China. Nov. 5-9.

Santini, D.J. (2010). *Highway Vehicle Electric Drive in the United States: 2009 Status and Issues.* Argonne National Laboratory Report ANL/ESD 10-9. Argonne, IL.

Santini, D.J. et al (2011). *Where Are the Market Niches for Electric Drive Passenger Cars?* Paper no. 11-3733, 90th Annual Meeting of the Transportation Research Board Washington, D.C. January 23-27.

Satyapal, S. and Aceves, S (2009). *Overview of Hydrogen and Fuel Cell Activities.* California Air Resources Board ZEV Symposium, Sacramento CA. Sept. 21.

Schamer, S., P. Lamp, and E. Hockinger (2010). *Development of Hybrid Vehicles at BMW and their Energy Storage Solutions.* Advanced Automotive Battery Conference, Orlando, FL. May 17-21.

Schiau et al (2009). *Impact of battery weight and charging patterns on the economic and environmental benefits of plug-in hybrid electric vehicles.* Energy Policy 37. pp. 2653- 2663.

Siry, D. (2008). *Mythbusters Part 2: The Tesla Roadster is not a Converted Lotus Elise* http://www.teslamotors.com/blog/mythbusters-part-2-tesla-roadster-not-converted-lotus-elise. accessed March 3.

Stephenson, R.R. (1975). *Should We Have a New Engine? An Automobile Power Systems Evaluation. Vols. 1 and 2.* Jet Propulsion Laboratory Report JPL SP43-17, V I&II. Pasadena, CA.

Suckow, T. (2009). *Hyundai Fuel Cell Vehicles.* California Air Resources Board ZEV Symposium, Sacramento CA. Sept. 21.

Sulzberger, C. (2004). *An Early Road Warrior: Electric Vehicles in the Early Years of the Automobile.* IEEE Power Engineering Society. May. http://www.ieee.org/organizations/pes/public/2004/may/peshistory.html

Tesla (2011). http://www.teslamotors.com/roadster. accessed May 19.

U.S. Department of Commerce. 1975. Historical Statistics of the United States: Colonial Times to 1970. Washington DC: GPO.

Vivian, D. (1994). Encyclopedia of American Cars. Brampton Books.

Vyas, A., D. Santini, and L. Johnson (2009). *Plug-In Hybrid Electric Vehicles' Potential for Petroleum Use Reduction: Issues Involved in Developing Reliable Estimates.* Transportation Research Record: Journal of the Transportation Research Board No. 2139. Transportation Research Board of the National Academies, Washington, DC, pp. 55-63.

Wikipedia (2011). *War of Currents.* http://en.wikipedia.org/wiki/War_of_Currents. accessed April 7, 2011.

Wing, G. (2010). *Mitsubishi Motors i-MiEV Fleet Test Experience.* Presentation at the 10th International Advanced Automotive Battery Conference, Orlando, FL. May 19-21.

Yokoyama, T. (2009). Progress and Challenges for Toyota's Fuel Cell Vehicle Development. California Air Resources Board ZEV Symposium, Sacramento CA. Sept. 21.

Zero to Sixty Times (2011). http://www.zeroto60times.com/. accessed April 7.

Chapter 4

WHAT IS THE SIGNIFICANCE OF ELECTRIC VEHICLES IN PROMOTING LOW CARBON TRANSPORTATION IN CHINA?

1. INTRODUCTION: AN OVERVIEW

In December 2009, the Chinese government formally established a voluntary quantitative objective to reduce carbon dioxide emissions, aiming to decrease carbon dioxide intensity (kg CO2 per GDP) by 40% to 45% by the year 2020, relative to 2005 levels. Transportation is a significant source of carbon dioxide emissions due to the global use of fossil fuels. In 2008, carbon dioxide emissions from vehicle fuel burning were around 8% of China's national total (Yang, 2011). This percentage lags far behind certain established economies, such as 33% in the United States in 2004 and 26% in Europe in 2004 (Wallington, 2008), among others. In both developing and developed nations, the road sector accounts for approximately 80% of total carbon dioxide emissions from transportation (Yang, 2011; Wallington, 2008), indicating that road transportation plays a crucial role in mitigating transportation-related carbon dioxide emissions presently and in the future. In 2008, the United States had 824 vehicles per 1,000 people, while Japan had 608 vehicles per 1,000 people in 2009; in contrast, China had approximately 68 vehicles per 1,000 people in 2010. It is evident that China's car population will double if vehicle ownership increases twofold while the national population remains stable. As an emerging economy, this scenario is likely to occur within the next 5 to 10 years. In the absence of a transformative overhaul of the transportation system, carbon dioxide emissions from road transportation are likely to increase. The transportation industry is anticipated to emerge as one of the most rapidly expanding sources of carbon dioxide emissions in China during the next several decades. A low carbon transport system is anticipated to be suggested imminently as a potential solution to reconcile the development of transportation and the economy with the mitigation of climate change. In response to apprehensions regarding the establishment of a low carbon transport system and the rising domestic petroleum consumption, there has been a significant increase in interest in the development of advanced vehicle technologies and alternative vehicle fuels over the past decade. The Chinese government supported numerous research and demonstration programs for various technologies, including light-duty vehicles (LDVs) utilizing methanol (M85) and ethanol (E10), buses and taxis employing liquefied petroleum gas (LPG), compressed natural gas (CNG), and liquefied natural gas (LNG), as well as passenger cars and buses utilizing dimethyl ether (DME) and diesel. Since 2003, ethanol gasoline (E10) has been mandated in five Chinese provinces: Jilin, Heilongjiang, Henan, Anhui, and Liaoning, as well as in several major cities within the provinces of Hubei, Shandong, Hebei, and Jiangsu. In recent years, China's strategy for the development of new vehicle technologies and alternative fuels has progressively transitioned from multiple pathways to a select few significant avenues, particularly focusing on electric vehicles, namely battery electric vehicles (BEVs), regular hybrid electric vehicles (HEVs), plug-in hybrid electric vehicles (PHEVs), and fuel cell vehicles (FCVs). The Ministry of Science and Technology integrated the research and development (R&D) of electric cars into the National High Technology Research and Development Program (863 Program). The most recent application guidelines for this program, released in October 2010, indicate that a total of 738 million RMB (approximately 113 million U.S. dollars) will be allocated to support laboratory research on critical technologies and system integration for electric vehicles (National High Technology Research and Development Program, 2010). The demonstration of several electric vehicles has commenced. During the 2008 Beijing Olympic Games, 370 battery electric vehicles (comprising 50 buses and 320 shuttles), 100 hybrid electric vehicles (including 25 buses and 75 passenger cars), and 23 fuel cell vehicles (consisting of 3 buses and 20 passenger cars) were utilized for transportation services. Two years later, 1,017 electric vehicles were exhibited at the 2010 World Expo in Shanghai, comprising 321 battery electric vehicles (181 buses and 140 shuttles), 500 hybrid electric vehicles (150 buses and 350 passenger cars), and 196 fuel cell vehicles (6 buses, 90 passenger cars, and 100 shuttles). The central government has implemented policies to encourage the proliferation of electric vehicles. In reaction to the profound worldwide economic slump triggered by the financial crisis in the United States in late 2008, the Chinese Automotive Industry Revitalization Plan, a crucial component of the national industry revitalization initiative, was released in March 2009. This three-year plan outlines China's objective to establish a production capacity of 500,000 "New Energy Vehicles" by 2011, encompassing battery electric vehicles, plug-in hybrid electric vehicles, and conventional hybrid electric vehicles. The plan established a target for 2011 to elevate the sales proportion of new energy vehicles to 5% of total passenger cars. In early 2009, a pilot project focused on energy saving and new energy vehicles was formally initiated in 13 cities, including Beijing and Shanghai, as per a circular released by the Ministry of Finance and the Ministry of Science and Technology. New energy vehicles were promoted for utilization in public transportation, taxis, postal services, sanitation, and other public services. The federal government offered a subsidy for car purchases, while the local government was tasked with infrastructure development, including the construction of charging stations for electric vehicles. It was reported that 12,000 new energy vehicles have been sold since the project's inception (Ministry of Science and Technology, 2010). In June 2010, the Ministry of Science and Technology and the Ministry of Finance initiated a subsidy scheme for the private acquisition of battery electric vehicles and plug-in hybrid electric vehicles in five cities: Shanghai, Changchun, Shenzhen, Hangzhou, and Hefei, extending through 2012. The subsidy is determined at 3,000 RMB (about 460 U.S. dollars) per kWh, with limits of 60,000 RMB (around 9,190 U.S. dollars) per car and 50,000 RMB (around 7,659 U.S. dollars) per vehicle for battery electric vehicles and plug-in hybrid electric vehicles, respectively. Numerous car models in the domestic automotive sector are anticipated to gain advantages from the program. An example is the BYD E6 model (battery electric car) priced at 270,000 RMB (about 41,000 U.S. dollars), of which 60,000 RMB will be subsidized by the central government. The BYD F3DM model, which operates in dual modes (battery electric and normal hybrid electric), is priced at 150,000 RMB (about 23,000 U.S. dollars), with 50,000 RMB subsidized by the central government. Furthermore, several major cities soon implemented their own regulations that provided an enhanced offer to consumers, including an additional subsidy of 50,000 RMB (about 7,659 U.S. dollars) or more. Private customers, however, have been mostly deterred by uncertainty surrounding the adoption of new technologies, resulting in suboptimal sales of new energy vehicles. For example, BYD Auto Company, a leader and the largest domestic manufacturer of electric vehicles, had sold only 480 vehicles (417 BYD F3DM and 63 BYD E6) by the end of 2010. The National People's Congress (NPC) of China approved the Twelfth Five-Year Plan for National Economic and Social Development (2011-2015) in March 2011. The new energy vehicle sector, recognized as one of the seven strategic industries, was incorporated into the national plan. The national strategy emphasized the extensive demonstration and eventual commercialization of plug-in hybrid electric vehicles and battery electric vehicles, signifying that electric vehicles will undergo a significant developmental phase in recent years.In this chapter, the title question was addressed by quantitatively analyzing the climate change impacts of electric vehicles in China. The circular life cycle energy consumption and greenhouse gas emissions (GHGs) of battery electric vehicles, fuel cell vehicles and conventional internal combustion engine vehicles (ICEVs) were calculated via well-to-wheel (WTW) method. An improved GREET (Greenhouse, Regulated Emissions and Energy use of Transportation) model was used in this study, inside which over 640 of total 730 parameters were updated with localized Chinese data by Shen (Shen, 2007; Shen & Zhang, 2008). Modeling studies indicated that battery electric vehicles significantly outperformed both traditional gasoline vehicles and fuel cell vehicles in terms of well-to-wheel fossil fuel consumption, petroleum consumption, and greenhouse gas emissions. Fuel cell vehicles were expected to assume a more significant role following advancements in hydrogen manufacturing technology. We additionally determined that electric vehicles will significantly enhance the future low-carbon transportation system. Furthermore, the market adoption of electric vehicles significantly diminished reliance on conventional petroleum. Electric vehicles offer a viable resolution to the challenges of transportation energy and climate change. Currently, electric vehicles in China confront several pressing challenges, including elevated purchasing costs, a lack of infrastructure, and complications related to battery disposal and recovery. Therefore, specific follow-up strategies must be implemented to further the commercialization of electric vehicles in China.

2. METHODS

The well-to-wheel technique is a particular life cycle assessment (LCA) applied to transportation fuels and automobiles. The energy consumption and greenhouse gas emissions associated with the fuel cycle constitute about 70% of the entire life cycle, which includes fuel production, vehicle production, and vehicle operating. Consequently, this study emphasizes the energy consumption and climate change implications of the fuel cycle rather than the vehicle cycle. The fuel cycle well-to-wheel analysis is often categorized into two phases: well-to-tank (WTT) and tank-to-wheel (TTW). The former denotes the upstream phase, encompassing the mining, processing, and transportation of feedstock, as well as the manufacturing, delivery, and storage of vehicle fuels. The latter is referred to as the downstream stage, specifically pertaining to vehicle functioning. This study utilized an enhanced GREET 1.7 model, in which the American database was ultimately substituted with a Chinese database. It may also be referred to as ChinaGREET, as 367 of 394 parameters for the feedstock and fuel production stage and 282 of 336 parameters for the transportation and distribution stage have been revised to reflect actual conditions in China.

2.1. Assumption

This study included 12 methods for the production and utilization of vehicle fuels, comprising a traditional gasoline vehicle track, a battery electric vehicle pathway, and ten fuel cell vehicle pathways (Table 1). Coal, natural gas, and water were identified potential hydrogen sources for fuel cell vehicles. The electricity utilized for vehicles was presumed to originate from the national electrical grid. Passenger vehicles were analyzed because of their greater potential for future growth relative to other types of vehicles.

Table 1. Feedstock, in-process product, fuel, and vehicle type of each pathway

Pathway name	Feedstock	In-process product (site)	Fuel	Vehicle
SI: 93# Gasoline	Petroleum	-	Gasoline 93#	Gasoline vehicle (spark injection)
FCV: MeOH-NG	Natural gas	Natural gas -"/ methonal -"/ hydrogen (on-board)	Gaseous hydrogen	Fuel cell vehicle
FCV: MeOH-Coal	Coal	Coal -"/ methonal -"/ hydrogen (on-board)	Gaseous hydrogen	Fuel cell vehicle
FCV: GH_2,RS,MeOH-NG	Natural gas	Natural gas -"/ methonal -"/ hydrogen (refill station)	Gaseous hydrogen	Fuel cell vehicle
FCV: GH_2,RS,MeOH-Coal	Coal	Natural gas -"/ methonal -"/ hydrogen (refill station)	Gaseous hydrogen	Fuel cell vehicle
FCV: GH_2,RS,Electrolysis	Water	Water -"/ hydrogen (refill station)	Gaseous hydrogen	Fuel cell vehicle
FCV: GH_2,CP,NG	Natural gas	Natrual gas -"/ hydrogen (central plant)	Gaseous hydrogen	Fuel cell vehicle
FCV: LH_2,RS,MeOH-NG	Natural gas	Natural gas-"/ methonal-"/ hydrogen (refill station)	Liquid hydrogen	Fuel cell vehicle
FCV: LH_2,RS,MeOH-Coal	Coal	Coal -"/ methonal -"/ hydrogen (refill station)	Liquid hydrogen	Fuel cell vehicle
FCV: LH_2,RS,Electrolysis	Water	Water -"/ hydrogen (refill station)	Liquid hydrogen	Fuel cell vehicle
FCV: LH_2,CP,NG	Natrual gas	Natrual gas -"/ hydrogen (central plant)	Liquid hydrogen	Fuel cell vehicle
EV	Various primary energy	Grid electricity	Electricity	Battery electric vehicle

2.2. Data

Data of coal-based, natural gas-based, and grid electricity pathways and data of vehicle stage were described below.

2.2.1. Coal-based pathways

Energy consumption of coal mining was mostly caused by mining equipments and boilers. The former mainly consumed electricity, and the latter basically used coal. Chinese raw coal was mostly provided by domestic coal mines since the country was rich in coal resources and the price was much lower than import coal. Therefore in this study we assumed that the coal used to generate hydrogen and electricity was produced in the country. According to the investigation of large national and local mines and the data of China Energy Statistical Yearbook, there were 34.4 kWh power and 26.7 kg raw coal would be used when 1 tonne coal was excavated in domestic coal mines. Coal chemical industry in China usually took washed coals as feedstock although they were only about 30% of raw coal output would be further washed. According to our investigation, 0.92 tonne coal equivalent (tce) raw coal, 3.0 kWh power, and 0.1 tonne water was consumed when 1 tonne coal was washed. Another issue that should draw our attention to was the release of absorbed gases from coal bed, such as methane and carbon dioxide. On considering current mining technology, we estimated that there were approximately 7~8 cubic meters methane, 6 cubic meters carbon dioxide, and a small quantity of sulphur dioxide and nitrous oxide that would be emitted when 1 tonne coal was excavated (Alternative Energy Program by National Development and Reform Commission, 2006).

It was known that coal resources were mainly located in the east and the north of China. Over 60%~70% of state coal reserves were found in Shanxi, Shaanxi, Inner Mongolia, and Xinjiang provinces. But end users of the energy were concentrated in north-eastern regions. So coal transportation from producing areas to consuming regions became a necessary and complicated work. Coal was usually delivered by rail, road, and water. The volume of coal transported and the average transferring distance by each means come from Year Book of China Transportation & Communications and China Energy Statistical Yearbook (Table 2). It can be found that sum of the share was over 100% because some coal was transported by more than one means which resulted in repeated calculation in statistics. Coal losses during transportation were assumed to be 0.5%~1.0% (Xiao, 2005).

Table 2. Share of coal volume and average transferring distance by each means

	Data source	Rail	Road	Water
Share of coal volume (%)	China Coal Research Institute (CCRI)	60%	30%	20%
	Ministry of Transport of China (MOT)	60%	10%	40%
	Assumption in this study	50%	30%	20%
Average transport distance (km)	Investigation	550~595	-	650
	Assumption in this study	600	80	650

Coal was initially gasified to generate syngas, a blend of carbon monoxide and hydrogen gas. Subsequently, syngas underwent conversion into high-purity hydrogen gas using decarbonization and desulfurization processes, specifically carbon monoxide conversion, low-temperature methanol washing, and solvent adsorption. The overall efficiency of hydrogen gas production from coal by the Chinese chemical industry was approximately 50%, which is inferior to that of international enterprises employing sophisticated technology. A large central facility for coal-based hydrogen production offers significant advantages over small-scale on-site plants. The resultant hydrogen product required further transportation to the refilling station. Hydrogen gas may be carried directly by vehicle or following liquefaction. In this study, gaseous hydrogen was supposed to be delivered via pipeline, whereas liquid hydrogen was delivered by tanker, owing to economic considerations. The average transport distance of gaseous hydrogen via pipeline was estimated to be 50 km, taking into account cost and energy efficiency. The average transport distance of liquid hydrogen by tanker was determined to be 110 kilometers.

2.2.2. Natural gas-based pathways

During natural gas extraction, 10% of the output was utilized as fuel for purification and separation processes, while around 0.4% of the output was unaccounted for. Natural gas was supplied via pipeline to a local chemical factory, some 50 to 100 kilometers distant, for methanol production. Methanol was transported via rail, road, and water to downstream facilities or refill stations for the production of hydrogen gas or liquid. The average distance from the hydrogen plant to local storage was estimated at 1000 km, whereas the distance from local storage to the refill station was approximately 50 km. Hydrogen derived from natural gas was transported using the same methods as hydrogen produced from coal (Chapter 2.2.1).

2.2.3. Grid electricity pathways

Grid electricity was utilized at the refill station for an electrolysis reaction to produce gaseous or liquid hydrogen. The electricity utilized for powering electric vehicles was sourced from the grid. Diverse main energy sources were combusted at the power plant to generate grid electricity. Approximately 80.4% of grid power was derived from coal, 1.0% from natural gas, 1.1% from oil, 1.9% from nuclear, 0.5% from biomass, 14.2% from hydro, 0.8% from wind, and 0.1% from solar. The energy efficiency of the thermal power plant was assessed at 360 grams of coal equivalent (gce) per kilowatt-hour (kWh) of electricity generated. Approximately 7% of power was lost during grid transfer.

2.2.4. Vehicle stage

The FOX 1.8MT passenger car made by Ford Motor Company was used to calculate the downstream energy consumption and greenhouse gas emissions. This model employed port injected spark ignition (PISI) technology and combusted gasoline that was labelled 93 (Research Octane Number, RON). Fuel efficiency of the car under urban condition was estimated to be 8.5L/100km (equal to 27.7mpg). For fuel cell vehicle, the fuel efficiency was assumed 80% higher than the above conventional gasoline vehicle. Electricity consumption of electric vehicles was assumed to be 22 kWh/100km.

3. RESULTS

This section presents and compares the well-to-wheel fossil energy consumption, petroleum consumption, and greenhouse gas emissions of coal-based, natural gas-based, and grid electricity paths with those of the traditional gasoline pathway.

3.1. Well-to-tank results

WTT fossil fuel consumption of each pathway was shown expect the 2 on-board hydrogen generation pathways (Figure 1). No pathway consumed less WTT fossil fuel than the conventional gasoline pathway when 1 MJ vehicle fuel was generated. The reason was that present overall energy efficiencies of hydrogen production or electricity generation from either coal or natural gas were between 40%~60%, which was much lower than the energy efficiency of petroleum refining process (over 90%). Large central plant of hydrogen production using natural gas as feedstock had the advantage of energy consumption by 350%~540% over other methods, indicating that this kind of central plant was likely a better choice to make hydrogen than refill station production or on-board generation waysThe fossil fuel need for generating grid energy was approximately 6.3 times greater than that needed for regular gasoline, mostly due to extensive coal usage in power plants.

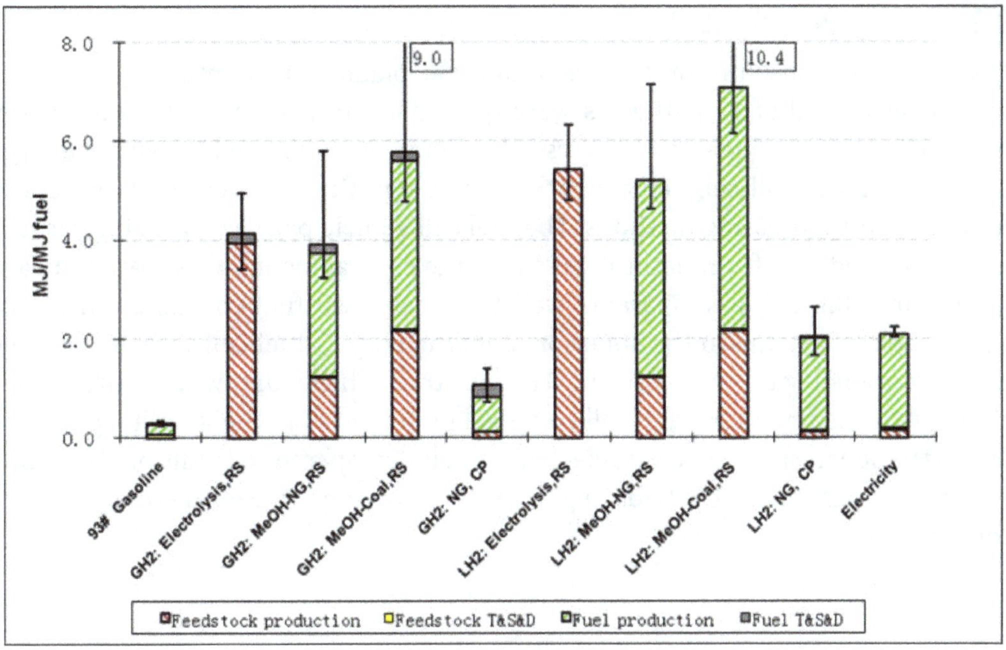

Figure 1. Comparison of WTT fossil fuel consumptions

WTT greenhouse gas emissions resulting from fossil fuel consumption of each pathway was presented expect 2 on-board hydrogen generation pathways (Figure 2). Greenhouse gas emitted during hydrogen and electricity generation was 5~35 times higher than gasoline production.

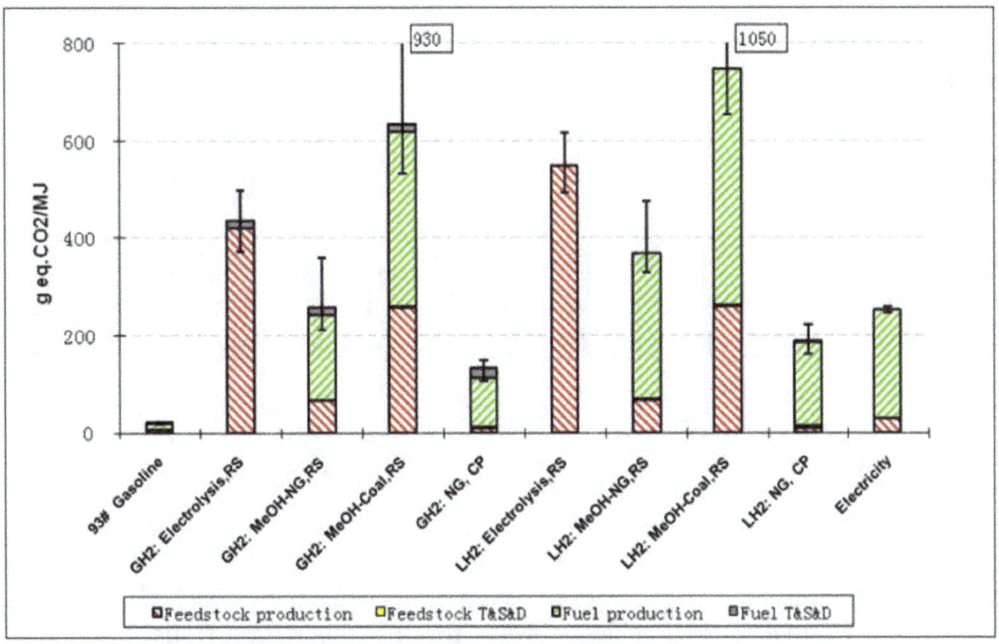

Figure 2. Comparison of WTT greenhouse gas emissions

3.2. Well-to-wheel results

The fossil fuel consumption (Figure 3) and petroleum consumption (Figure 4) of WTW over a total of 12 pathways were quantified in megajoules (MJ) of energy necessary for a vehicle to traverse 1 kilometer. The well-to-wheel energy consumptions of fuel cell vehicle paths varied significantly due to differences in feedstock and processing methods. The fuel cell vehicle pathway utilizing gaseous hydrogen produced from natural gas in a large central facility or via an on-board generator exhibited a comparable well-to-wheel fossil fuel consumption to the gasoline pathway, due to the superior fuel efficiency of fuel cell vehicles relative to conventional gasoline vehicles. The electric vehicle pathway utilizing grid electricity consumed 10% less well-to-wheel fossil fuel than the gasoline pathway, due to the superior efficiency of electric vehicles compared to traditional gasoline vehicles, significantly contributing to the reduction of well-to-wheel energy consumption.

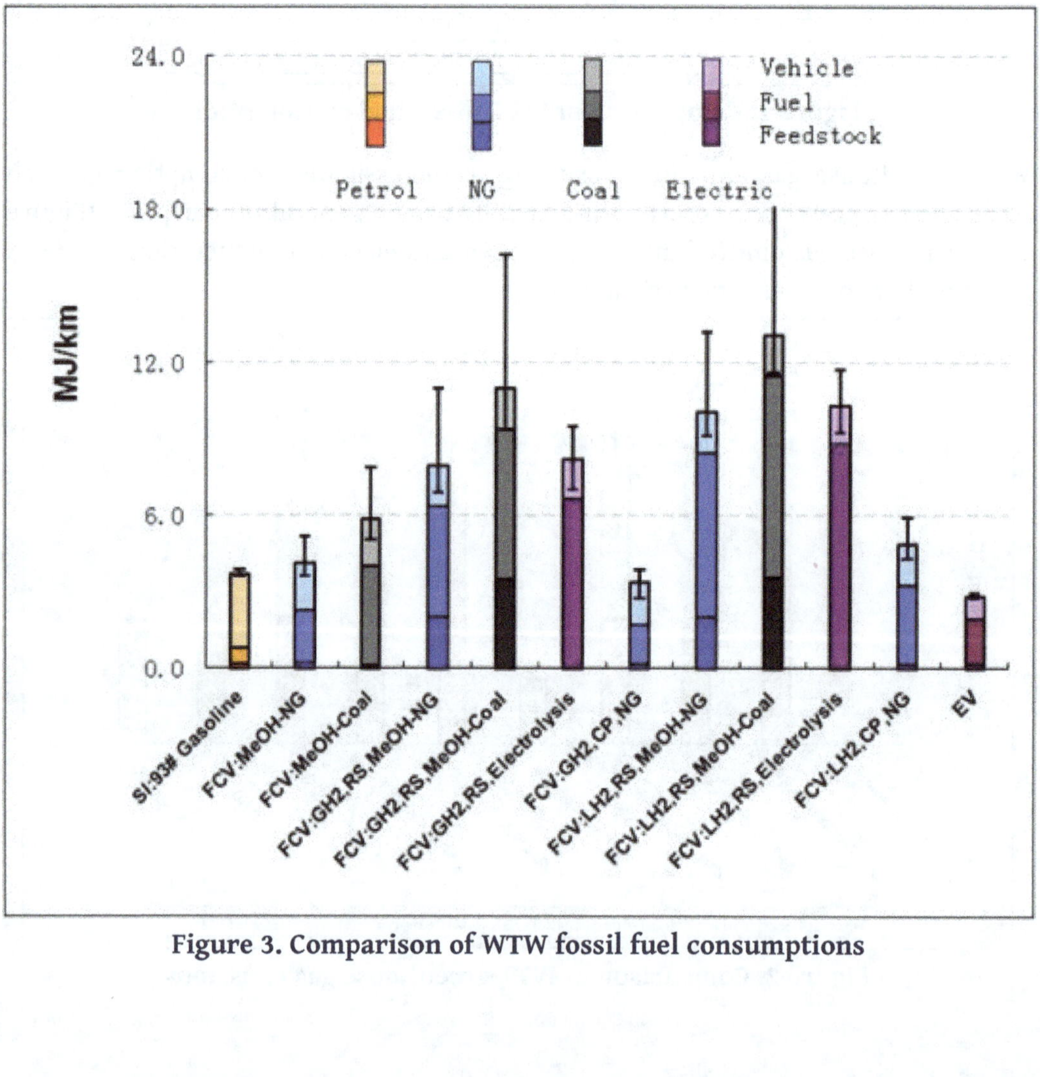

Figure 3. Comparison of WTW fossil fuel consumptions

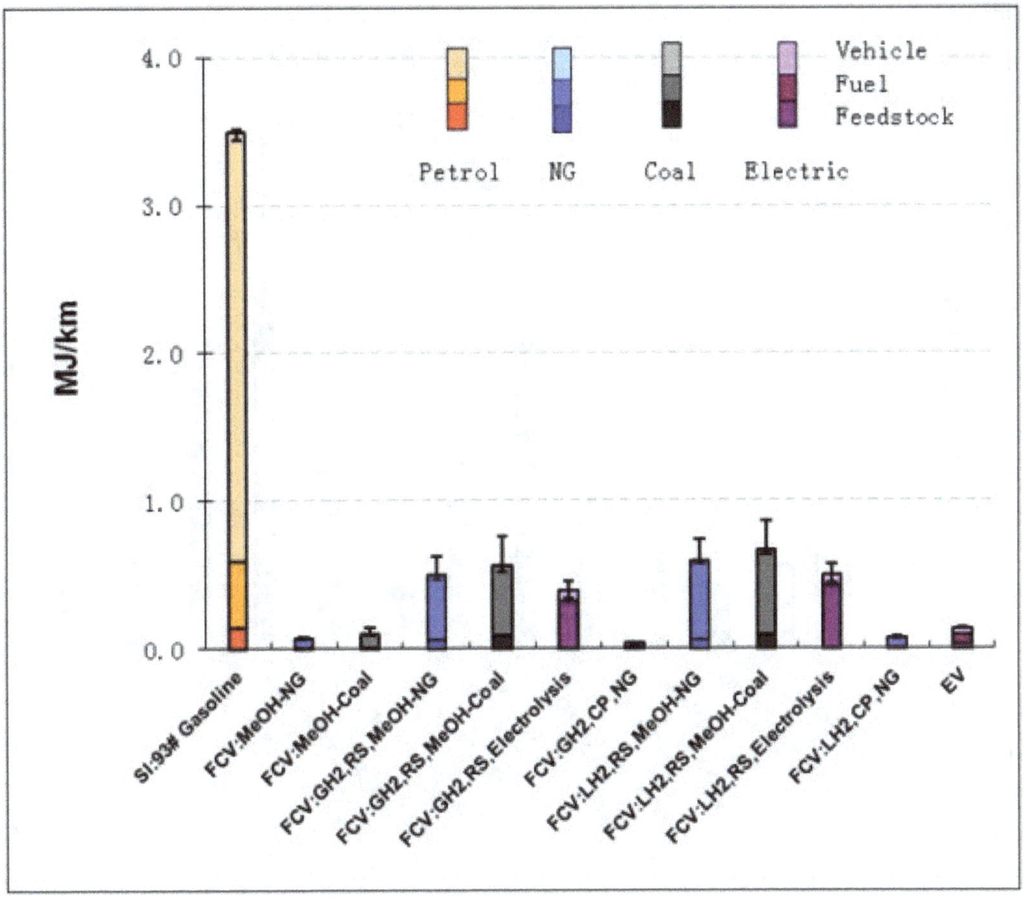

Figure 4. Comparison of WTW petroleum consumptions

Alternative fuels were able to largely substitute petroleum, and therefore import volume of petroleum would be reduced and energy security of the country would be strengthened. WTW petroleum consumption of fuel cell vehicle and electric vehicle pathways proved the above theory (Figure 4). All 11 alternative fuel pathways used less than 1/3 petroleum of the conventional gasoline pathway.

WTW greenhouse gas emissions of different pathways were described as how much grams of equivalent carbon dioxide (g eq. CO_2) emission was emitted when the car travelled 1 km (Figure 5). There were 2 fuel cell vehicle pathways that had lower WTW greenhouse gas emissions than the conventional gasoline pathway. One was fuel cell vehicle with hydrogen generated from natural gas by on-board generator (9% lower); the other one was fuel cell vehicle with gaseous hydrogen produced from natural gas in large central plant (23% lower). Besides, greenhouse gas emitted from electric vehicles using grid power was 12% less than that from the conventional gasoline vehicle.

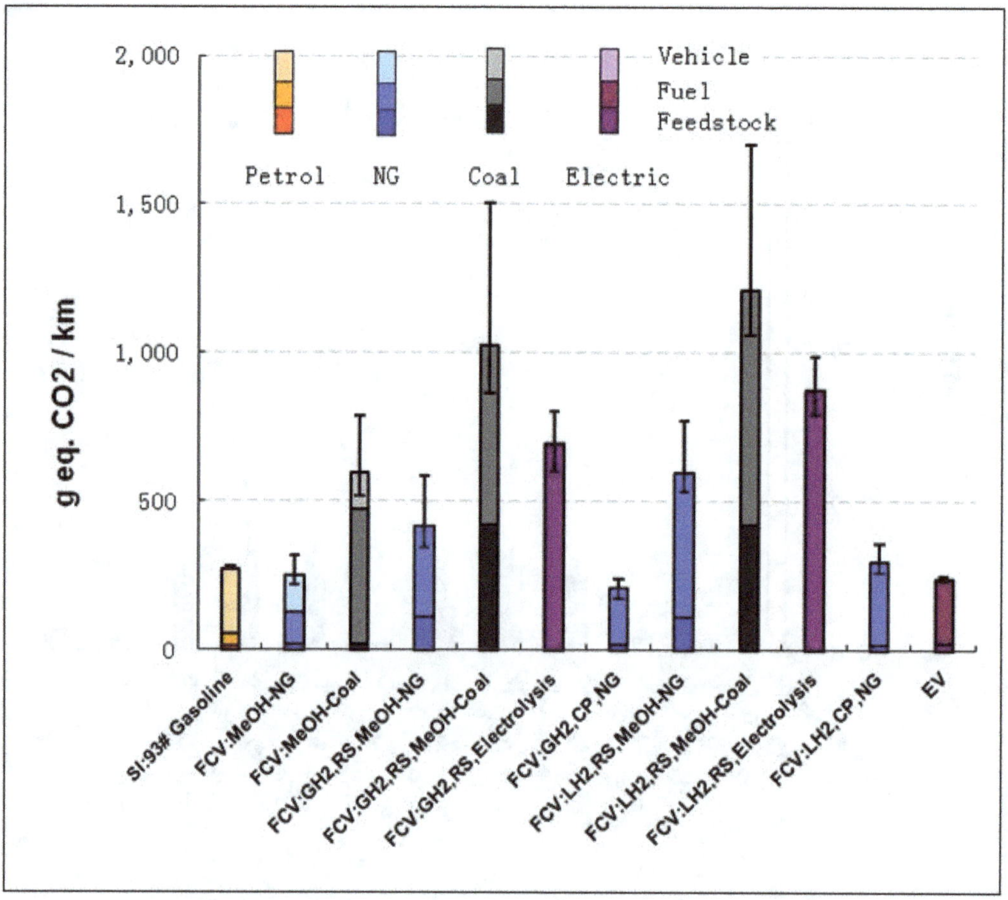

Figure 5. Comparison of WTW greenhouse gas emissions

4. CONCISE SUMMARY

The well-to wheel study revealed that the battery electric vehicle utilizing grid energy possesses certain advantages over both fossil fuel and petroleum use, as well as greenhouse gas emissions. It can be concluded that plug-in hybrid electric vehicles, which combine conventional gasoline vehicles and battery electric vehicles, likely possess similar advantages; 2) for fuel cell vehicles, there are limited pathways where the well-to-wheel energy consumption and greenhouse gas emissions are comparable to those of conventional gasoline vehicles. Consequently, fuel cell car paths today provide minimal advantages compared to both traditional gasoline vehicles and battery electric vehicles. Battery electric vehicles and plug-in electric vehicles should be prioritized in China's development of a low-carbon transportation system. Fuel cell vehicles are likely to emerge as a potential option in the future. Currently, electric vehicles in China encounter several significant challenges, including the elevated purchasing cost, the lack of an infrastructure network, and the disposal and recovery issues associated with batteries. Therefore, specific follow-up measures must be implemented to promote the commercialization of electric vehicles in China.

REFERENCES

1. Alternative Energy Program by National Development and Reform Commission.November 2006 Chinese Alternative Energy Research Report, National Energy Administration, Beijing, China
2. Ministry of Science and Technology.December 2010 New Energy Vehicles Been Put into Operation in 13 Demonstrating Chinese Cities, January 2011, Available from: http://www.gov.cn/jrzg/2010 12 /20/content_1769913.htm
3. National High Technology Research and Development Program (863 Program).October 2010 Application Guideline for Key Technology and System Integration of Electric Vehicle Program (Phase 1) in Modern Transportation Techonology Field, March 2011, Available from: http://www.863.gov.cn/news/1010/28/1010283763.htm
4. Shen W. . April 2007 Well-to-Wheel Analysis on Energy Use, GHG Emissions and Cost of Vehicle Fuels in Future China [Doctoral dissertation], Tsinghua University, Beijing, China
5. Shen W. Zhang A. Han W. Chai Q. . Octorber 2008 Life Cycle Assessment of Vehicle Fuels, Tsinghua University Press, 978-7-30218-281-8Beijing, China
6. Wallington T. Sullivan J. Hurley M. 2008Emissions of CO_2, CO, NO_x, HC, PM, HFC-134a, N_2O and CH_4 from the global light duty vehicle fleet. Meteorologische Zeitschrift, 17 2April 2008), 109 116 0941-2948
7. Xiao B. Chen G. Zhang A. 2005 Life cycle assessment report of clean coal power technology, Tsinghua University & China Coal Industry Clean Coal Technology Engineering Research Center, Beijing, China
8. Yang J. . April 2011 CO_2 reduction pathways of China transportation sectors under global stabilization target [Doctoral dissertation], Tsinghua University, Beijing, China

5 Chapter

PLUG-IN HYBRID AUTOMOBILES

1. INTRODUCTION: AN OVERVIEW

The plug-in hybrid vehicle (PHEV) exemplifies the automobile industry's response to green policies aimed at reducing pollution and fossil fuel consumption in transportation. The oil price is consistently increasing, and oil imports create an undesirable dependence of the national economy on unstable countries, as road transport is currently entirely reliant on oil fuels. The electric drive has been ready for vehicle implementation for many years; it provides optimal control and maximal efficiency. However, a sufficient battery for all-day energy supply is currently unavailable. However, the majority of vehicles in a household are often utilized for routine commuting, with an average daily distance of less than 50 kilometers, and are infrequently employed for extended journeys over weekends or holidays. The battery is accessible within an acceptable weight and price range. If users prefer not to maintain two vehicles, one electric for commuting and another petrol-powered for longer journeys, a Plug-in Hybrid Electric Vehicle (PHEV) presents an optimal solution. It integrates both powertrains, allowing for efficient collaboration between them, which can yield additional benefits. Furthermore, many materials and components for the second vehicle, such as the body, wheels, and suspension, can be conserved. Nonetheless, it should be noted that with superior battery technology (or comparable electrical energy storage systems), the inclusion of a generator and internal combustion engine (ICE) becomes superfluous, thereby simplifying the PHEV to a purely battery-operated vehicle (BEV). However, the operational engine generates some "free" heat, which can be advantageously utilized for air conditioning. The supply of green energy from renewable sources and nuclear power plants is increasing, and nighttime charging can mitigate the loss of oil fuels in road transport.

2. HISTORY OF EV AND HEV

Initially, electric drives were more successful than internal combustion engines in the realm of automotive mobility. The earlier steam engines, renowned for railway locomotives, were unsuitable for transportable lightweight applications due to their substantial mass and the continuous water loss in an open system without condensation. The electric motor (EM), powered by an inexpensive lead-acid battery and with straightforward speed control, was highly reliable and easy to operate for short-distance road passenger transport and sporting activities, given the low speeds associated with horse-drawn vehicles. Historical records indicate that the first vehicle to exceed a speed limit of 100 km/h was an electric car, and the number of registered electric vehicles was comparable to that of other types of propulsion systems. Only with Ford's widespread manufacture of affordable vehicles with internal combustion engines does the proportion of electric vehicles decline. Improved roads, increased speed, and extended operational range have rendered the low battery capacity and protracted charging process disadvantageous for electric vehicles in road transport competition. Only the cable supply was appropriate for urban transportation (trolleybus), while battery supply was limited to low-speed local transport, such as door-to-door milk and mail delivery, operations within production facilities, and at railway stations. In the past decade, golf carts and neighborhood electric vehicles have also emerged in the market.

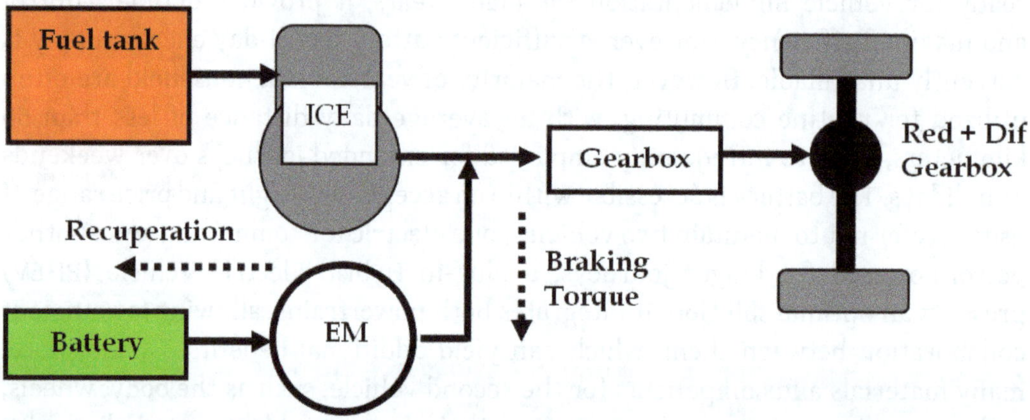

Figure 1. Typical configuration of common hybrid vehicle with parallel power flow

Hybrid electric vehicles (HEVs) integrate an internal combustion engine with an electric motor, reintroducing electric propulsion into vehicular traction. They have been on the market for almost ten years, and their second generation is now available. Initially, hybrid electric vehicles (HEVs) were exclusively produced by Japanese manufacturers; but, in recent years, every automotive manufacturing group has introduced at least one model featuring hybrid electric drive. This vehicle is undoubtedly more costly to produce; but, the benefit of a hybrid electric vehicle (HEV) lies in its diminished fuel consumption, particularly in urban settings characterized by low average speeds, where conventional internal combustion engine (ICE) vehicles exhibit greater fuel consumption (lower mileage) compared to high-speed land transport. Fuel savings are imperceptible outside the city.

3. HYBRID VEHICLES

The standard HEV (Fig. 1) possesses simply a low-power electric motor, which aids during vehicle acceleration and subsequently during braking, allowing for the recuperation of a portion of kinetic energy into the battery for future acceleration. The efficiency of this cycle (braking – acceleration) is suboptimal, with around 50% of energy dissipated; however, the frequent repetition of this cycle at traffic signals results in significant fuel savings. The energy produced in a single cycle is minimal; hence, only a tiny battery can be utilized. The battery's cycle life is crucial, as replacing it monthly is not feasible. The diminished depth of discharge (DOD) significantly prolongs the battery's lifespan, allowing a single battery pack to endure up to five years with a total cycle count exceeding 100,000, due to the low ratio between the energy of one cycle and the battery's total energy capacity. The HEV principal scheme can be followed in Fig.1, where it can be observed the parallel power ways from both torque sources to the wheel. The EM can work not only in motor run, when it produces the torque and mechanical power from electric energy, but it can be easily switched into generator run, when the mechanical power from kinetic energy of vehicle is changed into electric energy, recuperated back to battery. The advantage of EM presence is not only the energy recuperation, but also the torque production at any speed, like it was at old steam engines. No kind of ICE is able to produce the torque at zero speed and moreover there is some minimal value of crankshaft speed, called idle run, under which the ICE stops. To keep the ICE in the idle run needs also some fuel and in city transport the standing at the crossroads is very often and very long lasting. The engine stopping at any occasion and its automatic starting connected with touch of clutch pedal, known as STOP-START system is also effective, but it does not eliminate the energy wasting in brakes and the fuel consumption for acceleration. Also the clutch wear is much higher than in the case of vehicle accelerating by EM torque.

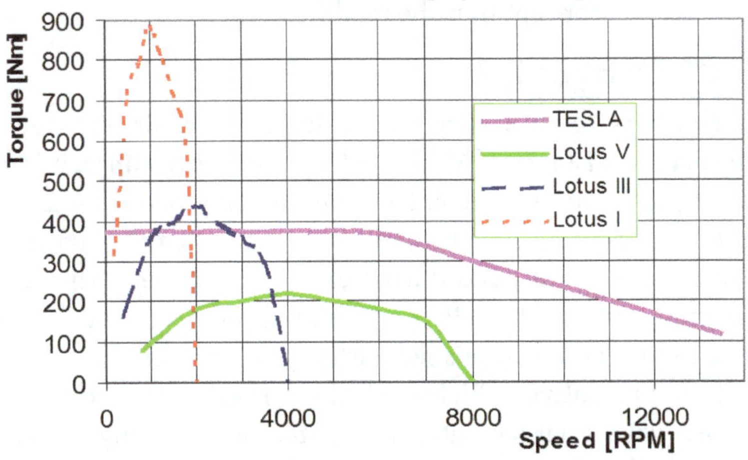

Figure 2. Mechanical characteristics of EM vs. ICE with gearbox (r_{III}=2, r_V=4)

3.1. EM advantages

Figure 2 illustrates the comparison of the mechanical characteristics of EM and ICE, indicating that any EM can exhibit identical characteristics when powered by an appropriate inverter. Each electric motor can be temporarily overloaded, as increased current results in augmented torque, which is available from zero speed. Each type of electromagnetic machine can recover energy while operating in generating mode, as the negative (braking) torque reverses the current back to the source. The velocity disparity between the zero and idle operational speeds of an internal combustion engine can be diminished through the utilization of a variable-ratio gearbox. In the gearbox, a reduction in speed results in an inverse increase in torque. A greater gear ratio results in reduced speed and increased torque while maintaining constant power (excluding losses). The mechanical power is given by:

$$P = T\,\omega \qquad (1)$$

where T is the torque in Newton-meters and ω is angular speed in radians per second. The common technical unit of speed n is revolve per minute, the transformation formula is:

$$\omega = (2\pi/60)\,n \approx n/10 \qquad (2)$$

The gear ratio, according to (1) gives:

$$r = n_2/n_1 = T_1/T_2 \qquad (3)$$

From this mathematics and from Fig.2 there is evident, that at high speed the ICE has always enough power, therefore it needs the help from EM2 primarily in the area of low speeds, where the power is also small as results from (1). Low power EM and low power battery are not able to drive the vehicle at speed over 10km/h. For fully electric drive the concept must be modified.

4. WHY PLUG-IN HYBRID?

Numerous car owners do not utilize their vehicles for business travel and do not exceed a daily driving distance of 50 kilometers. For this distance, it is unnecessary to consume fuel, as it can be efficiently covered using battery energy. However, a significant drawback of electric propulsion is that a “empty” battery cannot be filled in a matter of minutes, and during extended journeys, the reliability of a safe return is uncertain. Additionally, certain infrequent journeys during holidays, among others, cannot be accomplished with an electric vehicle, necessitating the acquisition of an alternative vehicles. All these issues are addressed by a serial hybrid equipped with a larger battery, capable of operating solely on battery power for the initial 50 kilometers. For extended journeys, the engine activates and functions at its optimal efficiency point, maintaining continuous power and speed. The produced electricity is utilized for motor supply or, in instances of low demand, concurrently stored in an unoccupied battery. The PHEV must operate solely in electric mode at any speed during short excursions under the daily limit. Consequently, it necessitates a sufficiently powerful electric motor (EM), leading to a series hybrid configuration where the internal combustion engine (ICE) is not mechanically linked to the wheels, as its assistance is unnecessary (Fig. 3).

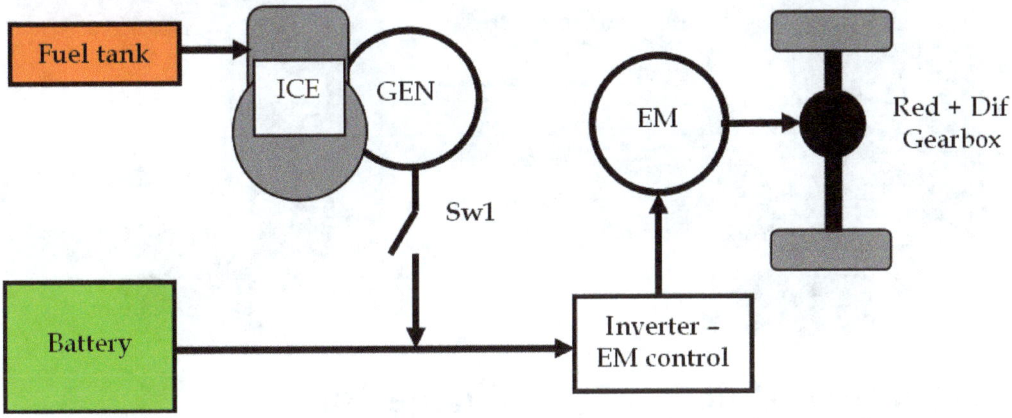

Figure 3. Typical configuration of PHEV with serial power flow

Omitting the generating unit in Fig.3, the PHEV is reduced into the simple BEV and only the parameters of battery determine the operating range of this vehicle. The idea of hybrid concept wants to eliminate the danger of empty battery in case of some complication in traffic like detour, lost way, waiting, etc. Because the battery charging is supposed from home plug during many hours and there are no charging stations in streets, the best way how to be mobile permanently is to have the energy source for charging on the board. The power of the charger does not have to be as big as is the EM power, because in the periods with full power both sources, generator and battery, work together.

All the mechanical energy output from ICE is converted by generator into electricity, which is typically divided between EM and battery, when EM does not work with full power. If the EM is loaded more than is the maximal power from the generator can be, then the battery must deliver the difference. It is typical in acceleration and in uphill slope, both lasts only very short time in second or minutes. The ICE has not to be so strong (maximal power) as it is in a petrol car and it can work here always near the optimal operating point (Fig.11) with maximal efficiency and minimal emissions.

It is not the new idea to have the Engine-Generator unit on board, but to develop and realise the mass production of PHEV it is a merit of American company Chevrolet (Fig.4). However their vehicle is about three or four years in prototype and they solve intensively the problem of optimal battery. The development and new inventions in this area are very fast and it is a problem to start the production of any battery, if tomorrow some principally better technology would appear.

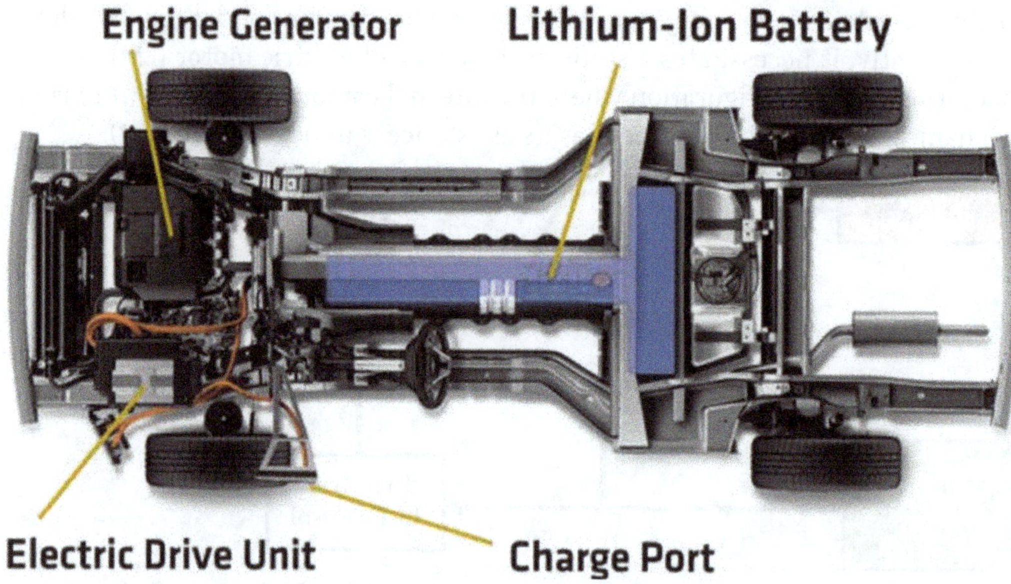

Figure 4. Chevrolet Volt Chassis Version 2008

But it is not only the batteries production technology, also the electric motor for traction manufacturing needs new knowhow in the automotive industry. Also the manufacturers of auxiliary components must prepare new products and some problem can create the dangerous voltage in the vehicle, because the battery voltage can reach up to 300V and the motor supply voltage AC up to 400V phase to phase or DC up to 600V. It is not the same situation as is in traditional 12 or 24V and the isolation check in metallic body must be perfect. But such systems are already developed and verified in trolleybuses e.g.

5. METHODS FOR SIZING PHEV COMPONENTS

The problem of this PHEV concept (Fig. 3) is how to reach high efficiency of all the drive train (ICE, Generator, Battery, Inverter, EM, Gear) for low power light vehicle with total mass about 1500kg. Its average power out of highway at 80 or 90km/h limit is only 5 - 10kW and the peaks are up to 100kW for dynamic drive in modern traffic. The situation in ICE cars is more dependent on the engine volume. The same model can be sold with three or more engines and each of them offers other dynamics. The power and the torque peaks are in case of ICE fix and they can only decrease due the wear, not optimal intake parameters or control, the EM is oppositely easily overload capable, of course for short time only, because the overload brings higher losses and the temperature rise consequently. Big traction machines have on their labels not only rated power but also one-hour power, which is 20 – 30% higher depending on the EM size. For vehicle acceleration in seconds the overload can be easily 100% or more, because after this period the torque and current falls down and the winding temperature also decreases back fast (with effective ventilation). From simulation in Fig.5 it can be seen, the differences between the power in acceleration and in constant speed drive for flat surface.

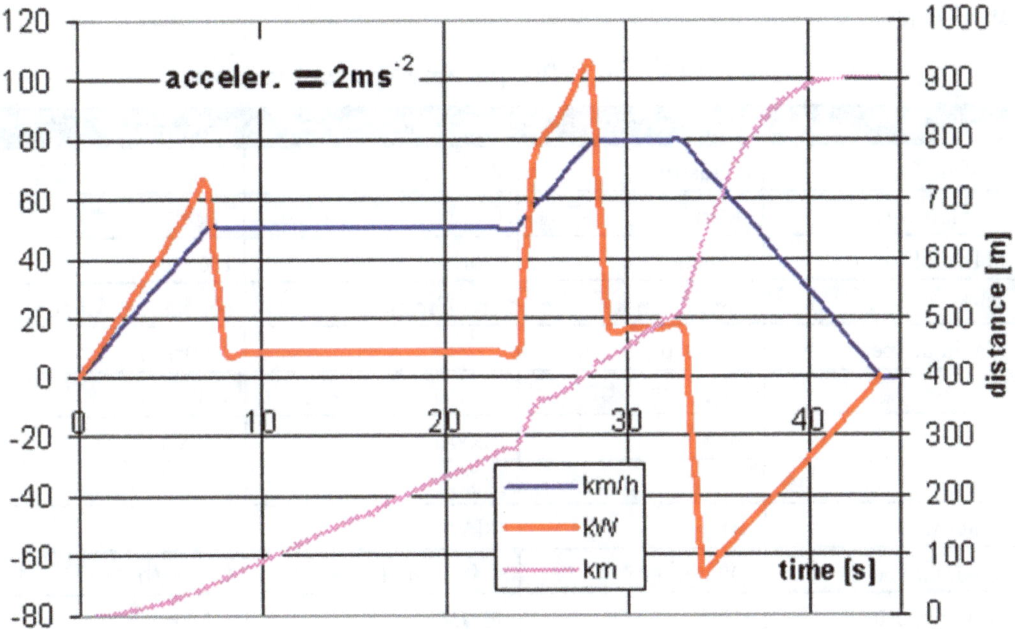

Figure 5. Time dependence of Power, Speed and Distance for exemplar cycle

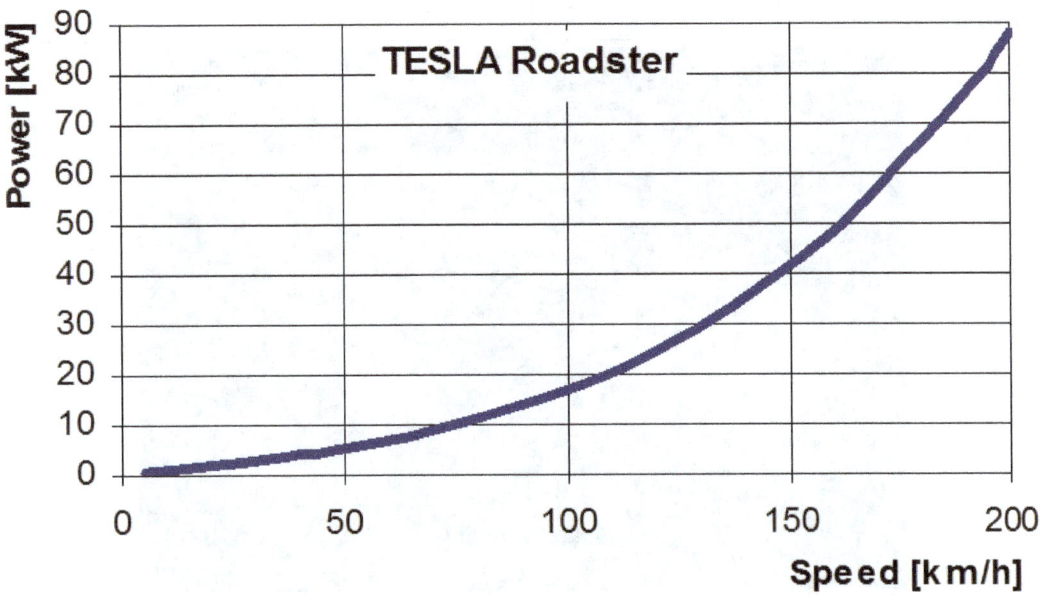

Figure 6. Power vs. speed for EV Tesla Roadster on the plane

The dependence of the power on the vehicle speed in steady state is in the Fig. 6 for strong sport car Lotus, which is professionally remade by TESLA Cars Company for the electric drive. Because at higher speed the aerodynamic drag represent the highest force, the power grow with the speed is nearly quadratic.-Tesla Roadster has no generator, but it can be interesting to compare it with two PHEV. The selected data of two typical PHEV are in Table 1 briefly compared with top power BEV Tesla.

Table 1. BEV and PHEV parameters comparison

Mark	Tesla	Chevrolet	Mitsubishi
Model	Roadster	Volt	iMiEV
AC Motor	185kW	120kW	47kW
Maximal torque	375Nm	320Nm	180Nm
	Asynchronous	Asynchronous	Permanent Magnets
Maximal speed	14.000rpm		7500rpm
Voltage	370V	320V	330V
Battery	56kWh	16kWh	16kWh
Mass		180kg	
Generator	NO	53kW	---
Maximal speed	200km/h	80/140km/h	130km/h
Maximal cruise	450km	64+960km	160 km
Vehicle mass			1080kg
80% SOC			30min

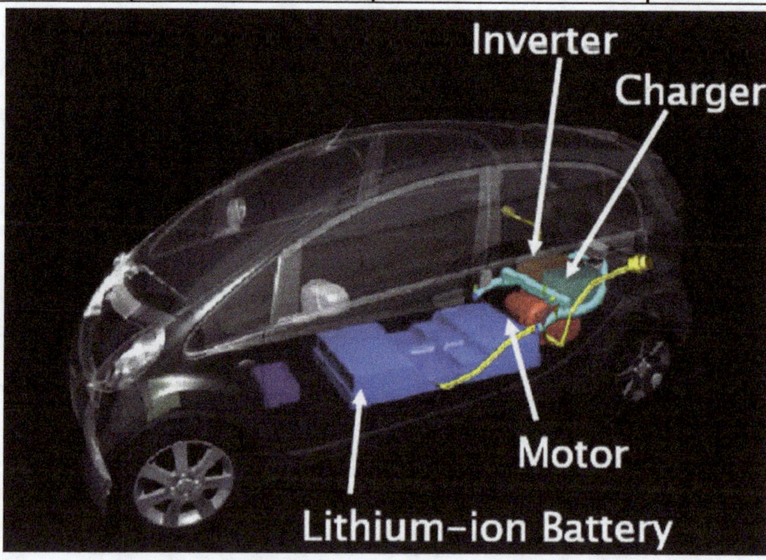

Figure 7. Mitsubishi iMiEV through view

Chevrolet's Volt is the first series hybrid concept car shown by a major manufacturer. Its 1,0-liter (1000ccm), 3-cylinder turbocharged engine runs an on-board 53kW generator that recharges a 16kWh lithium-ion battery made of 80 four-volt cells. Main components distribution on the chassis is in Fig.4. The engine-generator has typical handwrite of car engineers and not only its size, but also its location under front hood with long exhaust pipe. Guessingly 200kg mass can be replaced by another battery with more than twice longer range. The small generator for occasional long trips could be situated in backside trunk, or in tender, which could be hired. The fix mounting of ICE needs all the servicing as standard ICE vehicle, oppositely to BEV, which needs no regular service. Because of rapid evolution in battery technology GM newly opened advanced battery laboratories (3 000 square meters).

Opel Ampera is a Twin sister car to the Chevy Volt for European market. The other leading European producers are preparing their program in EV. New producers like Fiskers, Aptera, Th!nk and many other are only the small companies involved in EV development and they can bring some revolutionary solutions, because there are strong restrictions, in big companies, based on tradition.

Mitsubishi is leading Japan Company in preparing EV with lithium batteries. Its small car iMi with ICE has its twin with electric drive, which is presently (2010) long-term tested over all Japan. The main parts arrangement can be seen on Fig.7.

5.1. Motors

The EM choice is the most important part of PHEV design. As was mentioned above the EM is easily overloadable, its power P depends on the voltage U and current I from the battery:

$$P = \eta\, U\, I \tag{4}$$

where η describes the motor efficiency. This power is equal to that one from (1). Very simplified, it can be said, that the torque is given by the EM current and the EM speed is given by voltage:

$$T = K\, \Phi\, I_a \tag{5}$$

$$U_i = K\, \Phi\, \omega \tag{6}$$

These equation are exact for DC motor, where Φ is magnetic flux and I_a is the armature (or rotor) current and U_i is induced voltage in the armature, which is slightly different from the terminal voltage because of the voltage drop on the internal resistances. The AC motors are more complicated, they have more phases, in case of induction (or asynchronous) motor (AM) there is no separation of field circuit and armature circuit, but it is not the main goal of this book and more can be find in any electric machines textbook. The important for motor control is how to change the voltage, and in the case of AC motors the frequency must be also changed (together with voltage), because instead of mechanical current commutation in DC rotor, the switching technology must be used to create the three phase system in converter. The output frequency f gives the speed of AC rotor:

$$n_s = 2\pi f / p \tag{7}$$

where p is the number of pole pairs. The rotor of AM is slightly slower, that difference between the speed of the rotating field n_s and the real rotor speed n is called slip and its value is typically 3 – 5%, changing with the load.

DC motors are the first kind of traction motors, and in the period before power electronics their control was only by resistance controllers and serial-parallel switching. DC/DC inverters improved the efficiency of DC motor in controlled drive, it is cheaper than the DC/AC converter, but DC motor is not so robust and it needs often maintenance due to carbon brushes.

Brushless DC is in principle the DC with permanent magnets (PM) for field creation on the rotor and static electronic inverter, instead of collector with 3 segments on the rotor; it has 3 winding terminals on the stator, supplied from the 3-legs bridge, working as a commutator. Because the rare earth PM (REPM) are very expensive and very sensitive, namely to temperature and corrosion, they find place mostly in low power (0,2 – 2kW) EM for bikes, where they are in low volume only.

In the power range 20 – 200kW for passenger cars the **induction machine** offers its perfect robustness at low price and also the standard inverter supply, which widely used in industry and also in new trams.

The switched reluctance machine (SRM) has also very robust and simple construction with no rotor winding and it can be very prospective, but it is not yet widely used. It needs special inverter, which supposes some larger volume production for a good price.

Table 2. Typical Electric Motors 20 – 200kW Parameters

	DC Brush Type	Brushless DC (Permanent Magnet)	AC (Induction)
Peak efficiency	85 - 89	95 - 97	94 - 95
Efficiency at 10% Load	80 - 87	73 - 82	93 - 94
Max. RPM	4 000 - 6 000	4 000 - 10 000	9 000 – 15 000
Cost per shaft kW	\$120 - 200	\$120 - 180	\$60 - 100
Relative Cost of Controller	1	3 - 5	6 - 8

The survey of EM basic properties is in Table 2. It must be said here, that all parameters in this Table are valid for power range from 20 to 200kW and with growing power grows up also the efficiency and oppositely the maximal speed falls down.

Table 3. Electric Motors Properties Comparison

AC Motor	DC Motor
Single-speed transmission	Multi-speed transmission
Light weight	Heavier at equivalent power
Less expensive	More expensive
95% Efficiency at full load	85-95% Efficiency at full load
More expensive controller	Simple controller
Motor/controller/inverter more expensive	Motor/controller less expensive

Another EM properties comparison can be read in Table 3 from the drivetrain design point of view.

5.1.1. Motor volume

The volume and mass of electromagnetic (EM) entities are determined by their torque rather than their power. The vehicle bulk must be minimized, necessitating the design of the electric motor for maximum speed and least torque, while also considering efficiency and cooling capacity. Consequently, a direct drive without a gear is not desirable, and Figure 3 illustrates the reduction and differential gearbox between the electric motor and the wheel. Augmenting the velocity elevates the frequency, which must not surpass 400Hz. It is better to keep the frequency under 200Hz and for two pole AM it can give the speed from (7) n_s = 12 000rpm. For the vehicle speed 144km/h = 40m/s and the wheel circumference 2m its rotation speed n_W is:

$$n_W = 60 * 40 / 2 = 120\text{rpm} \tag{8}$$

Then the total ratio between the AM and the wheel must be from (3):

$$r_{Total} = 120 / 12\,000 = 1/100 \tag{9}$$

which can be realised by 3 pairs of cogwheels minimally.

5.1.2. Motor losses

Beside of typical mechanical losses due to friction and ventilation losses, which both are speed dependent, there are in the EM specific electrical losses and these can be divided into two groups. The current depend losses, or Joule losses grow up with the square of current:

$$\Delta P_J = R_a I_a^2 \tag{10}$$

and looking in (5), they grow up with square of torque.

The other group of losses has origin in the magnetic circuit (iron) due its alternating flux. These losses can be described by formula:

$$\Delta P_{Fe} = k\, m_{Fe}\, \Phi^2\, f^{1,6} \tag{11}$$

where m_{Fe} is the AC iron mass. The grow up with speed is more than linear, but if the EM has not PM field, the flux can be reduced when there is no need of full torque (5) to reduce the iron losses, but increased current results in the Joule losses grow up. The optimal flux at any speed and power can be estimated. The greatest advantage of controlled flux is at high speed and no torque run (by inertia or downhill), when the PM machine has high iron losses and they are supplied from kinetic energy of vehicle, which means they are braking the vehicle undesirably.

5.2. Battery and electrical energy storage

First EV has been built in the 1835 and 1836 respectively; the speed record 105km/h was also reached with EV in 1899 with lead acid battery. Edison tried to build EV with his Ni-Fe batteries, but without commercial success. From the year 1903 when Ford established 146km/h speed record, the petrol ruled the vehicles power supply, because of its very high energy density, which is about 36MJ/L = 10kWh/L, because the petrol density is only 0.72kg/L the mass density value is over 14kWh/kg. Also the charging power is enormous, if filling the tank by speed 2L/s the 60L tank can be full in 30s and “supplying power” is 72MW.

Comparing with liquid hydrocarbons, the best available batteries have only 0,2kWh in one kilogram. There are some projects of better electric storage devices based on electrostatic principle, but they are only in patents and no sample has been presented. In the electrochemical batteries, the most promising are lithium-air, which can change completely the electric vehicles during next ten years. Special problem of batteries with numerous cells is their cooling (and heating) system keeping optimal temperature in all the battery pack and the voltage distribution control (charge management) for in series cells avoiding overcharging which can damage the cells with danger of explosion and fire.

Two parameters must be watched if looking for optimal battery, the energy density and the power density. Survey of suitable batteries for EV is in Table 4.

Nearly all the 20th century there was no new chemistry in secondary cells (rechargeable) introduced and only the technologies were improved in two basic batteries.

Lead-acid is wide spread battery for engines starting and for emergency power supply, the Edison alkaline battery nickel - iron Ni-Fe was slightly improved by cadmium Ni-Cd. This kind of batteries was mostly used for railway vehicles and communication technologies. The silver based chemistry Ag-Zn was able to supply electric vehicle, but the silver is not widely available (precious metal) and such batteries can be used only in very special purposes for military or space technologies.

Table 4. Electrochemical Batteries Evaluation

	Lead-Acid	(Ni-MH)	Lithium-Ion
First Use (Commercial)	1859	1989	1991
Current Use(Automotive)	Traditional 12-volt batteries	For today's generation of HEV	Under development for PHEV and BEV
Strengths	Long proven in automotive use;Price	Twice the energy/weight as lead-acid	About twice the energy content of Ni-MH
Weaknesses	Heavy; low energy/ weight ratio for EV	High cost (four times the cost of lead-acid)	Expensive until production volume
Energy density	30 - 40Wh/kg	65 - 70Wh/kg	100 - 150Wh/kg
Recyclability	Excellent	Good	Very Good

Only the last decade of the 20th century and the new electronics devices, connected with communication and information technologies, bring the progress in the cell chemistry. The lithium ion and lithium polymer batteries replaced in few years the Ni-Mh in cellular mobile phones, notebooks and other audio and video portable players. The Ni-Cd has been also replaced in its last important area of use, which was hand-tools supply. This new batteries generation has, up to three times, higher energy density, then old chemistries, which give new possibilities for electric vehicle construction. Lithium is very promising and many new prospective chemistries with lithium are invented and developed, the survey is in Table 5.

Table 5. Lithium Battery Chemistry Survey

Chemistry	Company
Doped Lithium Nanophosphate	A123
Manganese Spinel	LG / NEC
Lithium Nickel Cobalt Aluminium Oxide	Panasonic
Lithium Manganese Oxide	Hitachi
Lithium Cobalt Oxide	Commercial offer
Lithium Titanate Spinel	Altair Nano

Chemistry	Company
Lithium Iron Phosphate	Lishen
Lithium Manganese Titanate	EnerDel

The greatest problem of any battery for electric vehicle is beside its lifetime in cycles, but also its low internal electric resistance, thermal stability between -40 to +60ºC, shock resistance, non toxicity, fire safety, but mostly its charging properties and its price, which can create important part of the vehicle price. For the car construction there is important also the volume and mass energy density. The cheapest chemistry is the lead acid, which his used in all cars for engine starting, but in electric vehicles only in old neighbourhood vehicles and golf cars. The modern hybrids use Nickel - Metal Hydride chemistry or Lithium, which is in strong development connected with communications and information technology produced therefore only in small cells about 2Wh. Tesla Car Company started the production of their BEV, which was designed few years ago with old technology lithium battery consisting from 6 831 small cells a little bigger than AA size (8Wh each).

5.2.1. Charging

The primary issue for electric vehicles is the protracted duration required to charge the battery, which is inconvenient for continuous transportation in business, especially when contrasted with gasoline refueling, which can be completed in under one minute. The calculation of the power of such filling results in megawatts. The fast charging brings not only the problems with battery cooling, but also problem with power peaks in grid, because there are no reservoirs for electrical energy (EE) similar to tanks for petrol and rapid charging results in high power peaks:

$$P_{Charge}\,[kW] = 60\ Ek\,[kWh]\ /\ t\,[min] \qquad (12)$$

To charge 10kWh (which is equivalent energy of one litre petrol) in one minute, from this formula, gives the charger power 600kW (!). Because the charging is not very efficient, and more than 100kW are the losses - the heat, such rapid charging connected with energy conversion is impossible. The only hope is here the storage of EE in electrostatic field, which is not connected with energy conversion.

The rapid charging can be more effective if there is not required full charging. The partial charging is described by SOC (state of charge) in percent of full capacity. Similarly is defined the DOD (depth of discharge), which define the percent of full charge, which was taken from battery and partial discharging extends the lifetime significantly.

The energy for charging is due to chemical processes much higher, than is the energy received during discharge. It can be demonstrated on the lead acid cell, which is discharged at 2V and for each 1Ah must be recharged 1,2Ah at voltage 2,4V. From this can be easily calculated the efficiency:

$$\eta = 2 * 1 / (2,4 * 1,2) = 0,70 = 70\% \qquad (12)$$

and similar value is valid also for the other chemistries.

5.2.2. Electrochemistry

Lead acid: The inaugural secondary battery (1859), widely manufactured for internal combustion engine vehicles, emergency power supply, and uninterruptible power supply (UPS). It is cost-effective, with approximately 400 full cycles; however, specialized designs for traction applications have enhanced the cycle count to 2000. The sulfuric acid electrolyte is highly corrosive, and lead is the only metal that withstands this environment. The sealed construction and gel electrolyte allow using this battery in any position without danger of stain or effusion. Its energy density is up to 40Wh/kg, due its price it is very popular for EV drives, but the operating range is very limited and the lifetime less than 5 or 10 years respectively. Another disadvantage is the danger of sulphatizing, which can be prevented by immediate charging after run that means the battery cannot be left discharged.

Nickel or Alkaline battery: The alkalic electrolyte with NaOH or KOH is less aggressive and Edison realized the first alkaline secondary cell with Nickel and Iron electrodes so called Ni-Fe battery. Nickel cadmium Ni-Cd is an improved chemistry with higher power density. The voltage of cell is only 1.2V, but the lifetime of such battery with minimal maintenance is about 20 years. It is mostly used in railway wagons and in hand tools. The toxic cadmium was about 20 years ago (1989) successfully substituted by metal hydrides in so-called Ni-MH cells, with better energy density. This chemistry is used in Japan HEV.

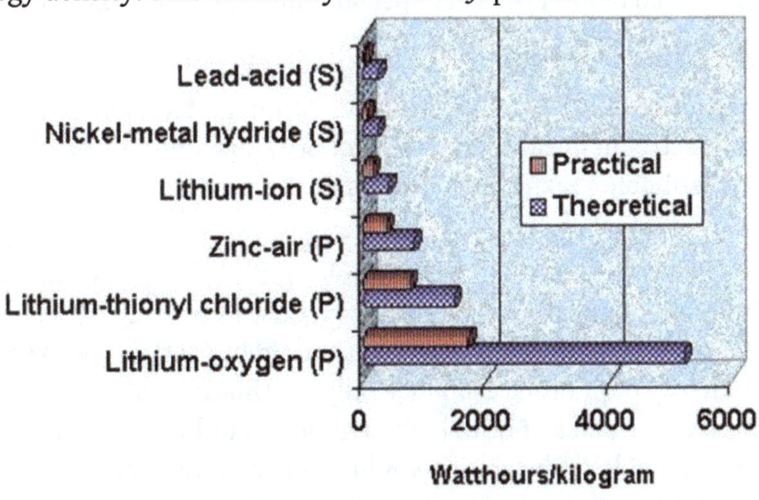

Figure 8. Energy density in various battery chemistries

Na-MCl2 chemistry has similar parameters with Ni-MH and also Ni-Zn is from the same family with similar gravimetric density but weaker volumetric density.

Lithium battery: Very reactive lithium has highest potential, but the technology was mastered only in 1991 (by Sony). The lithium ion cells LiCoO2 with rated voltage 3.75V are widely used in cellular phones and notebooks; the battery with serial connected cells must have electronic balancing system to avoid the overcharge at any cell. In last few years the big cells for traction are produced, with capacity up to 2 000Ah and new chemistry LiFePO4 (lithium iron phosphate) Table 5. Comparing to lead acid the fast discharge of lithium battery does not decrease the capacity, but the energy is lower due to joule losses in internal resistance.

The theoretical specific energy of lithium thionyl battery is 1 420Wh/L (explosive TNT has 1 920Wh/L) and the theoretical specific energy of lithium-oxygen is over 5 000Wh/kg, which gives more traction energy than the petrol of the same volume or weight, if the ICE efficiency is taken into account (Fig.8).

A lithium-titanate battery is a modified lithium-ion battery that uses lithium-titanate nano-crystals on the surface of its anode instead of carbon. This applied nanotechnology gives the anode a surface area of about 100 square meters per gram, compared with 3 square meters per gram for carbon, allowing electrons to enter and leave the anode quickly. This makes fast recharging possible and provides high currents when needed

Other chemistries: In last forty years after the renaissance of EV many new electrochemical batteries have been studied, but they did not convince. High temperature NaS has good parameters, but bad maintenance, Zn-Br needs two tanks for pumping electrolyte, which stores the energy similarly as the vanadium battery, suitable more for stationary applications. Special category is Ag-Zn chemistry with super performance, but due to limited silver cannot be widespread system and it is used only in the very special military or space applications.

5.2.3. Electrostatic storage

The super-capacitors (SC) are the first revolutionary technology, which can be compared to electrochemical batteries in energy density and have much better power density, suitable for the power peaks in short time.

Quantum Battery (QB) promises the surprising energy density, it is based on the discovery of quantum effect on TiO_2 sample, measured by Swiss inventor and described in patent application. The rutile crystals, 15nm long, absorb at 180V the energy with density 8 – 12MJ/kg. It is very optimistic, but without working prototype and with the theory of photon resonance only. Author describes cheap technology with possible market price 15USD/kWh. The predicted low self-discharge (about 6,3 % per month) and long durability would be optimal for EV.

EEStore from Cedar Park, TX, also announces the promising technology of high voltage solid dielectrics super-capacitor based on thin layer (nanometers) with high permittivity barium titanate composite. In the last period the power density 1200kJ/kg is referred, which is four times more than electrochemical battery, moreover without looses and with short recharging time less than one minute. But nanotechnologies in batteries with lithium chemistry can be also competitive, as is mentioned above.

5.3. PHEV auxiliary components

The aforementioned absence of an operational internal combustion engine in a plug-in hybrid electric vehicle results in a lack of thermal energy for cabin heating. The heating and cooling of the cabin air can be achieved using an electric-driven heat pump, which must be created for the next PHEV. The utilization of a heat pump can substantially reduce expenditures of constrained battery energy for non-traction applications. The modern vehicle lacks a mechanical drive for the power steering pump and power brakes when the internal combustion engine is not operational; these functionalities should be provided by localized electric drives. The lighting, dashboard, and cabin electronics require a low voltage supply, which must be provided by a power electronics DC/DC converter rather than a revolving generator. All components listed here that must be developed for PHEVs can subsequently be utilized for BEVs when superior batteries become available; however, the design, technology, and manufacturing of these components must be completed and proven prior to commencing mass production of PHEVs.

6. EFFICIENCY

The liquid hydrocarbons are optimal for transport due their extremely high energy density when in 50kg tank can be stored 500kWh, it is of course the thermal energy, but if the total efficiency (of all the energy conversion from fuel to heat, mechanical force on piston, torque from crankshaft, via gearbox and axis to the wheels) is only 16 - 24%, as is calculated in Table 6. Taking the middle value 20% there is the traction energy 100kWh to disposal. The common passenger car with such petrol can run between 500 and 1000km. It is about 0,2kWh/km and for 60km must be in the battery more than 12kWh.

In the Table 6 is the survey of all components of drive train with typical efficiencies and for more vehicles and operating modes is the total efficiency calculated. From first two rows is evident, that the classical petrol car in city transport has 50% increased fuel consumption, because its engine works with lower efficiency at low power (Fig.11). The new symbols in Table 6 are MGB for manual gearbox, REC for AC/DC inverter (rectifier), INV for DC/AC inverter and RDG for reduction and differential gearbox.

The PHEV without ICE has efficiency 77% if calculated from the battery energy, but only 54% if calculated from the plug. If the PHEV charges its battery from running ICE, its total efficiency can fall under the classic vehicle in the city traffic and if its ICE will be used only for EM supply its efficiency is still under the classic vehicle. Last example in the table is for Diesel – electric drive on big locomotive (Fig.9), where due better efficiencies of big power components is also the total efficiency satisfactory.

Table 6. Efficiency of Drive Train for Various Vehicles

	ICE	GEN	MGB	REC	BAT	INV	EM	RDG	TOTAL
Classic Car	0,30	---	0,85	1,00	---	---	---	0,95	**0,24**
Classic Car - City	0,20	---	0,85	1,00	---	---	---	0,95	**0,16**
PHEV - Electric						0,95	0,85	0,95	**0,77**
PHEV - Electric					0,70	0,95	0,85	0,95	**0,54**
PHEV – Charging BAT	0,30	0,85	---	0,95	0,70	0,95	0,85	0,95	**0,13**
PHEV – No Charging	0,30	0,85	---	0,95	1,00	0,95	0,85	0,95	**0,19**
Loco D-E - 1MW	0,38	0,90	---	0,95	1,00	0,95	0,90	0,95	**0,26**

6.1. Diesel-electric transmission

Diesel electric transmission is utilized in locomotives for rail transport; however, the powertrain lacks an energy accumulator, and the generator-inverter-motor system (Fig. 9) serves as an alternative to mechanical shafts and gearboxes, offering enhanced flexibility and high efficiency.. It is also the rated power of components, which gives the nice efficiency, because the motors are in the power 100 – 1000kW. The greater rated power the higher efficiency is standard property of every machine.

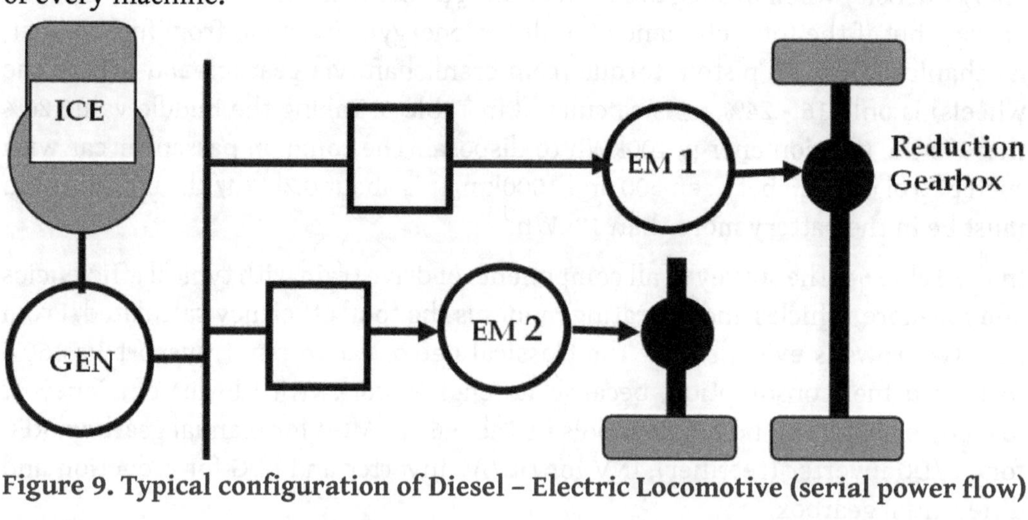

Figure 9. Typical configuration of Diesel – Electric Locomotive (serial power flow)

If the PHEV would be, after spending the energy accumulated in battery from the grid, operated similar way as this diesel - electric loco without charging the battery from engine and the battery would be used only in the same mode as is in actual HEV, that means for accumulating of kinetic energy during recuperative braking, the electric power train efficiency has not be so bad.

This locomotive may also replicate the multi-motor configuration, wherein each axle is equipped with an electric motor. Equipping each wheel of a road vehicle with its own engine can be advantageous; nevertheless, smaller motors tend to be less efficient. A potential solution is drive management that entails deactivating the motors at a constant speed, enabling efficient regenerative braking using individual electronic management for each wheel, while employing ABS control to prevent wheel lock, as torque and speed for each wheel may be regulated independently. The storage system can recapture energy from braking for subsequent acceleration, thereby decreasing energy consumption. This is akin to a standard hybrid system, where a larger battery capacity enables the storage of energy not only from a single acceleration-deceleration cycle but also, in hilly regions, the utilization of potential energy from downhill travel for subsequent uphill ascents.

6.2. HEV efficiency

What is the fuel savings composition in HEV is briefly explained in Fig.10, where the negative influence of increased vehicle mass (battery + EM) represents the first column. The next three columns are contributions from HEV technology given by no ICE idle run, ICE speed control and EE recuperation by electric braking.

The standard HEV as is in Fig.1 saves the fuel only in the city traffic. The ICE specific fuel consumption is in Fig.11 and it is evident, that for the torque under 15% of rated value, the fuel consumption grows more than twice.

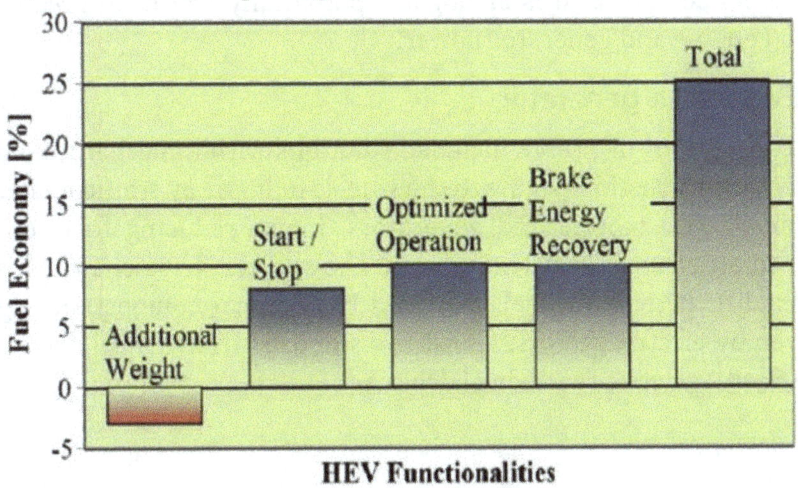

Figure 10. Fuel saving components of HEV in city transport

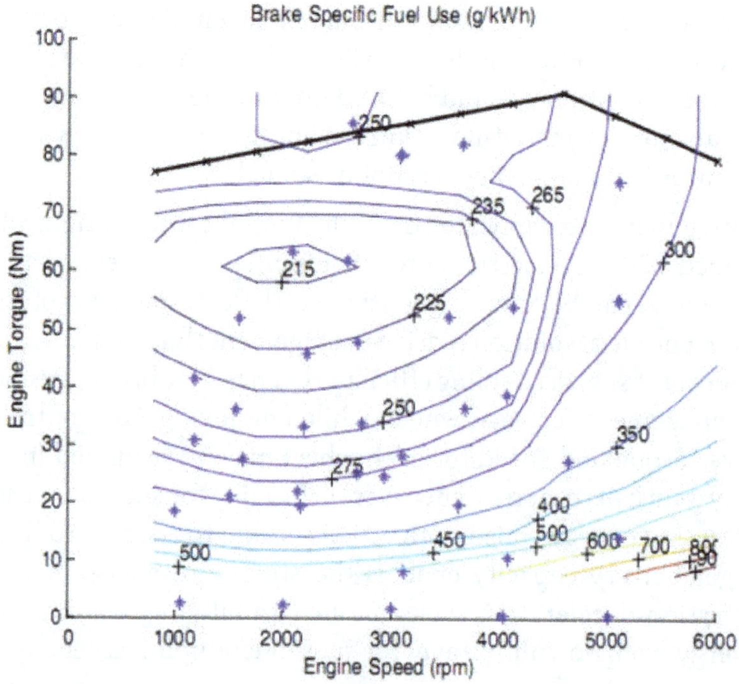

Figure 11. Specific Fuel Use for ICE in Honda Insight

7. OPTIONS

The serial PHEV serves as a viable alternative for military vehicles and other vehicles operating in remote areas, where the battery may alone be charged by a diesel generator. The benefit of night charging from the socket is not there; however, the generator with a battery can function as an autonomous power source for a local DC or AC grid via an inverter. It can be stated that this car is more plug-out than plug-in, yet its composition is analogous, potentially featuring greater internal combustion engine and generator power.

7.1. PHEV without generator

Previous conceptions of PHEVs included solutions with a mechanical connection of the internal combustion engine to the wheels in highway traffic mode, allowing for smaller electric motors and generators, hence reducing fuel consumption compared to an electric drivetrain due to the shorter drive chain. This concept may distinguish between wheels powered by internal combustion engines and those driven by electric motors, allowing a standard front-wheel drive vehicle to be supplied with electric rear-wheel drive.

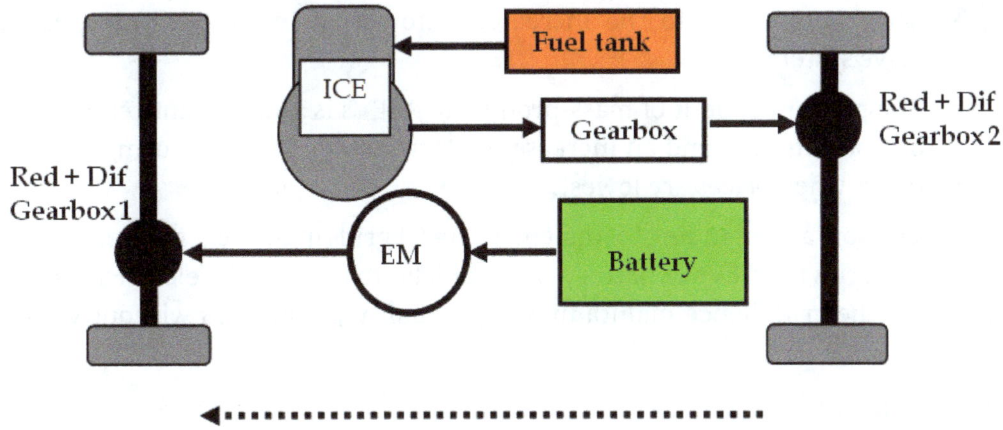

Figure 12. PHEV without Generator

Due to the inefficiency of charging from an internal combustion engine and its inability to conserve fuel, it is feasible to design a vehicle where each axle is powered by an individual motor (Fig.12). For short-distance journeys, Axis 1 is powered by electric motor energy sourced from the battery, which may also include energy recuperated from braking or descending slopes. During extended highway journeys, the internal combustion engine (ICE) powers axle 2, which is solely linked to the first axle via the road surface. The electric motor (EM) does not contribute to propulsion; however, it can recuperate energy. At low speeds, when the ICE operates below optimal efficiency, the driving torque from the ICE may exceed requirements, allowing the EM to function as a generator, converting excess energy into electrical energy (EE) to recharge the battery. In place of the generator in Fig. 3, there is the gearbox (manual or automatic) along with the secondary reduction and differential gearbox, both sourced from standard manufacture. It is an ideal amalgamation of two autonomous drives accessible in emergencies.

8. CONCISE SUMMARY

The rationale for PHEV includes ecological considerations, as electricity derived from renewable sources diminishes carbon emissions.

- Dependence on oil imports, as virtually all appropriate fuels for internal combustion engines are derived from oil.
- Coal hydrogenation was also advanced during the wartime period, and in certain tropical nations, such as Brazil, alcohol fuels are derived from sugar-containing plants. In recent years, plant-based oils have been manufactured to substitute mineral oils for diesel engines.
- Efficiency, particularly in urban transportation characterized by low average speeds, frequent stops, and traffic congestion, when combustion engines operate at minimal efficiency and substantial fuel is consumed during idling.

- Safety, attributed to the increased automation in electric transmission drive systems.
- The primary benefit of mass-producing PHEVs is a significant reduction in oil consumption and an increase in electricity generation during off-peak hours when prices are lowest.

The distributors are intending to implement smart grids in the near future, utilizing several batteries in PHEVs or battery-only vehicles to manage the electrical energy balance in the grid, hence maintaining high-quality parameters without voltage dips or sags.

REFERENCES

1. BršlicaV.2008Super-capacitor integration into hybrid vehicle power source, In Proceeding of International Conference on Renewable Energies and Power Quality (ICREPQ'08), 6978-8-46119-290-8Santander, Spain
2. BršlicaV. .Sept2005Co-generative Power Source for Electric Car, Proceeding of VPPC 2005, IEEE Cat. 05EX1117CCD-ROM) 0-78039-281-7IL USA
3. Altairnano, NanoSafe Battery Technology, ALTI 070404, pp. 4, In web Altairnano.com
4. Buchmann, I., How to prolong lithium-based batteries, BatteryUniversity. Com http://www.batteryuniversity.com/
5. StoberD.2007Nanowire battery can hold 10 times the charge of existing lithium-ion battery, Stanford News service,
6. FehrenbacherK.2009Reva to Boost Range with Lithium-Ion Battery, Earth2tech, http://earth2tech.com/2009/01/05/
7. Total Lithium-Ion Battery Sales Forecast to Double By2012to US$13.1B In Green car congress, http://www.greencarcongress.com/
8. CandaceK.et al.2007High-performance lithium battery anodes using silicon nanowires, Nature Nanotechnology 33135pp., December 16, 2007,
9. SoinoffN.Lithium Battery Power Delivers Electric Vehicles to Market, Scientific & Technical Information, http://www.sti.nasa.gov/tto/Spinoff2008/t_1.html
10. DeguzmanD.2009The race for car lithium battery is on, Green Chemicals, http://www.icis.com/blogs/green-chemicals/January 7, 2009
11. MillerC.2008Electric-Car Battery Makers Seek Federal Funds, December 26, 2008, http://bits.blogs.nytimes.com/2008/12/26/electric-car-battery-makers-seek-federal-funds/
12. ParkerR. .April2009Chevy Volt Battery Over-engineered Due To Unknowns, http://www.futurepundit.com/archives/cat_energy_batteries.html

13. Byoungwoo Kang & Gerbrand Ceder, Battery materials for ultrafast charging and discharging, Nature 458, pp.190193March 12, 2009

14. WeirR. D.et al.2008Utilization of poly(ethylene terephthalate) plastic and composition-modified barium titanate powders in a matrix that allows polarization and the use of integrated-circuit technologies for the production of lightweight ultrahigh electrical energy storage units (EESU), United States Patent 7,466,536, December 16, 2008

15. WeirR. D.et al.2006Electrical-energy-storage unit (EESU) utilizing ceramic and integrated circuit technologies for replacement of electrochemical batteries, United States Patent 7,033,406, April 25, 2006

16. EhrenberG.ScottG.et al.2008Nanoparticle ultracapacitor, United States Patent Application 20080316678 Kind Code A1, December 25, 2008

17. IlyanokM. A.2007Quantum Supercapacitor," United States Patent 7,193,261 B2, March 20, 2007

18. EisenringR.2008Method of storing electricity in quantum batteries, United States Patent Application 20080016681 Kind Code A1, January 24, 2008

19. Tesla Roadster Technical Specifications, 2009 http://www.teslamotors.com/performance/tech_specs.php

20. Chevy Volt: Reasons for Use and Cost of Operation http://gm-volt.com/chevy-volt-reasons-for-use-and-cost-of-operation/ EEStore Energy storage unit, http://bariumtitanate.blogspot.com/http://www.toyota.com/prius/ http://automobiles.honda.com/civic-hybrid/ http://www.chevrolet.com/electriccar/http://www.gm-volt.com/ http://www.ecom.cz/katalog_pdf/maxwell.pdf

21. HundT.2004Comparison Testing of Supercaps, Sandia National Laboratories, Albuquerque, NM, November 2004

22. United States Patent 7 033 406, http://www.freepatentsonline.com/7033406.html http://www.toshiba.co.jp/about/press/2007_12/1102/SCiB.pdf

23. SchindallJ.2007The Charge of the Ultra- Capacitors (Nanotechnology takes energy storage beyond batteries) http://www.spectrum.ieee.org/nov07/5636

24. Lockheed Martin Signs Agreement with EEStor, Inc.for Energy Storage Solutions, 10th January 2008 http://www.gm-volt.com/2008/01/10/lockheed-martin-signs-agreement-with-eestor/

25. EdgarJ.Brake Specific Fuel Consumption http://autospeed.com/cms/title_Brake-Specific-Fuel-Consumption/A_110216/article.html http://en.wikipedia.org/wiki/Brake_specific_fuel_consumption

6 Chapter ELECTRIC CARS WITH FUEL CELLS

1. INTRODUCTION: AN OVERVIEW

More than half of greenhouse gas emissions and a big chunk of air pollution emissions come from burning fuel for transportation. Over the next 100 years, even though technology will get better at using fuels more efficiently, greenhouse gas and air pollution emissions will rise because of rising demand, especially in developing countries. Conventional engines, called ICEs (internal combustion engines), which mostly run on petroleum fuels, cause the most problems. In this contest, different fuels and vehicles that are less harmful to the earth have been suggested. Advanced internal combustion engines (ICE), spark-ignition (SI) or compression ignition (CI) engines, hybrid electric vehicles (ICE/HEVs), battery-powered electric vehicles, and fuel cell vehicles (FCVs) are some types of truck technologies. Using hydrogen in fuel cell cars could lead to lower emissions than other options and the ability to use a different main fuel type (Ogden, 2005) (Fig. 1).

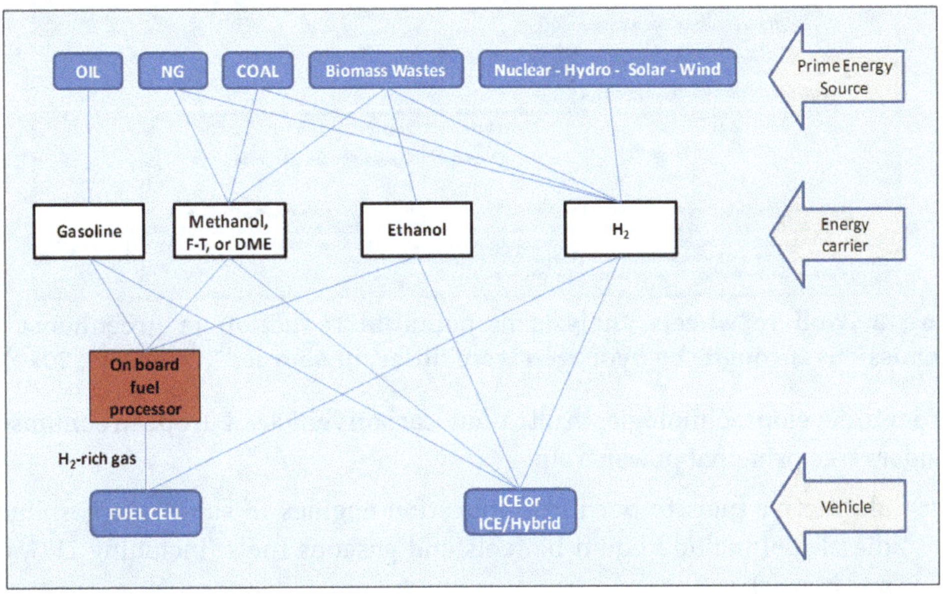

Figure 1. Alternative fuel vehicle pathways.

There are "zero emissions" from fuel cell cars that run on pure hydrogen. In fact, the only local emissions are water vapor. But in this case, it's important to look at the "well-to wheels" emissions, which include emissions from making the fuel, transporting it, and delivering it. The primary source for making hydrogen is very important for how well cars do in terms of the environment. If hydrogen made from renewable energy sources like solar or wind power is linked to an electrolysis process and then used in fuel cells, it can cut emissions by a large amount.Recent research on alternative fuels has found that hydrogen-powered fuel cell cars are the most promising technology when it comes to fuel cycle emissions. The next figure shows a study of how different hydrogen FCV pathways can help cut down on emissions and oil use.

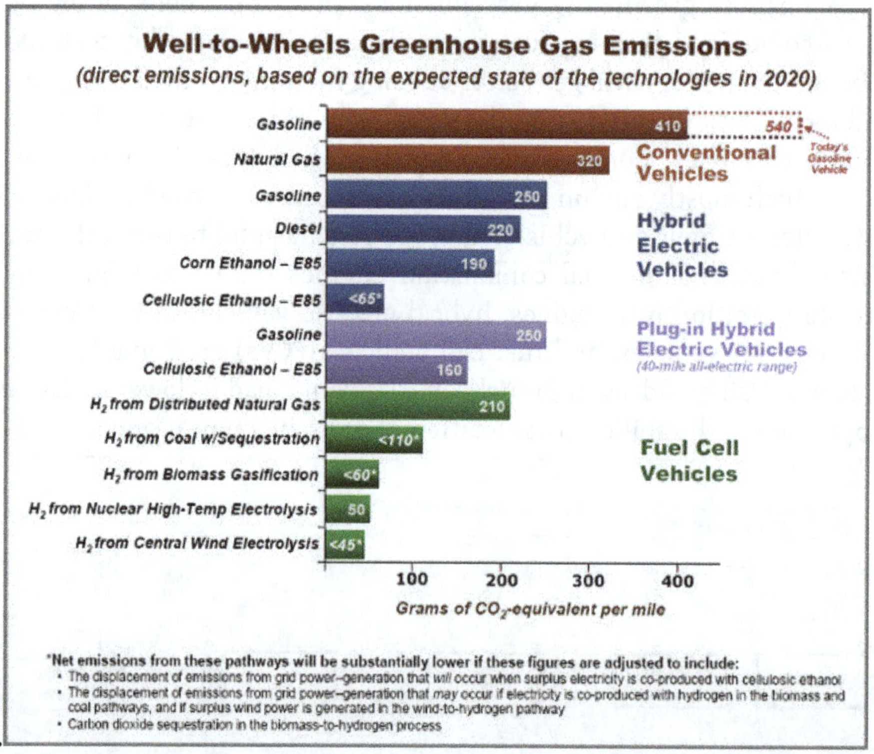

Figure 2. Well to wheels analysis of potential reduction in greenhouse gas emissions through the hydrogen from different sources. (DOE 2009, 2010)

In order to develop technologies in ultra-low-carbon vehicles, European Commission considers tree principal power train:

- alternative fuels to burn in combustion engines to substitute gasoline or diesel fuel include liquid biofuels and gaseous fuels (including LPG, CNG and biogas);
- Electric vehicles;
- Hydrogen fuel cell vehicles.

Advanced vehicles with internal combustion engines may not achieved full decarbonisation alone (Mc Kinsey & Company 2010). It is therefore important to develop different technologies to ensure the long-term sustainability of mobility in Europe.

According with this strategy, hydrogen fuel cell vehicles and battery electric vehicles have similar environmental benefits (European Commission COM(2010)186).

Today, in the light of numerous tests in a customer environmental (500 passenger cars - both large and small - covering over 15 million kilometres and undergoing 90,000 refuellings, Mc Kinsey & Company, 2010) FCVs may be considered technologically ready. Moreover, they are still expensive and further research is needed to bring costs down. To became competitive with today's engine technologies, FCVs must reach large enough markets to reduce the cost via mass production. The figure 3 reports the most important technological challenges of FCVs for commercialization.

Despite great improvements in automotive fuel cell system of last years, significant issues must be still resolved. These challenges include:

- Development and cost of hydrogen refuelling infrastructures for direct-hydrogen FCVs;
- Storage systems for hydrogen simultaneously safe, compact and inexpensive;
- Cost reduction in fuel cell stack and durability;

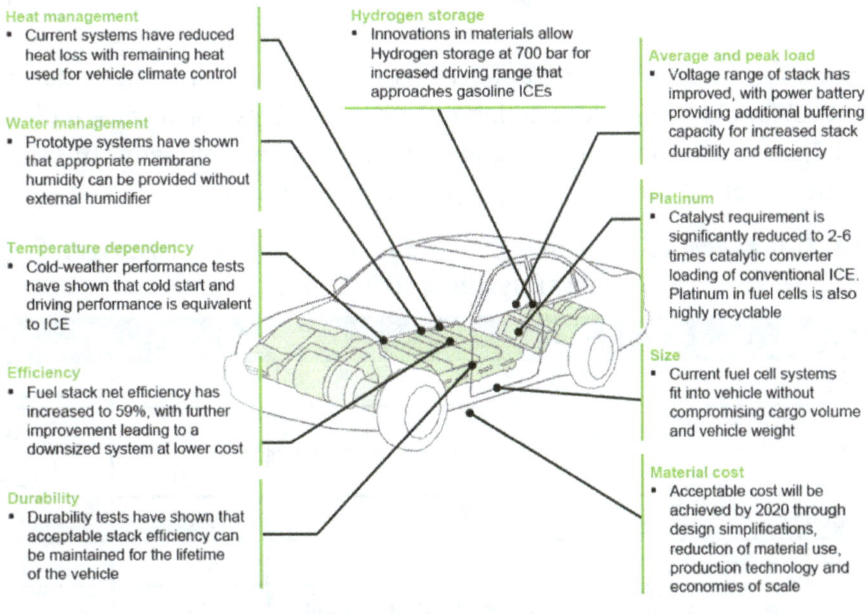

Figure 3. FCVs: from demonstration to commercial deployment (Mc Kinsey & Company, 2010).

The U.S. Department of Energy (DOE) is working on activities that will help get hydrogen and fuel cell technologies developed and used. These activities will address both technological and non-technological problems that are getting in the way. The next picture shows the actions that the program did to get around all of the problems that were stopping hydrogen and fuel cells from being used in businesses.

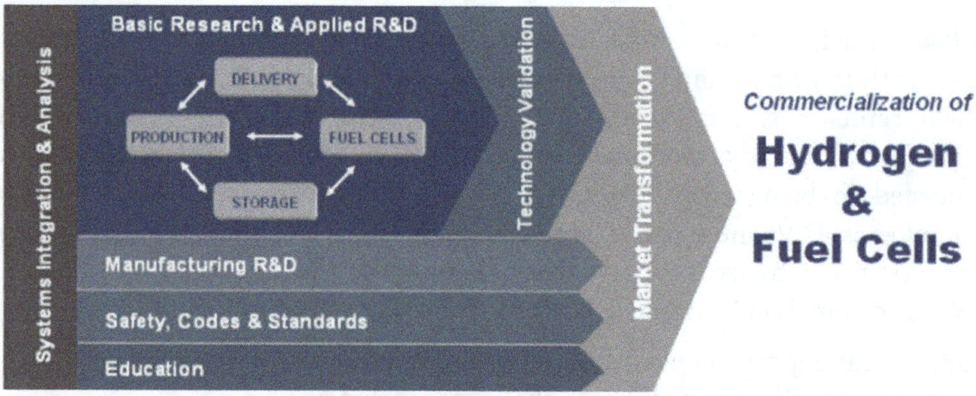

Figure 4. The DOE program's activities for fuel cell commercialization.

Regarding the stacks, the targets are to develop a fuel cell system with a 60 percent of efficiency and able to reach a 5000-hours lifespan, corresponding to 240000 km at a cost of $30/kW (at large manufacturing volumes) by 2015 (fig. 4.). The Program is also conducting RD&D efforts on small solid-oxide fuel cell (SOFC) systems in the 1-to 10-kW range, with possible applications in the markets for auxiliary propulsion units (APUs).

DOE targets for transportation applications were derived with information from FreedomCAR and Partnership, a collaborative technology organization of Chrysler Group LLC, Ford Motor Company and General Motors Company. In table 1 are showed the targets of direct hydrogen fuel cell power systems.

Table 1. DOE targets for automotive application of direct hydrogen fuel cell power systems (DOE, 2010).

Characteristic	Units	Target 2015[a]
Energy efficiency @ 25% of rated power [b]	%	60
Energy efficiency @ rated power	%	50
Power density	W/L	650
Specific power	W/kg	650
Cost [c]	$/We	30
Transient response (time from 10% to 90% of rated power)	s	1

Cold start up time to 50% of rated power @–20°C ambient temperature @+20°C ambient temperature	s s	30 5
Start up and shut down energy [d] from -20°C ambient temperature from +20°C ambient temperature	MJ MJ	5 1
Durability with cycling	hours	5,000 [e]
Unassisted start from low temperatures [f]	°C	-40

[a] *Targets exclude hydrogen storage, power electronics and electric drive.*

[b] *Ratio of DC output energy to the lower heating value (LHV) of the input fuel (hydrogen). Peak efficiency occurs at about 25% rated power.*

[c] *Based on 2002 dollars and cost projected to high-volume production (500,000 systems per year).*

[d] *Includes electrical energy and the hydrogen used during the start-up and shut-down procedures.*

[e] *Based on test protocols in Appendix D.*

[f] *8-hour soak at stated temperature must not impact subsequent achievement of targets.*

An other important issue in fuel cell vehicles commercialization is hydrogen storage. Currently, compressed hydrogen is the principal technology used on board but the research is addressed towards a advanced materials able to store hydrogen at lower pressures and near ambient temperature, in compact and light weight systems (metal hydrides, chemical hydrogen storage and hydrogen sorption).

In this chapter the prospects of fuel cell in transport application will be discussed and particular attention will be paid to the CNR ITAE experiences. CNR ITAE is the National Council Research of Italy that studies advanced technologies for energy. The Institute is involved in different demonstration projects regarding the development of fuel cell hybrid electric vehicles (FCHEVs) and in particular minibus, citycar, bicycle and tractor. Some kind of projects are addressed to different markets, in particular the so-called "early markets" are deal with. In this case the powertrain is electric and hybrid because it is composed by known technologies, like batteries, but also by supercaps and fuel cells that are innovative technologies. Fuel cells have a small size because are used like on board batteries recharge, "range extender" configuration, allowing to increase the range of traditional electric vehicles. The lower fuel cell power means a reduction in terms of stack size then a less cost of it as well as hydrogen storage amount.

Other one kind of projects is instead addressed to a future market. The configuration used is the "full power fuel cell", in which FCVs have a big size of power close the electric motor power. The full power fuel cell vehicles are provided with innovative components such as radio systems (information technology systems - ITS) able to broadcast with other similar vehicles and fleet managing station. They represent a new concept of vehicle because they are a high-tech products, equipped with hardware and chassis made with new light materials and with a platform having interchangeable upper bodies.

2. FUEL CELL TECHNIQUE FOR USE IN TRANSPORTATION

Proton Exchange Membrane Fuel Cells (PEMFC) are the most used technology in FCVs. In part, this dominance is due to large number of companies interested in PEMFC development. In technical terms, PEM fuel cells have high power density, required to meet the space constraints in vehicles, and a working temperature of about 70 ºC allowing a rapid start-up. The electric efficiency is usually 40-60% and the output power can be changed in order to meet quickly demanded load. Other characteristics of PEMFC systems are compactness and lightness. As a result of these characteristics, PEMFC are considered the best candidates for mobile applications. The disadvantages of this technology are sensitive to fuel CO impurities and expensive catalyst, higher CO levels result in loss of fuel cell performance. Furthermore, the electrolyte must be saturated with water and the control of the anode and cathode streams therefore becomes an important issue. In transport applications this technology is used in hybrid configuration with electricity storage devices, such as batteries or super capacitors.

Today real competitors in transport market are SOFC (Solid Oxide Fuel Cell) systems, particularly suited for auxiliary power unit (APU) such as heating, air conditioning, etc (heating, air-condition etc..). SOFCs are characterised by their high working temperature of 800-1000ºC. There are two configuration of stack, tubular and planar. The tubular concept is suitable for large-scale stationary applications while the planar concept is preferred for transport application tanks to the higher power density. The SOFC applications in vehicles are limited to APU rule due to long start-up time and slow dynamic behaviour caused by high temperature operation. However, it is also considered an important option for auxiliary power units on board of vehicles in the 5 kW range. The power density of the SOFC is in the range of 0.15-0.7 W/cm^2 but high temperature corrosion is a problem that requires the use of expensive materials. Delphi automotive and BMW companies have already been examined this technology in prototype vehicles.

Other different typologies of fuel cells used in transport are the AFC (Alkaline Fuel Cell). The use of this kind of FC is, today, limited if compared with other FC technologies. Several units are installed in niche transport sectors such as motorbikes, forklift trucks, marine and space applications. Several installations (80%) were introduced before 1990 and used in space applications especially. The rest were installed in transportation development and demonstration vehicles. After 1990 some units were installed in light duty, portable and small stationary end-use. When PEM units were introduced in the 1980s, the interest was shifted to this fuel cell alternative, particularly for the transport sector. Recently, some companies have been considered AFC technology for operation in stationary and portable application. The main problem of this technology is the carbon dioxide poisoning: small amounts of CO_2 reduce the conductivity of electrolyte. As consequence of this, pure hydrogen must be used. Besides, air needs to be cleaned from CO_2, which limits the application for terrestrial applications considerably.

Finally, DMFC (Direct Methanol Fuel Cell) technology is used to power portable devices and some types of transportation, like boats, motorcycles, and APU. Ballard and Daimler Chrysler put a DMFC system on a light duty truck in 2000, but since then, no other cars have been made. A few years ago, DMFC was seen as a hopeful technology because all refueling infrastructures could be kept up with methanol, which is a liquid fuel. The DMFC power density is lower than that of PEMFC, but the high energy density of the fuel (methanol) means that micro fuel cell systems could be used instead of batteries.

Table 2. Fuel Cell technologies for transport applications.

FC technology	Working temperature	Efficiency	Automotive applications	Advantages	Disadvantages
PEM	70-90°C	50-60%	Buses, Niche transport, light duty vehicles, APU (niche transport vehicles)	High power density rapid start-up capacity to meet quickly demand load Solid electrolyte	sensitive to fuel CO impurities expensive catalysts
SOFC	700-1000 °C	50-60%	APU ((niche transport vehicles)	Tolerance to fuel CO impurities Fuel flexibility Solid electrolyte	Long start-up Slow dynamic load behaviour High temperature corrosion of components
DMFC	60-130°C	40%	APU (niche transport vehicles)	Storage of liquid fuel (methanol)	Low power density high noble metal loadings
AFC	90-100 °C	50-60%	APU (niche transport vehicles)	Low cost components	Sensitive to CO2 in fuel and air

Table 2 summarizes fuel cell technologies for transport sector by application. The used technologies are PEM, SOFC, AFC and DMFC while PAFC (Phosphoric Acid

Fuel Cell) and MCFC (Molten Carbonate Fuel Cell) systems are generally suitable to provide stationary power and generation of heat for residential and industrial applications.

3. THE CURRENT MARKET OF FCVS

Fuel cell vehicles are still in development and demonstration phase. All automakers have substantial development programs underway. Most attention is focused on the use of PEMFCs for transportation applications. Actually, PEM technology is used in different application as shown in figure 5. The majority of units installed globally are used for portable applications. Niche transportation, light duty vehicles and buses are only around 15% of total installed units because the request is low compared to the other markets.

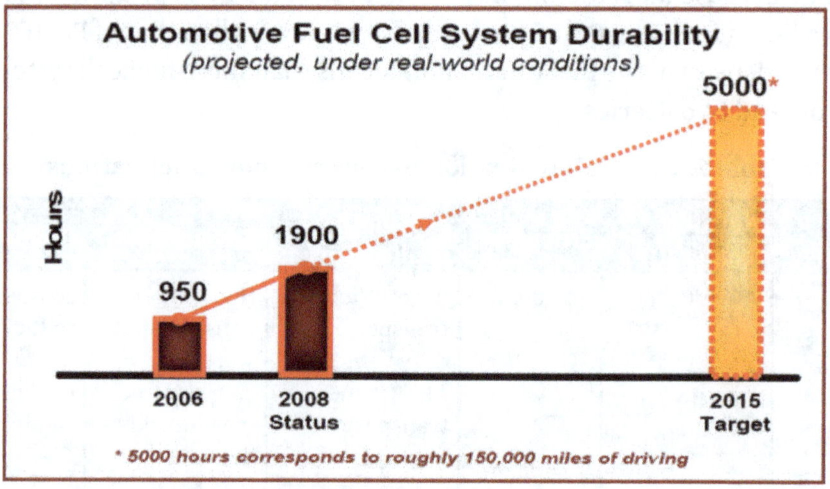

Figure 5. Target of durability of FCVs in order to reach 240000 km (150000 miles). (DOE 2009).

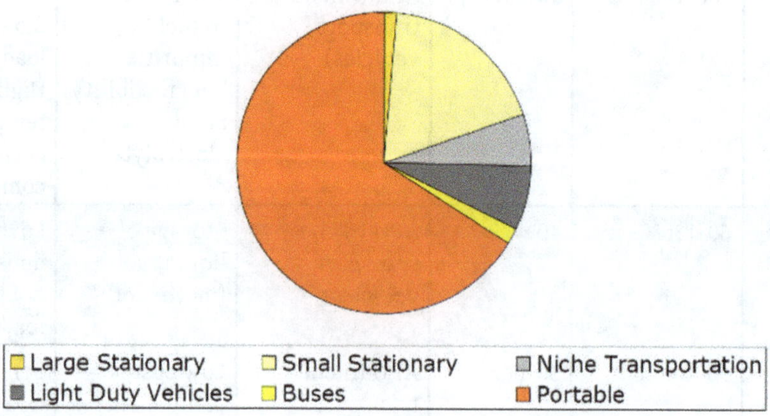

Figure 6. Percentage of PEM units installed by application (Gemma Crawley, 2006).

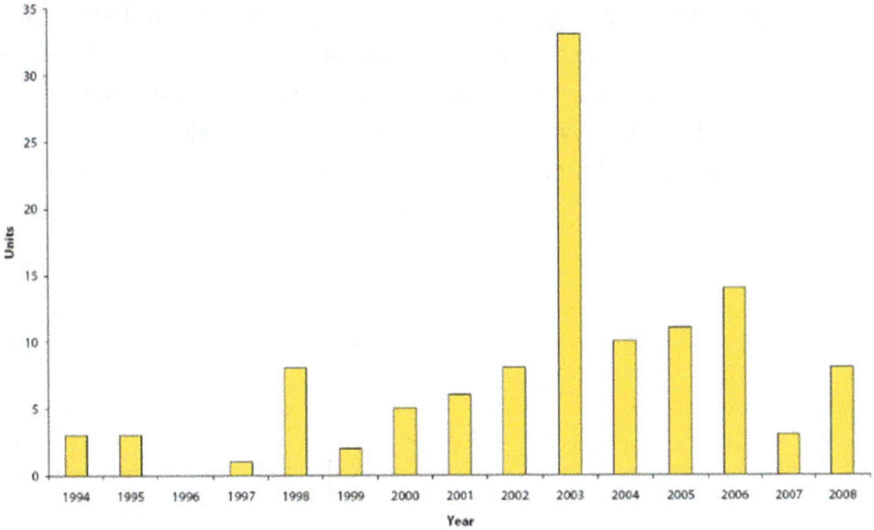

Figure 7. Buses fuel cells units produced from 1994 to 2008 (Lisa Callaghan Jerram, 2008).

The PEFC systems was chosen for providing primary power train for buses involved in the clean urban transport for Europe (CUTE) (2003-2006). A total of 27 Mercedes-Benz Citaro buses, equipped with fuel cell power train, were used on three continents. Figure 7 shows the buses units number produced per years, the peak number is in 2003 when started the CUTE project. In term of fuel cell bus deployment, Europe is leader with 53% of total deployments and 17 cities involved in demonstration projects. Asia's projects are focused in Japan, China and South Korea, in USA several activities are presented in California (fig. 8).

With regard to the regions of manufacture the situation mirrors the deployments with Europe 67% of total. The reason of that is the CUTE project, supported by Mercedes Benz fuel cell Citaros, and a Belgian fuel cell buses company, Van Hool.

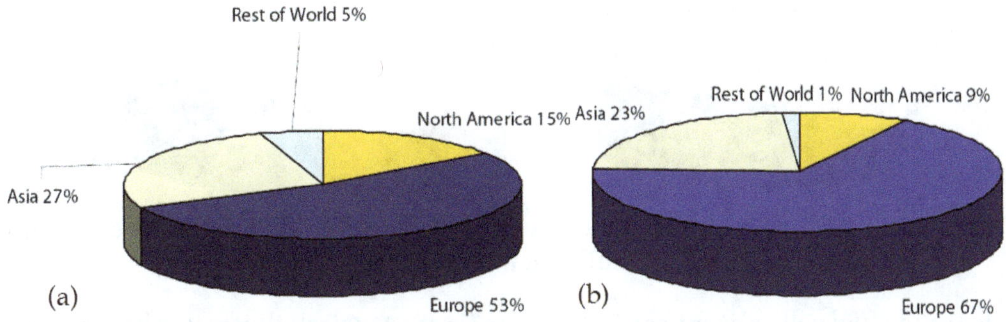

Figure 8. Fuel Cell bus deployment (a) and region of manufacture (b) from 2003 to 2008 (Lisa Callaghan Jerram, 2008).

In the bus sector, the technology selection for duty vehicles is PEM. The yearly distribution of units is variable and indicative of the pre-commercial characteristics of this market (fig. 9). Furthermore, the objectives from previous years have been postponed until subsequent years. The principal automakers engaged in this industry are Honda, General Motors, Nissan, Hyundai-Kia, and Toyota.

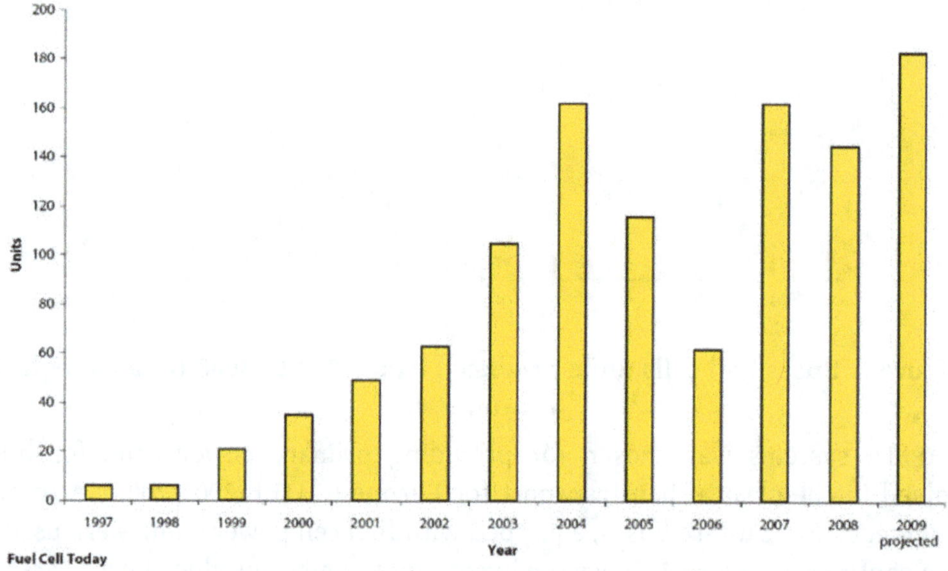

Figure 9. Annual New Light Duty Vehicle (Lisa Callaghan Jerram, May 2009).

Figure 10 reports the manufacture and deployment percentage by region. For 2007-2009 (projected) Asia and North America have become the major areas of manufacture. California, with the presence of infrastructures and the ZEV mandate, is a leading market for fuel cell vehicles. Germany, due to government programs, is a promising country for fuel cell market. With regard to Asia, fuel cell vehicles are used as small fleets leased to government officials.

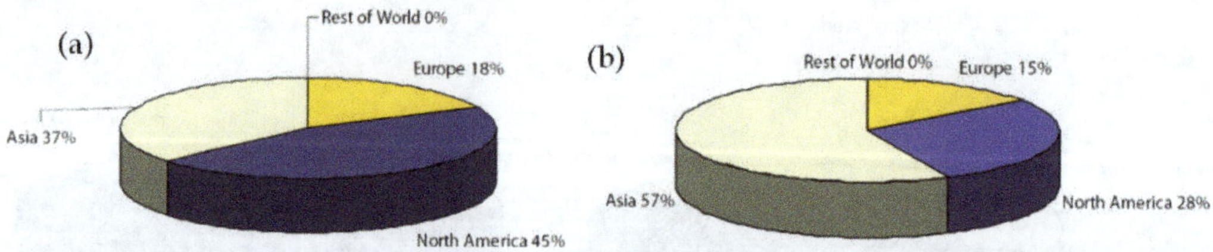

Figure 10. Light Duty Vehicle deployment (a) and region of manufacture (b) 2007-2009 (Lisa Callaghan Jerram, May 2009).

The Honda fuel cell concept car is shown in figure 11 witch illustrates how modern fuel cell systems can be packaged into a small light-duty vehicles. Like most FC cars, the vehicle is equipped with a compressed hydrogen storage system.

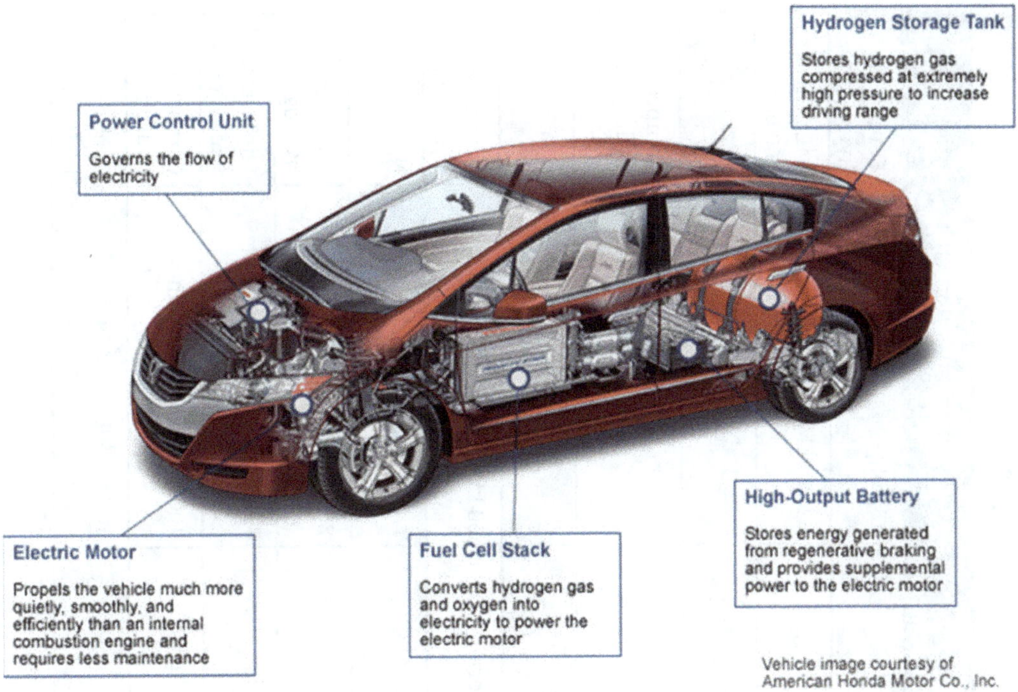

Figure 11. Fuel cell concept car (manufacturer Honda, type FCX Clarity).

In table 3 the last fuel cell vehicles produced by some carmakers are listed. FC manufacture, range and fuel type are reported.

Transport sector comprises applications as aircraft and aerospace, scooters, motorbikes and other two- and three-wheeled vehicles, materials handling vehicles such as forklift trucks, trains and the 'other' category, including such applications as wheelchairs and mobility assistance vehicles. Annual growth from 2005 through 2008 is showed in figure 12. The units installed in 2008 regard principally materials handling vehicles, scooters and motorbikes and the 'other' category including mobility assistance vehicles, each of which saw tens to hundreds of units deployed.

Table 3. Some recent fuel cell Vehicles prototype by automaker (Fuel cell 2000, 2011).

Automaker	Vehicle type	Year	Engine type	FC manufacturer	FC size/type	Range (mi/km)	Fuel type
Audi	Q5 HFC	2010	FC/battery hybrid	N/a	131 bhp	N/a	Compress. hydrogen
AVL list GmbH	AVL FCC: 4-5	2010	EV with FC range extender	N/a	3 kW PEM	150 km	34L CGH @ 200 bar
BMW	FC/hybrid electric 1-Series	2009	FC/battery hybrid	UTC Power	N/a	N/a	N/a
Daimler	Mercedes-Benz F 800	2010	FC/battery hybrid	N/a	N/a	18 mi battery/ plus 375 mi hydrogen	Compress. hydrogen
Fiat/Alfa Romeo	MiTo	2010	N/a	N/a	N/a	N/a	N/a
Fiat	Panda	2007	Fuel Cell	Nuvera	60 kW PEM	200 kW	Compress. hydrogen
Ford Motor Company	HySeries Edge	2007	Fuel Cell plug-in hybrid	Ballard	HySeries Drive	491 km	N/a
GM	Provoq	2008	FC/battery hybrid	GM	88 kW PEM	483 km	Compress. hydrogen
Honda	Fc Sport	2008	Fuel Cell	Honda	PEM	N/a	Compress. hydrogen
Honda	FCX Clarity	2007	Fuel Cell	Honda	100 kW/PEM	570 km	Compress. hydrogen
Hyundai	Tucson ix35 FCEV	2010	FC/supercap hybrid	N/a	100 kW/PEM	650 km	Compress. hydrogen

Automaker	Vehicle type	Year	Engine type	FC manufacturer	FC size/type	Range (mi/ km)	Fuel type
Kia	Borrego/Majoave FCEV	2008	FC/battery hybrid	Ballard	115 kW/PEM	685 km	Compress. hydrogen
Microcab Industries Limited	Microcab	2008	Fuel Cell	N/a	N/a	160 km	Compress. hydrogen
Morgan	LIFECar	2008	Fuel Cell	QinetQ	22 kW/PEM	402 km	N/a
Pininfarina	Sintesi	2008	FC/battery hybrid	Nuvera	Four 20 kW PEM	N/a	N/a
PSA Peugeot Citroen	FiSyPAC	2009	FC range extender	N/a	N/a	496 km	N/a
Renault	Scenic FCV H2	2008	FC/battery hybrid	Nissan	90 kW	240 km	N/a
Shanghai Automoti-ve Industry Corp.	Shangha	2007	Fuel Cell	ShenLi	60 kW/PEM	N/a	N/a
Suzuki	SX4-FCV	2008	Fuel cell	GM	80kW/ PEM	250km	Compress. hydrogen
Tecnalia	H2CAR	2008	Fuel cell/ battery hybrid	N/a	5kW/ PEM	N/a	Compress. hydrogen
Toyota	FCHV-adv	2008	Fuel cell/ battery hybrid	Toyota FC stack	N/a	830km	N/a
Volvo	C30	2010	Fuel cell/ battery hybrid	Powercell Sweden	N/a	400 km	Hydrogen (reformed onboard from gasoline)
VW	Passat Lingyu	2008	Fuel cell/battery	SAIC (Shanghai VW parent co.)	55kW/ PEM	300km	N/a

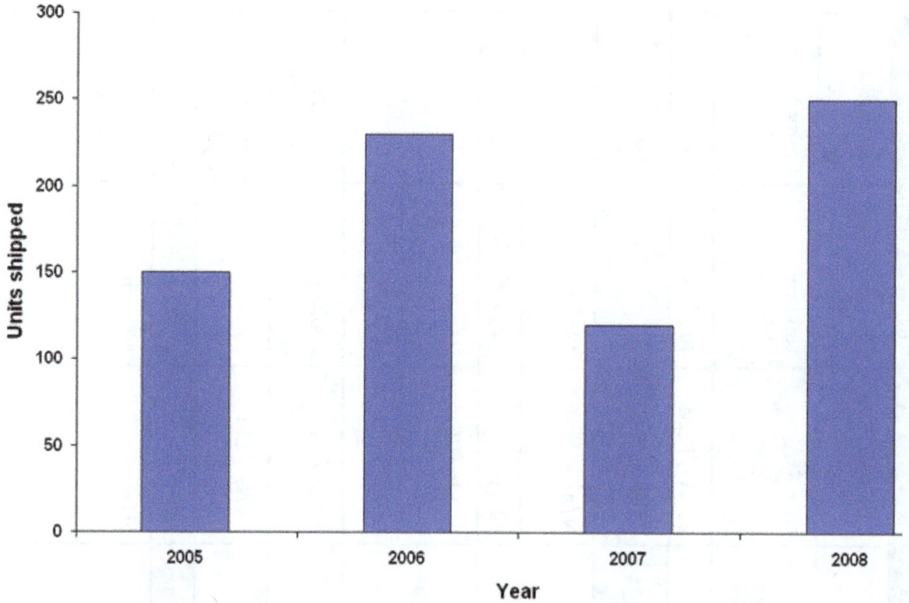

Figure 12. Niche transportation: annual growth (Jonathan Butler, July 2008).

The technologies used for this application are PEM and DMFC, very little units installed are SOFC (fig. 13). In particular, there is most units PEM in aerospace and aircraft sector, a two thirds to one third split between PEM and DMFC in the scooters and motorbike market. In the materials handling market, almost exclusively PEM units are used. In the 'other' category, there are roughly six times the number of DMFC units compared with PEM units.

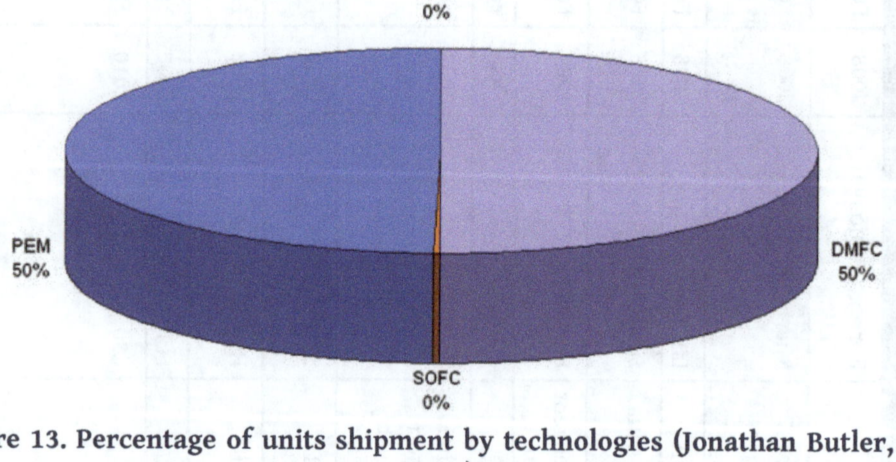

Figure 13. Percentage of units shipment by technologies (Jonathan Butler, July 2008).

Marine and APU market is another interesting application where fuel cells have about 7000 units installed into 2008. Starting with very low units installed in 2005 the numbers increased during 2007/2008.

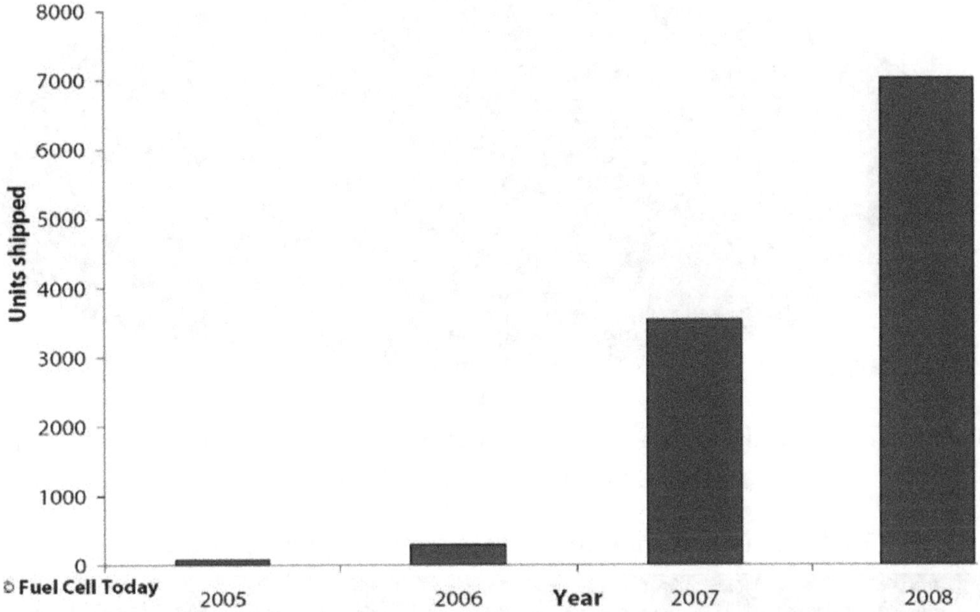

Figure 14. Fuel cells units installed in niche transportation sector (APU and marine) from 2005 through 2008 (Dr. Jonathan Butler, 2008).

In terms of technology, DMFC is the principal chose thanks to the liquid fuel and flexibility of refilling. The followed figure shows the percentage by technology. Number of PEM units are limited to APU applications for on board yachts due to their silent operation.

Very interesting is the SOFC technology used mainly for APU demonstration units for road vehicles and marine vessels. In fact, units installed in this transport sector are higher than in large stationary applications (Gemma Crawley, 2007). In this sector SOFC units are used as APUs to supply auxiliary power to selected vehicles. Companies involved in development of automotive system for fuel cell, as Delphi Automotive Systems, believe in SOFC technology for the high efficiency, simply reforming technology and less stringent fuel requirements. The majority fuel cell manufacturer in transportation sector are indicated in table 4.

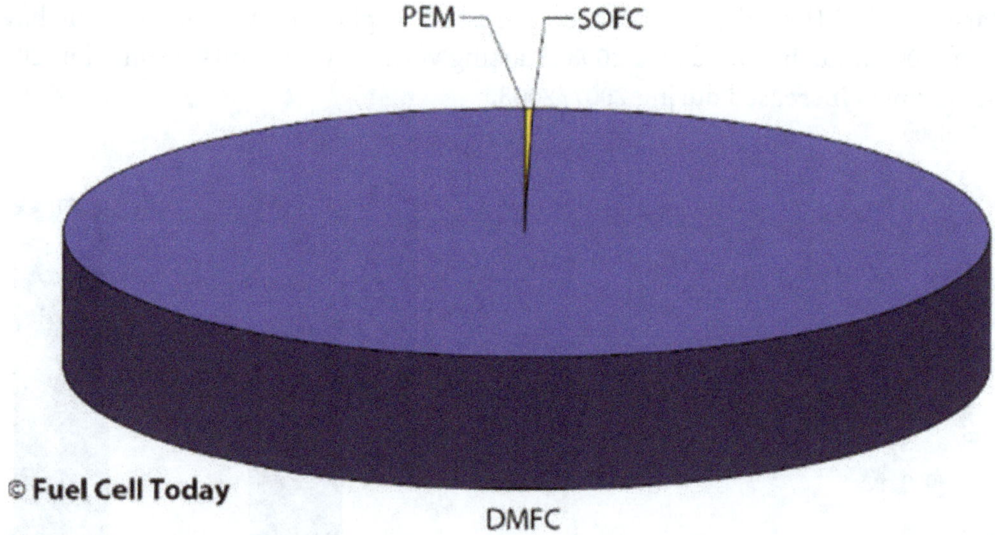

Figure 15. APU and Marine sectors: percentage by technology 2007/2008 (Dr. Jonathan Butler, 2008).

Table 4. Fuel cell manufacturers for transport applications.

Manufacturer	Application	size	Technology
Ballard	Transportation/bus/light mobility	4 - 85 kW	PEM
General Hydrogen	Small/medium/large trucks	/	PEM
SFC Smart Fuel Cell	Mobile/niche transport	600 - 1600 Wh/day	DMFC
UTC Power	Transportation	120 kW	PEM
Hydrogenics	Mobility application	4 – 65 kW	PEM
Nuvera FC	APU, transportation	5 – 82 kW	PEM
Delphi Automotive System	APU	1-5 kW	SOFC

4. VEHICLES CONFIGURATION

Fuel cell vehicles are electric vehicles that utilize batteries and fuel cells for power. Various configurations exist for fuel cell hybrid electric vehicles (FCHEVs). The arrangement specifically relies on the intended hybridization level and the regulations governing fuel cells and batteries. In normal electric vehicles, batteries supply power to the electric motor; in fuel cell hybrid electric vehicles (FCHEVs), batteries and fuel cells are interconnected in a parallel system to collectively generate electricity. Figure 16 illustrates a fuel cell stack linked to a DC/DC converter required to deliver a regulated output voltage. The battery pack is connected to auxiliary devices and the fuel cell. The DC/AC inverter transforms direct current

(DC) into alternating current (AC) to supply an electric motor. The engine can recuperate a portion of energy typically dissipated during braking (regenerative braking). The recovered energy is utilized to recharge batteries.

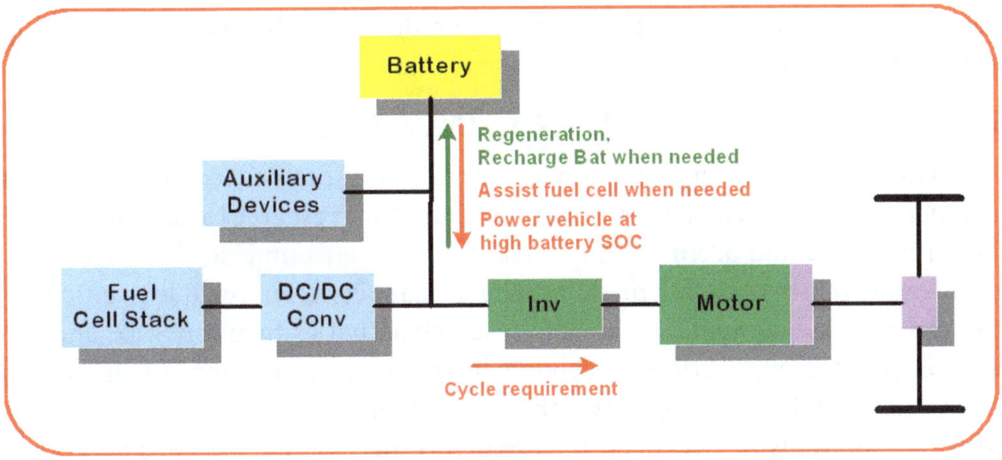

Figure 16. Fuel Cell Hybrid Electric Vehicle configuration

The dimensions of batteries and fuel cells determine the level of hybridization and the configuration. A traditional electric car (fully charged) has inherent limitations such as range (dependent on battery capacity) and charging duration (about 6-8 hours), which may hinder its utilization. FCHEVs enhance range, measured by operational hours or distance, as it depends on the onboard stored hydrogen, and the hydrogen refueling duration is not equivalent to battery recharge time. In the initial form, termed "total fuel cell" or "full power fuel cell," the electric drive motor is entirely powered by the fuel cell, with a minor battery perhaps included solely for vehicle startup or peak power demands. In this instance, the power of the fuel cell is comparable to that of the electric motor. A comparable architecture, featuring a large fuel cell, entails a substantial volume of stored hydrogen (also contingent upon the desired range) and elevated expenses.Another configuration consists of an architecture in which the fuel cell is used as APU (auxiliary power unit) and provides the electrical power required by the auxiliary devices. In this case the fuel cell size is very small and its function is essentially addressed to cover small loads like air conditioning, electric windows, lights, etc.

Finally, the "*range extender*" configuration is characterized by a small size fuel cell used like on board batteries recharge. This solution, depending on the on board stored hydrogen, allows to increase the range of traditional electric vehicles. Using this configuration it is possible to define a specific batteries recharge strategy; in particular the batteries can be recharged when the electric motor doesn't require load, i.e. during the stops and at the terminus. In some case the fuel cell can contribute to the electric traction providing energy when the vehicle runs also.

In this way the fuel cell works in optimal operation conditions at a fixed power, avoiding the load following operation that could cause thermal and mechanical stress of materials.

Moreover, the lower fuel cell power means a reduction in terms of stack size then a less cost of it as well as hydrogen storage amount.

5. CNR ITAE CHALLENGES AND ACTIVITIES

CNR-ITAE's automotive approach focuses on examining fuel cell technology to assess stack performance and its system integration, directing research towards optimizing system interface and architectural trade-offs. This encompasses hydrogen stack modules and reformate stack modules. The primary concerns include to efficiency, cost, durability, and manufacture. The research activities focus on SOFC and PEM technologies for applications involving hydrogen and reformed hydrocarbons (NG, GPL). Despite the long-term objective of utilizing hydrogen as fuel, the existing limitations in hydrogen infrastructure necessitate the consideration of alternative fuels, such as Reformed Natural Gas, as a short-term solution (figure 17). The initiatives in which CNR ITAE participates pertain to the development of electric vehicles powered by electric motors. This type of endeavor targets various markets, specifically addressing the so-called "early markets." The powertrain in this instance is both electric and hybrid, comprising established technology like as batteries, alongside novel technologies like supercapacitors and fuel cells. Fuel cells are compact in size as they function similarly to onboard batteries in a "range extender" configuration, so enhancing the range of conventional electric vehicles. This method facilitates the progressive introduction of FC technology due to its reduced power requirements and expenses.

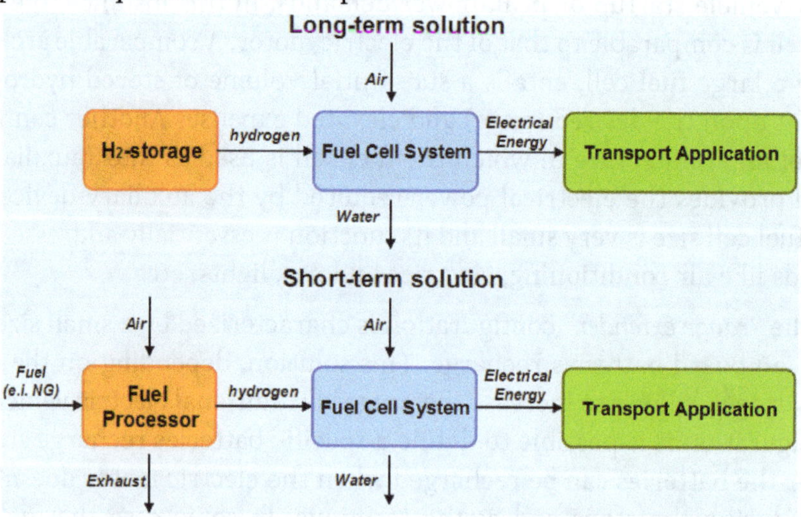

Figure 17. Fuel options for fuel cell power generation.

The other one kind of projects is instead addressed to a future market. The configuration used is the "full power fuel cell", in which FCs have a big size of power close the electric motor power. The full power fuel cell vehicles are provided with innovative components such as radio systems (information technology systems - ITS) able to broadcast with other similar vehicles and fleet managing station. They represent a new concept of vehicle because they are a high-tech products, equipped with hardware and chassis made with new light materials and with a platform having interchangeable upper bodies.

Demonstration projects regarding the development of fuel cell hybrid electric vehicles (FCHEVs) and in particular minibus, citycar, bicycle, tractor and airplane. The tractor projects intends to demonstrate that the fuel cell can be applied in the farm context because hydrogen could be produced on site using different methods (biomass, wind energy, photovoltaic). In this case the hurdle of hydrogen distribution is avoided.

The research activities of CNR ITAE are supported by numerous partners involved in fuel cell development. This includes collaborations and with fuel cell developers like Nuvera Fuel Cell, SOFCPower and with industrial partners.

The most of important projects in automotive sector which CNR ITAE is involved are "Meccano", "BHYKE", "HY-TRACTOR" and "H-BUS".

The project, called "MECCANO", is to develop a highly evolved concept vehicle which offers competitive advantages in terms of optimized ergonomics, low running costs, high levels of safety, modularity and low environmental impact. This new product wants to meet societies demand for reduced congestions, low road-space occupation and improved intermodality with public transportation systems. In this context, the vehicle is characterized by the following features (fig. 18):

- Very highly efficient propulsion system: the powertrain configurations include full-power fuel cell, plug-in battery electric, battery electric with auxiliary motor-generator (series hybrid), and parallel hybrid with methane fuelled internal combustion engine.
- Compact body and short vehicle length (approx. 3m) with high vehicle habitability.
- Advanced technologies for integrated preventive, active and passive safety, in order to attain the highest levels of Euro-NCAP consumer ratings.
- Latest solutions for human-machine and machine-infrastructure interactions and communication personalized depending on the user and on the specific application.

- This economic and ecological urban vehicle aspires to become an ideal mode of transport for environment-aware individuals and municipal authorities. This dual-use concept (ie. individuals as conventional cars or as a means of personalised public transport) introduces radically new opportunities for vehicle design: a) the development of a platform: chassis and low part of the vehicle can be configured in a highly flexible manner in order to accommodate the different propulsions listed above; b) the design of two vehicle bodies and their relative interiors offering different styles and appropriate technologies.

In MECCANO project several automotive companies are directly involved (FIAT, Michelin, Magneti Marelli, Marangoni, ecc.) in conjunction with research institutes.

Figure 18. Meccano project.

The bicycle project "BHYKE" is the study of an innovative electric bicycle in joint venture with an Italian company, called TRE S.p.A. (Tozzi Renewable Energy). The bike, having pedal assistance, is provided with a 250 W fuel cell and a hydrogen solid state storage cylinder of 900 Sl at 12 bar. The targets for this project are: a range of 130-150 km, a maximum speed of 35 km/h and total weight of 30 kg (fig. 13). The aim of the project is to realize, through a new concept of bike sharing service, a representative sample of field test of hydrogen refuelling station from renewable energy (photovoltaic and wind).

Bike	Technical Characteristic
Max. Power Output	250 W
Max. torque	15 Nm@66 rpm
Motor Voltage range	24V Regulated DC
Hydrogen Storage	Solid State
Hydrogen capacity	900 Sl@12bar
maximum range	150 km@pedal assist
Total Weight	26 Kg

Figure 19. Hydrogen bicycle and technical characteristics.

In order to demonstrate that fuel cell technology can be used also in farm sector "Hy-Tractor" project wants to develop a fuel cell tractor fed by hydrogen. In farm sector the hydrogen distribution is not a problem because hydrogen can be produced on site using the available renewable energies: wind, photovoltaic, biomass. The main activities are:

- Development of a hydrogen production and storage system based on: 1) photovoltaic and electrolyzer (fig. 14), 2) biomass, 3) low temperature thermolysis, 4) high temperature pyrolysis;
- Design and development of tractor equipped with fuel cell powertrain, on board hydrogen storage system and other needed auxiliary subsystems.
- Development of energy saving systems for efficiency increase. Some of these are: photovoltaic roof, high efficiency air-conditioning and external lights, hydraulic systems and power take-off (PTO) with electric drive.
- Replacement of hydraulic drive with electric drive, avoiding oil (that is a polluting substances) and increasing the check.
- Design of a *Multi-Power Testing-Trailer* able to carry out simultaneous tests on the traction, hydraulic system and electric devices.
- Field test of the FC tractor during operation both in external sites and inside places (hayloft).

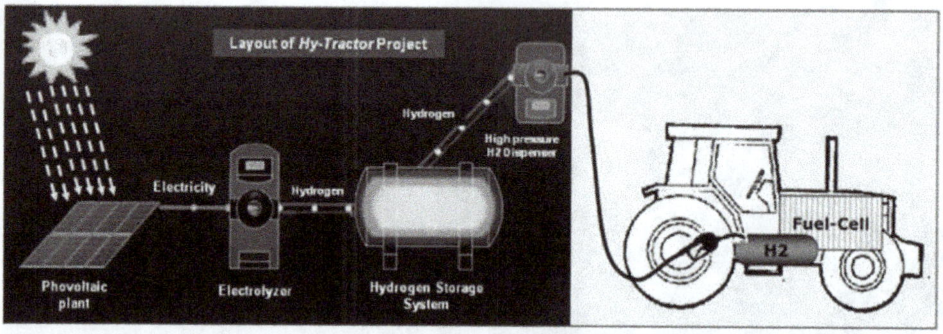

Figure 20. "Hy-Tractor": Project layout with photovoltaic plant.

The H-BUS is a joint project of National research Council of Italy and two supplier companies to develop a range extender Fuel Cell/Battery Hybrid Electric city bus. The aim of H-BUS project is to realize a pre-commercial Fuel Cell/Battery HEV able to increase the range (at least 30%) with respect to same bus in a standard electric configuration, using a small size of fuel cell that works as batteries recharge on board. Within the project, CNR TAE Institute is involved in determining the optimal level of hybridization assessing all boundary conditions (mission, performances, hydrogen consumption, range, etc...). The bus selected for the prototype realization is an electric vehicle having an 85 kW rated power of electric drive motor and a capacity of 44 passengers (Fig.21).

Figure 21. The selected bus for the H-BUS project

5.1. Fuel cell systems development

The CNR ITAE collaborations with fuel cell developers are focused on improving durability, architecture and cost reduction of fuel cell systems and stacks. As above said, in automotive sector, PEMFC and SOFC are the principal technologies studied. The development of PEM fuel cell systems is summarized in table 5, all devises are fed by pure hydrogen. Gen 3 is a hybrid system composed by a stack of 5 kWe and a battery pack with a power output of 4 kW. Besides, this system is equipped with a new kind of hydrogen recirculation system which increases stack durability up to 10000 hr.

A fuel cell system is composed by fuel cell stack and the linked ancillaries: a blower for the air, a pump for the water and a fan for the cooling circuit (Fig 22). Dedicated micro–computer and software are used for the management of the entire system in terms of operation and safety

The stack is the core componentof a fuel cell system but, for the electrical energy production, hydrogen and air have to be fed into the stack. Excess heat must be removed through a cooling system. The operational characteristic curve of a stack (polarization curve) illustrates the device's performance unambiguously. The experimental curve of the fuel cell PEM system is is shown in Fig. 23a.It demonstrates that the stack works in a defined range of voltage of 0.65-1Vcell. In this range of voltage it is possible to obtain high performance in terms of efficiency and to limit the materials stress in order to assure a long durability. The figure also reports cell voltage of stack (average voltage of two contiguous cells) at different power levels (fig.23b). The stack is composed by 40 cells.

An important issue in automotive sector is the response time of system. For this reason start-up/warm-up times have been evaluated at different temperatures in order to determine system limitations and the best operative conditions. The aim was to minimize the battery pack that supply the load and the FC system ancillaries at the same time. The first remark is that batteries cannot be completely eliminated, due to start-up operations. In fact, during the start-up, system drains an average current of 13.5 A (P = 648W), from an external power supply (Fig. 18). The minimum time needed by the FC system to generate power is ever 7 seconds (FC system software setting), but its value never reaches the maximum value (5 kW) before the warm-up.

Table 5. PEM Fuel Cell Systems development.

Rated Power (kW)	5	5	5 kW FC + 4 kW batteries
Number of cells	40	40	40
Temperature (°C)	80	80	80
Active area (cm²)	500	500	500
Efficiency (%)	52	54	54
Durability (hr)	1500	3000	10000

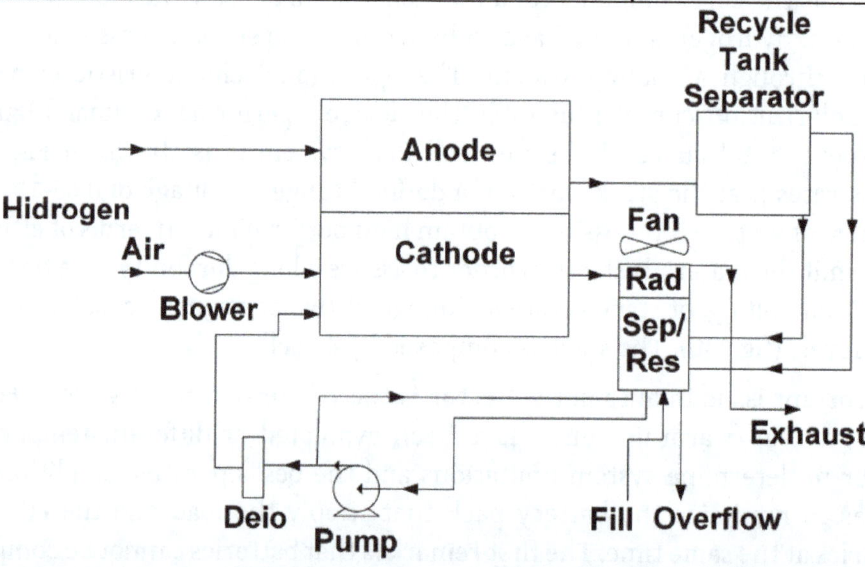

Figure 22. Schematic diagram of the Fuel Cells System.

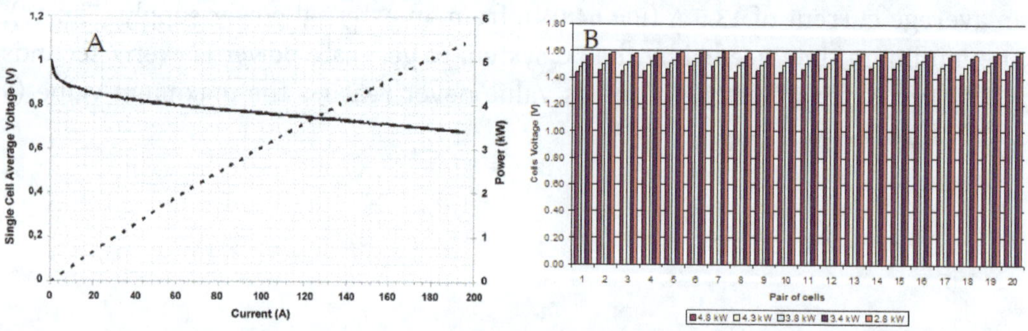

Figure 23. Polarization curve (A) and voltage distribution (B) for a stack of 40 cells PEM.

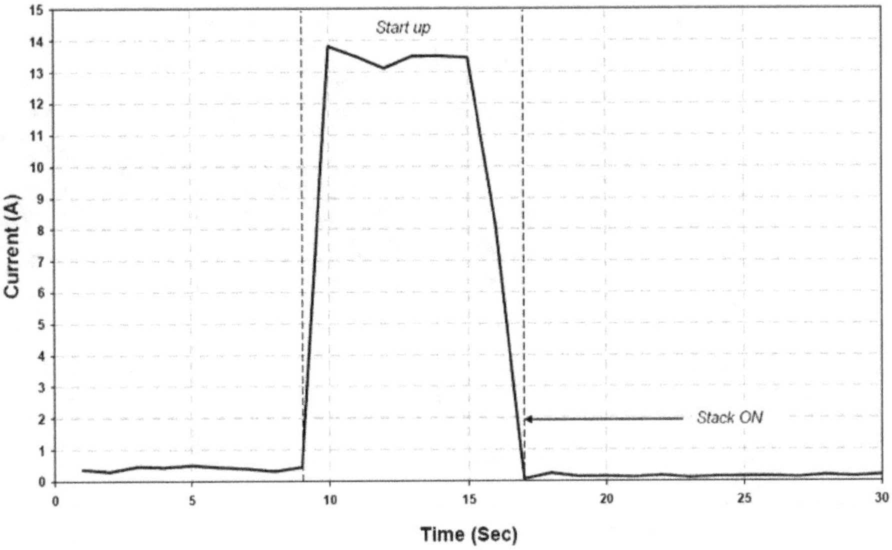

Figure 24. Current demand from external 48V power supply by FC system during then start-up.

The minimum time needed by the FC system to generate power is ever 7 seconds (FC system software setting), but its value never reaches the maximum value (5 kW) before the warm-up. FC system produces the best response when it starts to run at the nominal temperature as shown in the following figure 19, where is reported the start-up/warm-up time depending on the different initial FC system temperature. At the nominal temperature, FC system generates maximum power after 76 seconds during start-up routine runs (7 seconds) and FC stack is warmed-up (69 seconds) at the nominal temperature.

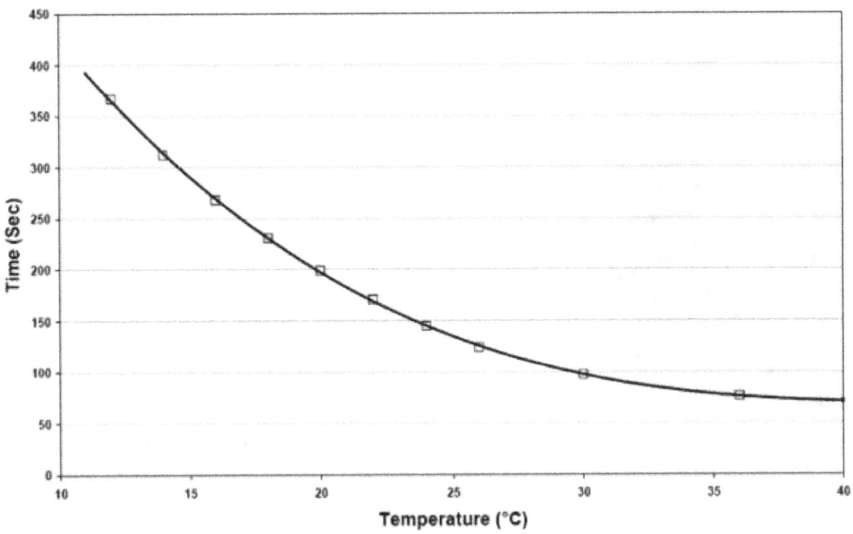

Figure 25. Start-up times at different initial FC system temperatures.

Table 6. Features of the three SOFC stacks.

		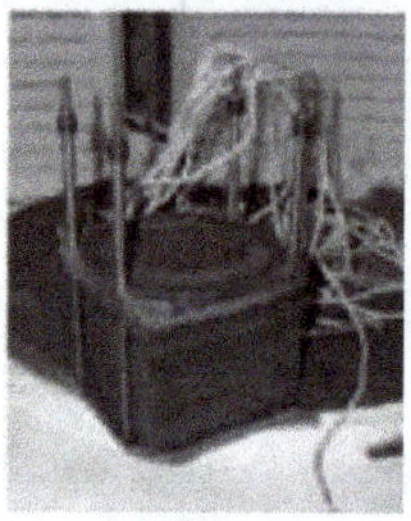	
Rated Power (kW)	1.248	1.811	1.096
Number of cells	75	10	32
Temperature (°C)	650	500	750
Active area (cm^2)	92	360	50
NG reforming	Internal with pre-reforming	Internal	Internal

Among Fuel Cells, SOFCs show the great advantage of working with more flexible gas than than polymer electrolyte fuel cells. The table 6 reports the performance of three intermediate temperature SOFC stacks. The aim is to build a complete SOFC power generation system around the stack. All these three stacks have planar bipolar plate, but they are arranged with different technical solutions: active areas, volumes, dimensions, ect.

A polarization curve of a SOFC system tested is showed in figure 26. The stack power output is 500 W and 55% of electric efficiency is expected. The working temperature is about 750 ºC.

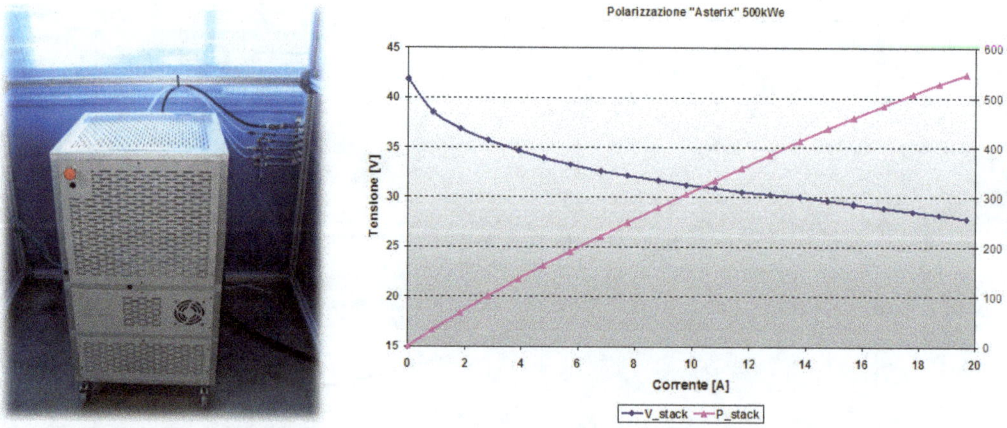

Figure 26. Polarization curve of a SOFC system with a power of 500 W.

These systems are suitable for small recreation vehicles (i.e. motor cycles, golf car), utility vehicles (i.e. fork-lift trucks) and hybrid vehicles in range extended configuration.

5.2. Investigations of hybrid powertrains

CNR ITAE has assessed several powertrain configurations regarding energy flows and the dimensions of system components over the years. This document reports various architectures selected for hybrid powertrains utilized in small cars and buses. The configuration of a hybrid powertrain for a golf cart is identical to that depicted in Figure 16. The hybrid powertrain consists of the following primary components: fuel cell power source, battery pack, and static power converter (positioned between the fuel cell and load diode, which is situated between the static power converter and load). The fuel cell system is a compact power module with a nominal output of 5 kWe, developed in collaboration with Nuvera Fuel Cells. The lightweight vehicle was sufficiently equipped for data collecting through the implementation of a speed transducer and voltage and current sensors (fig. 27); it underwent a work cycle under heavy load conditions, both on-road and in a laboratory setting replicated by electric load. In this latter configuration, the fuel cell is used as main power source for the powertrain, also providing battery charge. The battery has the role to provide peak power during the start up of the vehicle and to supply the necessary energy to the fuel cell system during the start up. The hybrid powertrain has shown a fast response even at extreme and impulsive loads and a wider range compared to a battery vehicle, without compromising the weight limitations on the vehicles.

The figure 28 shows the response of the battery and the fuel cell system during a rising transient. The behaviour of starting batteries is characterized by a short delay in the load response when rising transient begins. This phenomenon is due to a small power inlet from fuel cell to batteries. The batteries package is connected directly to the electronic load and, in correspondence of the power demand, voltage decreases. As a consequence the recharging current of the batteries increases, since the voltage difference between PowerFlow and batteries is higher than the pre-fixed control value. During this very short time (0.1 s) the fuel cell tries to recharge the batteries even if the demand is higher than its rated power. This delay occurs every time the load changes. Moreover, the load response is slightly lower than the electronic load demand.

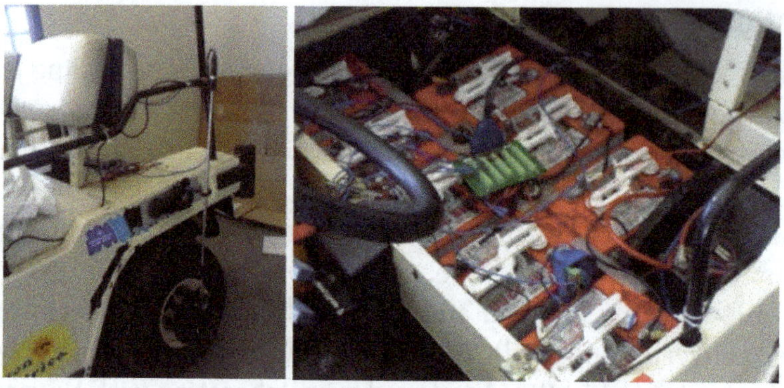

Figure 27. Golf car Hybrid Powertrain.

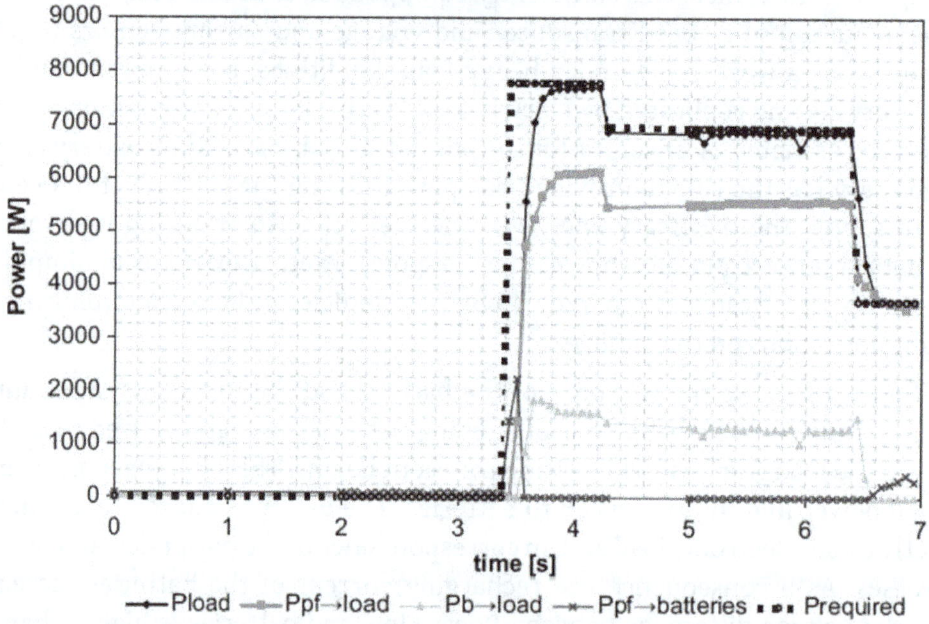

Figure 28. Response of the battery and the fuel cell system during a rising transient.

An important instrument to identify the most favourable vehicle configuration in specified operating conditions is the computer simulations. Figure 29 shows a power train simulation for a bus in range extended configuration. A range-extender HEV is essentially an EV with an on-board charging system(Suppes GJ et al., 2004). Simulation studies have been performed to evaluate the potential SoC saving and autonomy increase with respect of pure battery EV bus. The simulation models have been developed in the Matlab® Simulink® environment utilizing the SimPowerSystems tool.

In the proposed configuration FC system works as batteries recharge that provides, following an identified strategy, the necessary power to the driving cycle to increase the autonomy of the vehicle. The storage system (traction batteries) provides, however, the energy required to satisfy the peak power demand. PEM Fuel Cell and ZEBRA® (Zero Emission Battery Research Activities) technologies have been selected for the fuel cell system and batteries, respectively.

The study has demonstrated that a power train with 6 ZEBRA® batteries connected with 5 kW FC system appears as the best solution. This configuration allows to increase the range of about 40% as shown Figure 30.

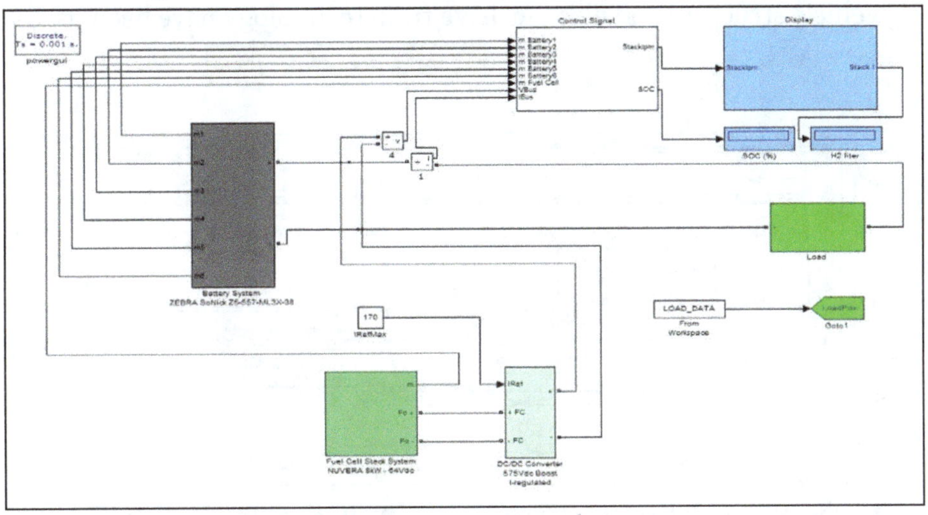

Figure 29. Simulink® model of the powertrain for bus application.

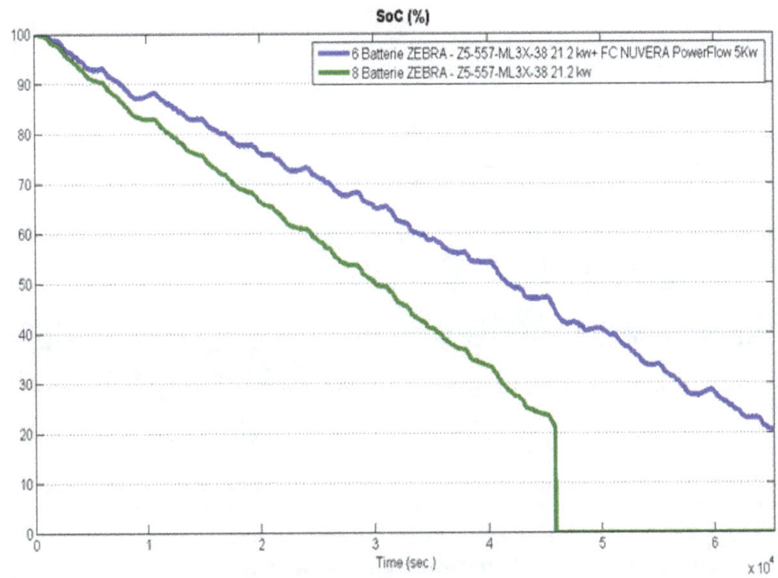

Figure 30. SoC (%) analysis: Comparison of proposed HEV (blue) and pure battery EV (green).

The obtained results show that Fuel Cells and Batteries achieve an optimal synergy because their combination provides better performance and lower costs than batteries or total fuel cells vehicles.

With regard to the integration of fuel cell in the vehicles, the figure 31 shows the layout bus for the project "H-Bus". The fuel cell system and hydrogen storage are assembled on the top of the vehicle in substitution of Nº1 batteries box. TIn order to reduce costs and improve the fuel cell system technological development the exiting vehicle structure and electric drive train technology have been used.

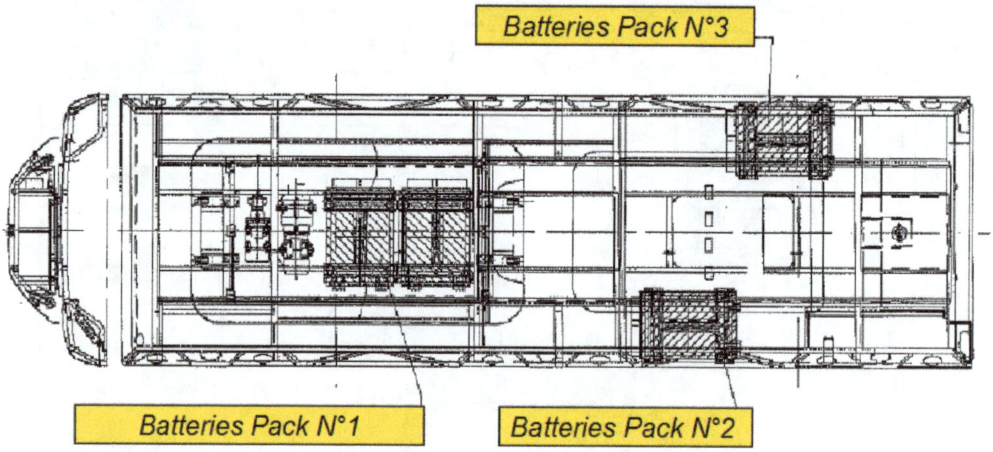

Figure 31. Position of batteries packs on the top of the electric Bus version (only battery electric vehicle).

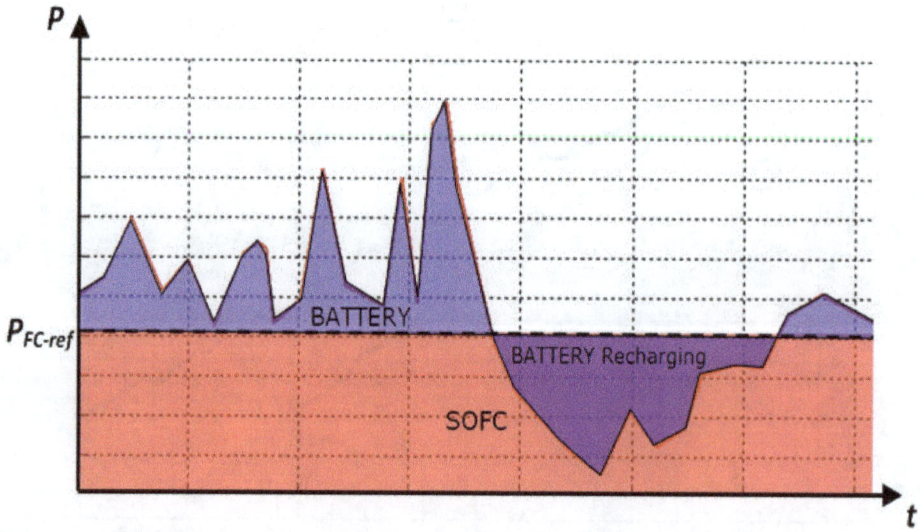

Figure 32. Example of distribution of the power between SOFC and battery.

Certain studies concentrate on Solid Oxide Fuel Cell (SOFC) technology, primarily for Auxiliary Power Unit (APU) demonstration units in hybrid-configured road cars (Battery and Fuel Cell). This study pertains to the integration of a small 500 W SOFC system with a battery. A unique control algorithm was devised to employ the SOFC system as the primary power source, with the battery serving as a supplementary source (Fig.32). In contrast to PEM technology, SOFC devices cannot accommodate rapid and significant fluctuations in load due to their elevated operating temperature. The objective is to create an efficient hybrid system capable of meeting power demands, integrating energy storage, and ensuring long-term functionality. To derive advantages from a hybrid system, the power flows must be meticulously orchestrated and governed in alignment with a suitable energy strategy to maximize overall efficiency and protect the devices from stress that could diminish their lifespan. This 500 W research can be scaled and optimized for particular settings.

REFERENCES

1. World Resources Institute, Sustainable Development Information Service, The global commons: Proceed with caution: Growth in the global motor vehicle fleet.
2. Ogden J. M. 2005Alternative fuels and prospects-Overview, In: Handbook of Fuel Cell. Wolf Vielstich Arnold Lamm Hubert A. Gasteiger, 3 24Wiley, England.
3. European CommissionCOM(2010April 2010. A European strategy on clean and energy efficient vehicles. Available from: < http://eur-lex.europa.eu/LexUriServ/LexUriServ.do?uri=COM:2010:0186:FIN:EN:PDF>.
4. McKinsey & Company on behalf of a consortium of 31 public and private companies.November 2010 A Portfolio of Powertrains for Europe: A Fact Based Analysis: The role of Battery Electric Vehicles, Plug-in Hybrids and Fuel Cell Electric Vehicles. Available from :< www.zeroemissionvehicles.eu
5. DOE, (January 2009 Hydrogen and Fuel Cell Activities, Progress, and Plans. Report to Congress. Available from: < http://www.hydrogen.energy.gov/pdfs/epact_report_sec811.pdf
6. DOE, 2010 Well-to-Wheels Greenhouse Gas Emissions and Petroleum Use for Mid-Size Light-Duty Vehicles. Available from: < www.hydrogen.energy.gov/program_records.html
7. Lisa Callaghan Jerram. 2008 Bus Survey, Available from < http://www.fuelcelltoday.com/
8. Gemma Crawley. . March 2006 Proton Exchange Membrane (PEM) Fuel Cells, Available from < http://www.fuelcelltoday.com/

9. Gemma Crawley. . August 2007 Direct Methanol Fuel Cells (DMFC). Available from < http://www.fuelcelltoday.com/

10. DOE 2010 Research and Development of Fuel Cells for Stationary and Transportation. Available from: < www.energy.gov/>.

11. Lisa Callaghan Jerram (May, 2009 Light Duty Vehicle Survey. Available from < http://www.fuelcelltoday.com/

12. Dr. Jonathan Butler 2008Niche Transport 2Available from< http://www.fuelcelltoday.com/

13. Gemma Crawley (2007 Solid Oxide Fuel Cells (SOFC). Available from< http://www.fuelcelltoday.com/

14. Fuel Cell 2000 2011Available from< http://www.fuelcells.org/info/charts/carchart.pdf>.

15. Dr.Jonathan Butler (July 2008Niche Transport Survey 1Available from < http://www.fuelcelltoday.com/

16. Suppes G. J. Lopes S. Chiu C. W. 2004 Plug-in fuel cell, hybridsas transition technology to hydrogen infrastructure". Int. J Hydrogen Energy 29 369 374

7 Chapter SUPERCAPACITORS AS A POWER SOURCE FOR ELECTRIC VEHICLES

1. INTRODUCTION: AN OVERVIEW

Accelerated advancements in science and technology enhance human existence while simultaneously generating new crises. Humanity faces unprecedented perils throughout its history. Global warming is a quintessential example. While the majority of climate change experts assert that global warming is a reality caused by human activity, there are many scientists that question these assertions. One of the primary issues associated with critical situations is the matter of accountability. World governments must confer with experts and accurately assess when to declare a hazardous situation. Robust political effort is essential to address significant ecological issues, such as global warming. The global political accords reached thus far under the Kyoto Protocol are insufficient to mitigate this issue. This convention mandates an eight percent reduction in greenhouse gas emissions in European Union countries, seven percent in the United States, and six percent in Japan during the initial years of the 20th century. Even assuming that all signatories of the protocol adhere to the agreement, global warming will persist unless there is a significant reduction in gas emissions. Conversely, industrial behemoths such as General Electric, Toyota, and Sharp, together with financial firms like Goldman Sachs, are allocating billions of dollars towards clean technologies. The development of clean technology is not solely a social issue managed by individuals focused on environmental protection; it also represents a viable economic opportunity that generates revenue and gradually integrates into commercial sectors. The economy faces additional problems stemming from elevated oil prices, resource constraints, global environmental issues, and security threats. Clean technologies, designed to deliver enhanced performance at reduced costs while minimizing losses compared to traditional alternatives, have significant potential to drive future economic growth. Science, undoubtedly, addresses all issues pertaining to planetary survival and life therein. It endeavors to address these issues and achieves success to the extent feasible in reality, considering political, social, economic, and technological variables. Electric drive vehicles represent a significant technological advancement in addressing the proliferation of this type of environmental pollution. Recently, there has been heightened global interest in hybrid automobiles, which exhibit reduced fuel usage and significantly lower emissions of pollutants compared to conventional vehicles. Hybrid vehicles can be broadly defined as vehicles that employ a combination of energy production and storage (Van Voorden et al., 2007). The advantageous characteristics of conventional vehicles, such as extended range, rapid acceleration, and an extensive supply network, are integrated with those of electric vehicles, including zero emissions, silent operation, and regenerative braking energy utilization. Two types of these vehicles are under consideration: parallel hybrids and series hybrids. In parallel hybrids, a connection exists between the power generator and the driving wheels, whereas in series hybrids, such a connection is absent. Series hybrids possess significant advantages over parallel hybrids owing to their mechanical simplicity, design flexibility, and capacity for the integration of novel technologies (Stević et al., 2002). In the next five years, electric and hybrid vehicles may become viable alternatives to conventional vehicles in major cities, as indicated by study conducted by the consulting firm McKinsey & Company. In 2015, New York aimed to have approximately 70,000 electric vehicles, constituting 16 percent of newly registered automobiles. Paris is aiming for 60,000 such vehicles, while Shanghai is planned for 25,000. Research indicated that establishing a network of charging stations is unnecessary to boost electric vehicle sales, as owners are willing to charge their vehicle batteries at home. Supermarkets, parking facilities, and restaurants might potentially provide charging stations for their customers as a further step. A highly effective option is to exchange depleted batteries with charged accumulators at specialized stations. A significant aspect influencing political decisions is public opinion. It is crucial to enhance global ecological consciousness and broaden public education accordingly. The objective of this chapter is to familiarize the reader with innovative motor technologies aimed at environmental and ecological conservation.

2. SUPERCAPACITORS VS. ACCUMULATOR BATTERIES AND FUEL CELLS

Supercapacitors are relatively new type of capacitors distinguished by phenomenon of electrochemical double-layer, diffusion and large effective area, which leads to extremely large capacitance per unit of geometrical area (in order of multiple times compared to conventional capacitors). They are taking place in the area in-between lead batteries and conventional capacitors. In terms of specific energy (accumulated energy per mass unity or volume) and in terms of specific power (power per mass unity or volume) they take place in the area that covers the order of several magnitudes. Supercapacitors fulfill a very wide area between accumulator batteries and conventional capacitors taking into account specific energy and specific power (Conway, 1999). Batteries and fuel cells are typical devices of small specific power, while conventional capacitors can have specific power higher than 1MW/dm3, but at a very low specific energy. Electrochemical capacitors improve batteries characteristics considering specific power or improve capacitors characteristics considering specific energy in combination with them. In relation to other capacitor types, supercapacitors offer much higher capacitance and specific energies, as illustrated in Figure 1.

Accumulator batteries and low temperature fuel cells are typical devices with low specific power, where conventional capacitors may have specific power over 1MW/dm3, but at very low specific energy. Electrochemical capacitor can improve characteristics of batteries in terms of specific power and improve properties of capacitors in terms of specific energy when they are combined with them (Stević, 2001).

Supercapacitors, ultracapacitors (commercial denominations given originally by its manufactures Nippon Electric Company, NEC, in Japan, and by Pinnacle Research Institute, PRI, in USA) or electrochemical double-layer capacitor (EDLC, technical name) are devices that can be used as energy storage systems, that have high energy and power densities, a high efficiency, nearly 95% and a large life expectancy (Guerrero at al. 2009).

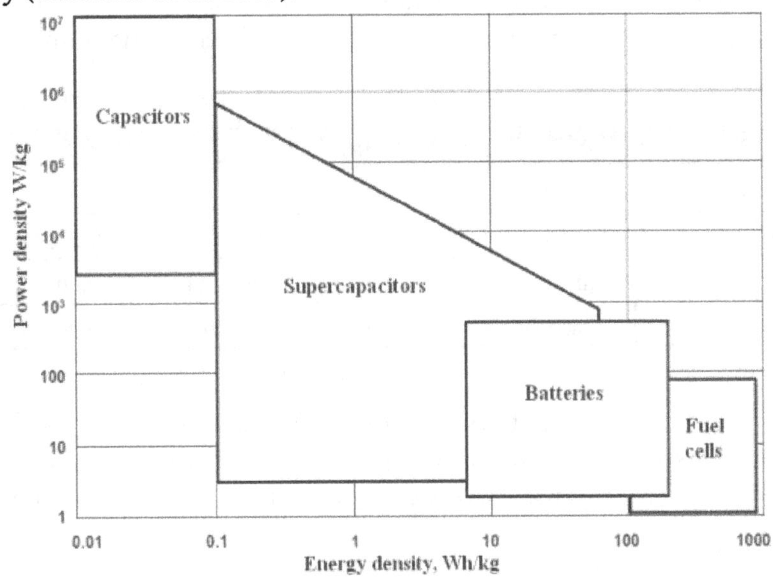

Fig. 1. Area diagram for various energy storage systems

Supercapacitors store charge in a similar way to conventional capacitors, but the charge does not accumulate in two conductors, but in the interface between the surface of a conductor and an electrolytic solution. Devices consist of two electrodes which allow a potential to be applied across the cell, therefore they present two double-layers, one at each electrode/electrolyte interface (Bugarinović at al., 2007).

An ion-permeable separator is placed between the electrodes in order to prevent electrical contact, but still allows ions from the electrolyte to pass through. The electrodes are made with high effective surface materials, such as porous carbon or carbon aerogel (Bugarinović at al., 2008). Two principal technologies are used (Stević at al., 2010): aqueous (maximum voltage of 1.2 V and work voltage of 0.9 V) and organic (voltage near 3 V but with a much higher series resistance).

The principal supercapacitor characteristic that makes it suitable for using in energy storage systems (ESS), is the possibility of fast charge and discharge without lost of efficiency, for thousands of cycles. This is because they store electrical energy directly. Supercapacitors can recharge in a very short time having a great facility to supply high and frequent power demand peaks (Kotz & Carlen, 2000).

Supercapacitor can be manufactured in any size because they do not need a dielectric, form high capacitance supercondensators for hybrid vehicles, to small capacitance ones to be used in low power applications such as wireless systems.

Data given in Table 1 clearly show supercapacitor characteristics that make those devices adequate for purposes requiring great specific energy and great specific power combination or long lifetime denoted by charging and discharging number of cycles In other words, capacitors have retained classical capacitors positive property to achieve almost unlimited charging and discharging number of cycles.

Table 1. Capacitor, supercapacitor and accumulator basic characteristics

Characteristic	Classical capacitor	Supercapacitor	Accumulator
Discharging time	µs – ms	ms – weeks	min - months
Charging time	µs – ms	ms – minutes	hours
Specific energy	< 0,01 Wh/dm3	0,5 – 5 Wh/dm3	< 500 Wh/dm3
Specific power	> 104 W/dm3	(1-3) 103 W/dm3	< 500 W/dm3
Cycles number	106 - 108 (unlimited)	106 - 108	200 - 1000

Literature offers data on two basic kinds of supercapacitors with different ways for energy storing (Arbizzani at al., 2001). New trends in electrochemical supercapacitors. *Journal of Power Sources,* Vol.100, No1-2, (November 2001):

1. double layer capacitors
2. redox supercapacitors.

Capacitance of the first kind is electrostatic by its nature, taking into account that distance between quasi electrodes is extremely short, and electrode material has highly developed surface. Typical examples are Faraday inactive coal powders including both assumptions. The name electrochemical double layer capacitor describes the elementary principle of energy storing at those devices. Principally, energy storing at double layer capacitor is a result of electrical charge separation at the interface between electrode as electronic conductor and electrolyte as ionic conductor. Capacitance created at that interface is the double layer capacitance. Owing to the large specific area, coal is one of the best examples of materials for double layer capacitors (Stević & Rajčić-Vujasinović, 2006). Electrochemical processes taking place on double layer capacitors can be described in the following way (Zheng at al. 1997):

Positive electrode:

$$E_s + A^- \underset{\text{Discharging}}{\overset{\text{Charging}}{\rightleftarrows}} Es^+ // A^- + e^-$$

Negative electrode:

$$E_s + C^+ + e^- \underset{\text{Discharging}}{\overset{\text{Charging}}{\rightleftarrows}} E // C$$

Overall reaction:

$$E_s + E_s + C^+ A^- \underset{\text{Discharging}}{\overset{\text{Charging}}{\rightleftarrows}} E // A + E // C$$

where E_s represents carbon electrode surface, // presents the double layer in where charges are accumulated on the two sides of the double layer, while C+ and A- represent the cation and the anion of the electrolyte. From above given equations it can be concluded that during the charge electrons are forward from the positive electrode to the negative electrode through an external power source, while positive and negative ions are separated from the bulk electrolyte and moved to the electrode surfaces. During the discharge, electrons move from the negative electrode to the positive electrode through the load, and ions are released from the electrode surface and moved back into the bulk of the electrolyte. As it can be seen from overall reaction, the salt (C+A-) in the electrolyte is consumed during charge, so the electrolyte could be considered as an active material. During the charge and discharge, the charge density at the interface electrode– electrolyte changes as well as concentration and electrolyte conductivity (Stević at al., 2002).

Figure 2 presents the scheme of a symmetrical double layer elemental cell as well as illustration of interface on which electrochemical double layer and potential drop appear (Rajčić-Vujasinović at al., 1999).

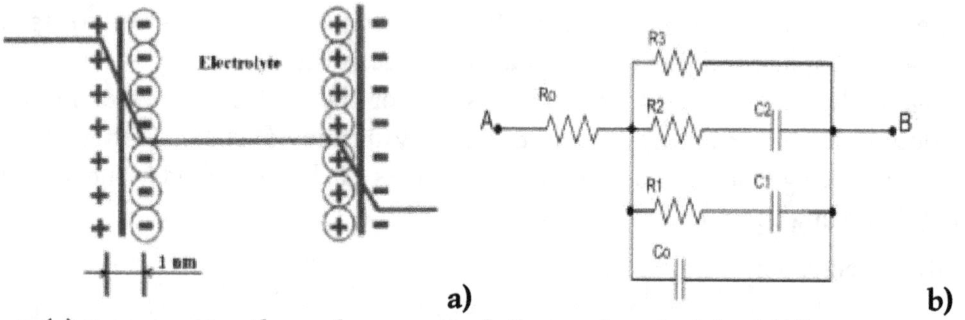

Fig. 2. (a) Cross section through a practical electrochemical double layer capacitor and profiles of the potential across an electrochemical capacitor; (b) Simplified equivalent electric circuit for an electrode – electrolyte interface

In the case of redox capacitors, in the course of electricity pass, Faraday processes are occurring as at batteries and a new phenomenon, called pseudocapacitance is appearing (Ardizzone at al., 1990). That is the reason why this kind of supercapacitors has got the name pseudocapacitors. Typical representatives of the supercapacitors type are RuO2, Co3O4, NiO2 and IrO2 in acid water solutions. Investigation of materials by cyclic voltammetry method showed extremely capacitance behavior that can be explained only by double layer capacitance. The latest researches in this field are focused to the development of supercapacitors based at fast reversible Faraday reactions. Specific capacitances realized at those supercapacitors are from 50 to 64 F/g, specific energy from 25 to 40 kJ/kg and specific power 4 to 17 kW/kg.

Pseudocapacitors have rather greater values of specific energy comparing to carbon double layer capacitors. Besides, electrical conductivity of metal oxide (RuO2) is extremely higher than at carbons and all together lead to greater specific power or, in other words, to less RC (resistance – capacitance) value of time constant. These pseudocapacitors advantages are decreased by their high price compared to carbon. However, advantages realized with carbon materials can be combined with advantages achieved with transient metals oxides, so it led to the new class of electrochemical capacitors. Development of combined double layer Faraday pseudocapacitors as a result may have the use of the advantages of metal oxide Faraday capacitance and carbon material double layer capacitance (Miller at al. 1997). Investigations of behavior of copper sulphide minerals during their anodic polarization and modeling of these reactions were performed at Bor Technical faculty. On this occasion, it was established that equivalent electrical circuit must contain very high capacitances, indicating possibility of copper sulphide mineral use as potential material for electrochemical supercapacitor electrodes (Stević at al., 2003).

3. CHARACTERIZATION OF SUPERCAPACITORS

Electrochemical investigation methods are widely used for characterization of different kinds of materials, as well as of the processes in systems where the electrochemical reactions take part. There is a series of well known methods, but some new methods from electrotechnical area have been introduced (Stević at al., 2008). So, first of all it was given an overview of the standard electrochemical methods and parameters, beginning with potential measurement and simple methods such as chronopotentiometry and chronoamperometry, till electrochemical impedance spectroscopy (Stević at al., 2010). The last named method is adapted for systems containing large capacitances that became actually with appearance of electrochemical supercapacitors. New methods are Dirac voltage excitation and Dirac current excitation. Measurement system described here allows application of electrochemical methods, as follows: measuring open circuit potential, chronopotentiometry, chronoamperometry, galvanostatic method, potentiostatic method, Dirac voltage excitation, galvanodynamic method, cyclic voltammetry and electrochemical impedance spectroscopy.

For signal generation and data acquisition it was developed a measuring and control system based on PC Pentium 4. Beside PC, hardware consists of ADDA converter and external interface for analog signals conditioning. ADDA conversion is performed using commercially available converter NI 6251 from National Instruments, M series high-speed multifunction data acquisition. They have an onboard NI-PGIA2 amplifier designed for fast settling times and high scanning rates, ensuring 16-bit accuracy even when measuring all channels at maximum speeds.

Measurement interface has been designed for the needs of the electrochemical investigations by controlled current or voltage excitation and response registration.

The software platform for predicted measurement methods was National Instruments Lab VIEW package, which is regarded as a high standard in the area of modern virtual instruments. In Lab VIEW, one builds a user interface by using a set of tools and objects. The user interface is known as the front panel. As an example, the description, as well as Front Panel has been showed for the Electrochemical Impedance Spectroscopy (Stević at al., 2004).

3.1 Electrochemical impedance spectroscopy

In this method differs, before any measurement has been done, on the outlet there has been put the direct (DC) voltage to last long enough in order to achieve a stationary regime, and then the current measurement can be started as a response to a complex excitation (superimposed alternating voltage of an order of magnitude milivolts on DC level up to 10V). The assigned parameters are: open circuit potential, Eoc, direct voltage, EDC, and time of the regime duration, to (Stević at al., 2009).

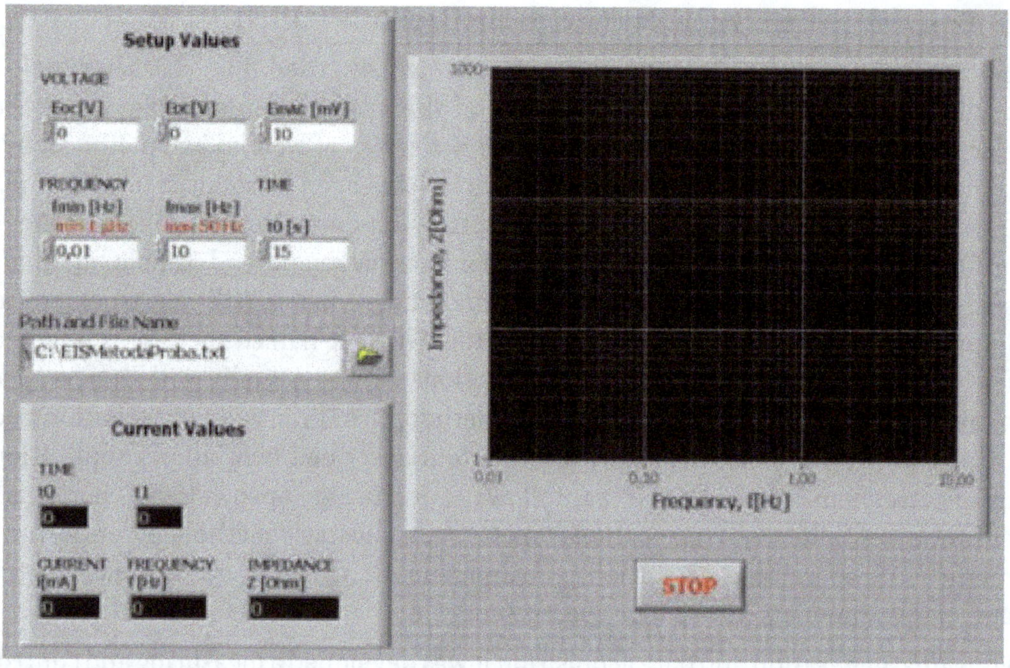

Fig. 3. Front panel of the measuring system for electrochemical impedance spectroscopy

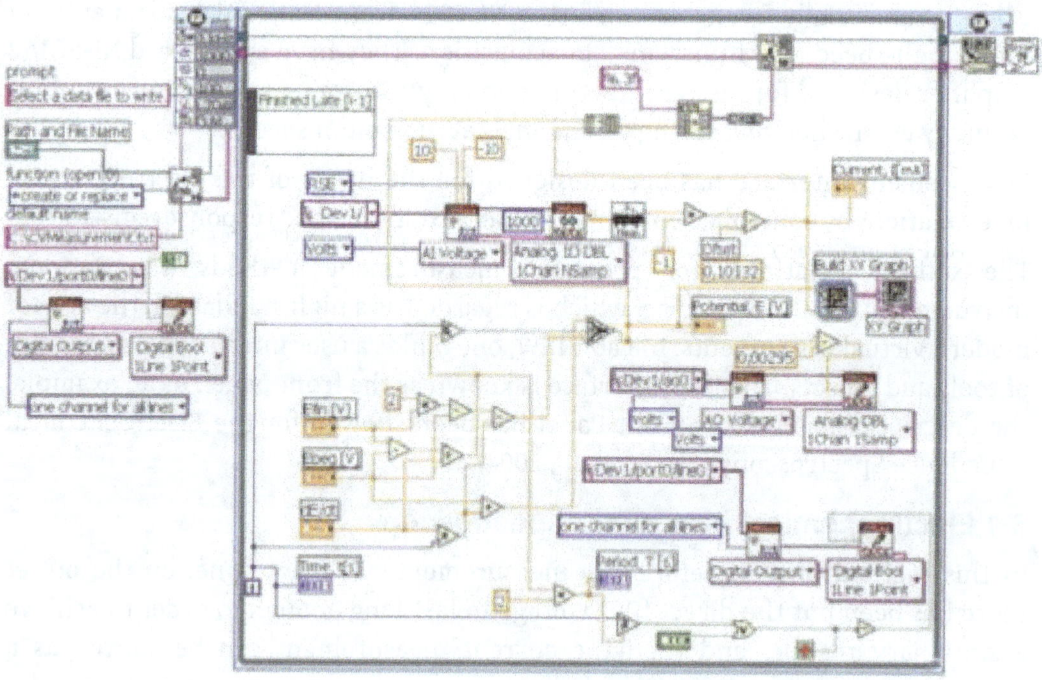

Fig. 4. Electrochemical impedance spectroscopy diagram

After the DC regime expires, the initial WHILE loop starts, in which the current value of the alternate voltage of the assigned amplitude EmAC and the calculated frequencies (block DAQmx Write), the channel (AOCH1) have been generated. The frequency is calculated in the external loop according to octaves, in relation to the assigned values starting from fmin to fmax. The lowest frequency can be 1μHz necessary for systems with high capacitances.

The generated alternating voltage is being superimposed with already installed DC voltage and they both make together the excitement of the electrochemical system.

The current response, which in the pseudo stationary regime has also got the sinusoidal shape with a DC component, is measured on the analog input channel, being averaged, converted into an array and led in a block Array Max&Min. On the base of this block it is possible to calculate the average maximal value of the superimposed component of the voltage excitation so that the outcome is the module of system impedance which is memorized and graphically displayed as a function of the logarithm of the frequency – Bode diagram of an impedance (Fig.3 and 4).

4. APPLICATIONS OF SUPERCAPACITORS

Considering applying, there are four groups of supercapacitors. Depending on applying place, different characteristics of supercapacitors can be more or less taken into account. Some of them are of crucial importance for capacitor choice, and some of them can be of no importance at all. The most important catalogue data for different commercial supercapacitor types are given in Table 2.

First group includes capacitors for supplying electronic consumers of small power and very low voltage (CMOS memories, watches, micro controllers, intelligent sensors etc.). They are most frequently produced as a miniature cell of great capacitance. Crucial role at choice have cell voltage, capacitance and self-discharging current. Inner (linear) resistance is resistance of less importance.

Second group includes filter capacitors for obtaining "ideally" filtered one-way voltages. Apart from great capacitance, cell voltage is of great importance for them, and for achieving working voltage, it most frequently requires linear connection of several cells. Self- discharging current and linear resistance are practically of no importance.

Third group of supercapacitors in future will have great application in energetic electronics complexes of medium power, using for electrical energy reservoirs in transient regime. There is an actual possibility to replace soon massive inductivities, which are at the same time great sources of electromagnetic disturbances. Used for those purposes, supercapacitor must have great capacitance and relatively great working voltage (implicating linear cell connection and all problems related to that). Inner resistance has to be rather small, and leakage current is not of greater importance.

Most strict requirements are related to capacitors of fourth group applying in electric haulage, i.e. for vehicles of the future. Nowadays, batteries of several hundreds farad capacitance are with working voltage of several hundred volts have been produced. Beside great capacitance and relatively high working voltage, these capacitors must have great specific energy and power (because of limited space in vehicle). Considering their specific power, they have great advantage in relation to accumulator batteries, but, on the other side, they are incomparably weaker considering specific energy. Hence, ideal combination is parallel connection of accumulator and condenser batteries. In an established regime (normal drawing) vehicle engine is supplied from accu-battery, and in the case of rapidly speeding, from supercapacitor. Very important is the fact that in the case of abrupt breaking, complete mechanical energy could be taken back to system by converting into electrical energy only in presence of supercapacitor with great specific power. Because of mentioned reasons, inner resistance of these supercapacitors has to be extremely small. Leakage current is not of essential importance. Vehicles with such drive are not still in wide use, and the reasons for that are for sure economic.

Table 2. Some of the most important characteristics of different supercapacitor types

Parameter / Model	Pulse fast supercapacitors		Supercapacitors for applying: Continually Reserve			
Dimensions, mm	55 x 35 x 4	27 x 17 x 2,0	φ24 / h5		φ21 / h15	
Nominal voltage, V	4,5	4,5	3,6	5,5	3,6	5,5
Nominal capacitance, mF	200 – 400	50 –80	1000	400	2400	1000
Leakage current, μA	< 50	< 20	< 3			
Working/storage temperature, o C	- 20 do + 60	- 20 do + 60	- 25 do + 70			

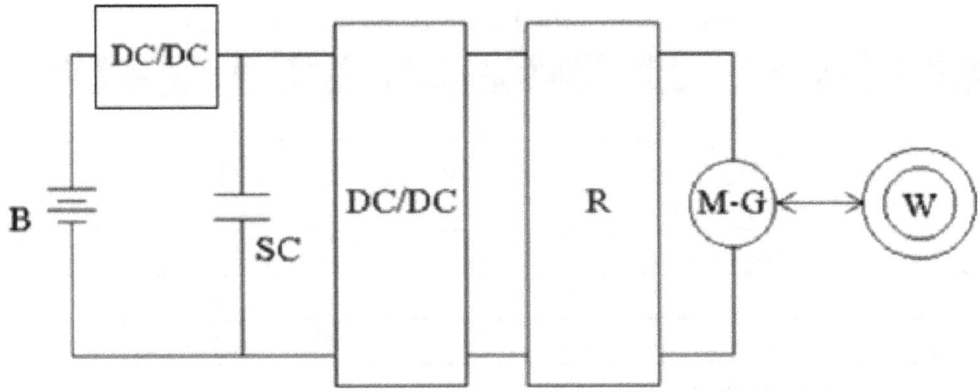

Fig. 5. Scheme of electrical drive vehicle with supercapacitor with possibitlity for using breaking energy; B – one-way voltage source, SC – supercapacitor; DC/DC – direct voltage converter; R – regulator; M-G – engine – generator (depending on working regime; W – drive wheels

Vehicles with electrical drive would present one of the most significant ecological advantages taking into account diffusion of this kind of nature pollution. There is an increased interesting for so-called hybrid vehicles in the world lately, with less fuel consumption and relatively less harmful product emission. Generally, hybrid vehicles could be described as vehicles using combination of technologies for energy production and storage. Two types of the vehicles are in consideration – so called parallel and linear hybrids. Parallel type possesses mechanical connection between power generator and drive wheels, while in linear one such connection does not exist. Serial hybrids have significant advantages in relation to parallel ones because of their mechanical simplicity, design flexibility and possibility for simple incorporation of new technologies.

In Figure 5 the scheme of an electrical drive vehicle in which supercapacitor is used for energy storage and so-called regenerative breaking is presented.

Critical component at each hybrid or pure electrical vehicle presents electrical storage. Supercapacitors are nowadays the only available technology, which can provide great specific power (over 1 kW/kg) and great number of cycles at reasonable price and save and reliable work. Supercapacitors have other characteristics that make them attractive in hybrid vehicles such as possibility of complete energy using (so called regenerative breaking) for increasing energy efficiency, with no additional maintenance, great recovery of electrical energy, little toxicity and easy disposal after usage.

The greatest advance is expected at hybrid vehicles in city traffic. More variants of vehicle prototypes are tested in real conditions in heavy traffic of the largest world cities. Data on installed supercapacitors in a vehicle are given in Table 3.

Table 3. Data on installed supercapacitors in a vehicle

Characteristic	Value
Voltage maximum, V	400
Current maximum, A	1400
Average current, A	178
Capacitance, F	4
ESR, W	0.143
Mass, kg	44.6
Average power, kW	51
Power maximum, kW	286
Average effficiency, %	86
Total energy, kJ	300
Useable energy, kJ	226
Energy density, Wh/kg	1.4
Power density, kW/kg	6.4

5. SUPERCAPACITORS IN REGULATED ELECTRIC PROPULSION SYSTEMS

Regulated electrical drives are nowadays 20-25% of all electric drives. They are developing quickly and present to constructors stricter and stricter speed regulation (and position) and torque. From energy point of view it is desirable their more participation, since optimal speed setting or required can lead to reduction of used energy.

Typical regulated speed electrical drive is comprised of:

- converter located between source and motor that adjusts typical source values: frequency, voltage, current, number fo phases to required by motor. Dosing this energy also achives motor governing.
- Mechanical transmission between motor and load that adjusts motor speed and torque to torque of the working mechanism (load)
- Data from all elements (source, converter, motor, transmission, load) are collected by regulator (controller), which based on given (required) parameters performs automatic drive control.

DC source voltage is performed by means of DC-DC converter (chopper) combined with non regulated rectifier. Figure 6 shows principal scheme of such a system.

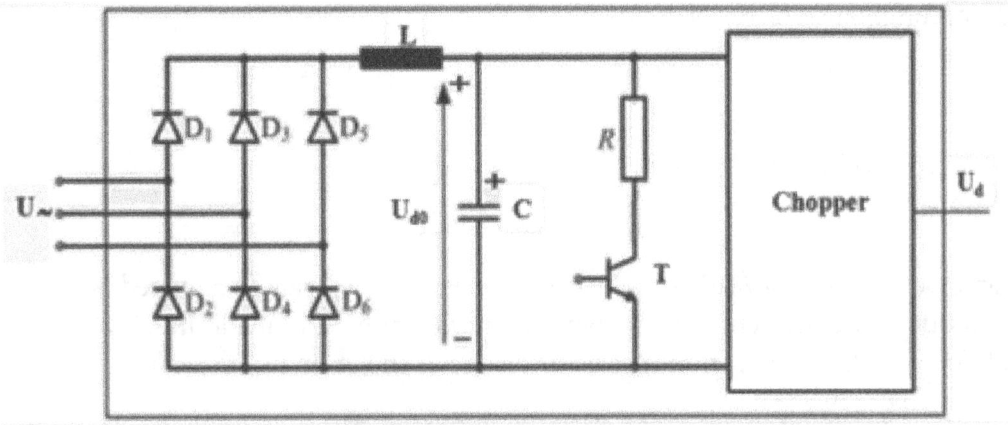

Fig. 6. Principal scheme of chopper supply

To provide breaking, or to dissipate braking energy that cannot be returned to the network through diode rectifier, it is required to have braking device with transistor T and resistor R. Rectified voltage from rectifier (D1 – D6) is filtered by simple LC filter and brought to the chopper input that regulates mean value of output voltage Ud.

For motor supply there are mostly used chopper voltage reducers, so they will be considered here. Figure 7 shows simplified presentation of the chopper supply of a DC motor. Chopper is shown as ideal breaker controlled by voltage (Uup), so it can control switching on (TON) and switch-off (Toff) exiting voltage (Udo).

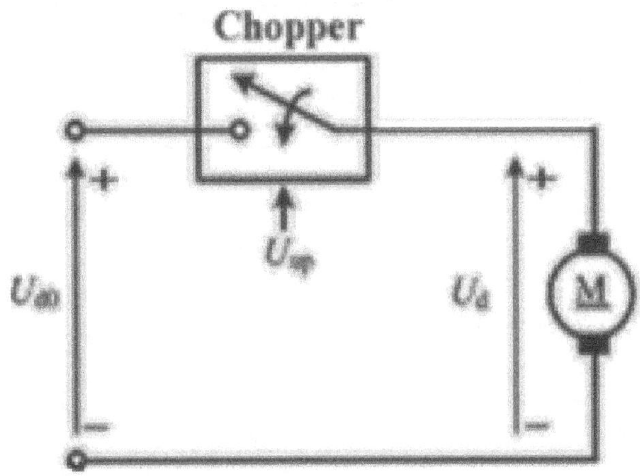

Fig. 7. Simplified presentation of the chopper supply of a DC motor

Chopping frequency is by the definition:

$$f_c = \frac{1}{T_{on} + T_{\text{off}}} = \frac{1}{T}$$

Also, compliance factor is defined:

$$d = \frac{T_{on}}{T},$$

So, output voltage is:

$$U_d = d\,U_{do}$$

So, by changing Ton/Toff ratio, the output voltage Ud can be adjusted between 0 and Udo. Simplified chopper shown can provide only first quadrant operation. For all four quadrant operation transistor bridge as shown in figure 8 can be used:

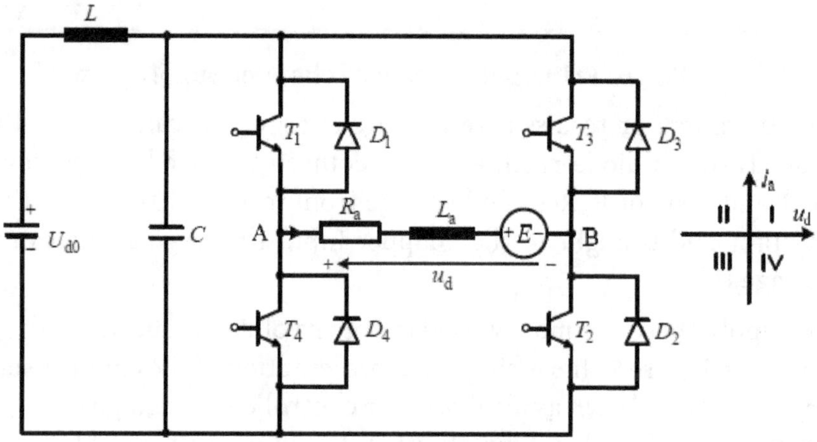

Fig. 8. Transistor bridge

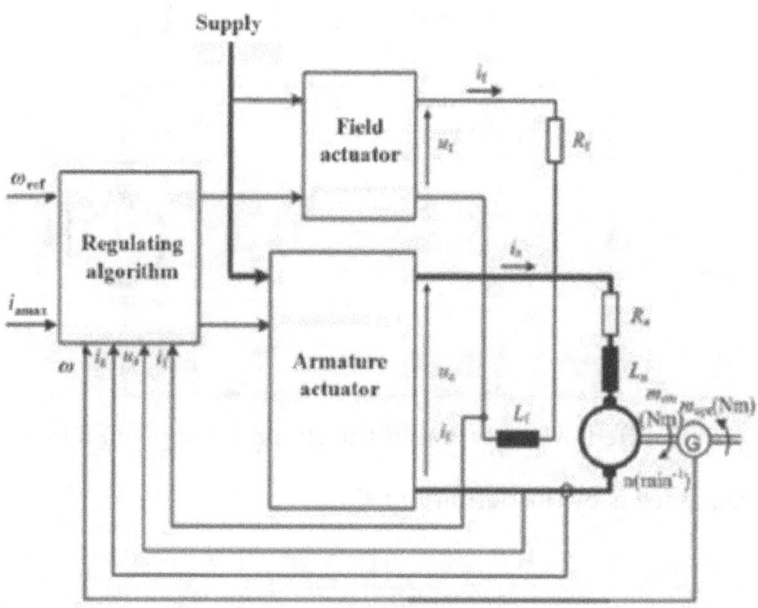

Fig. 9. Circuit for speed regulation of DC motor with independent field

By switching on transistor pairs T_1-T_2 **or T_3-T_4 positive or negative polarity of motor voltage u_d is provided. To close motor current at null or reverse polarization, diodes D_1 to D_4 are provided.**

General modern circuit for speed regulation of DC motor is shown in figure 9. Reference rotary speed Wref is set and also maximum armature current Iamax and their actual values are monitored and also brought into regularotr which outputs present command values for excitation actuators and inductor.

Out of base range (for speeds above nominal) method of reduced field is used so among basic values excitation current, if, is monitored.

Apart from classic PID action, regulating algorithm comprises other tasks (actuator command input adaptation, change of regulating method in accordance with the given speed, alarms etc.). Standard way of regulating DC drives, cascade regulation, consists of two feedbacks: internal - current and external - speed.

Asynchronous motor at constant frequency and amplitude of supply voltage rotor speed depends of load torque, which requires complicated governing algorithms in case when precise speed control and/or position. This phenomenon is a consequence of principle of asynchronous motor, and it is electromagnetic induction, which requires difference in between rotor speed and rotary magnetic field generated by stator to create electromagnetic torque. Electronics that creates algorithms mentioned was expensive earlier and such a use of asynchronous motors was difficult, but today with cheaper electronics components and use of microprocessors for regulating algorithms they are more often used.

Depending on use and requirements, some of feedbacks and regulators can be left out. Power block (converter + motor) has two input and five output values. Input (command) parameters are effective polyphase supply voltage U_d **and frequency W_s.** Output (regulated) values are motor current Is, flux w, position O, rotary frequency w and torque me. Each of those has proper regulator in negative feedback, in order as shown in figure 10.

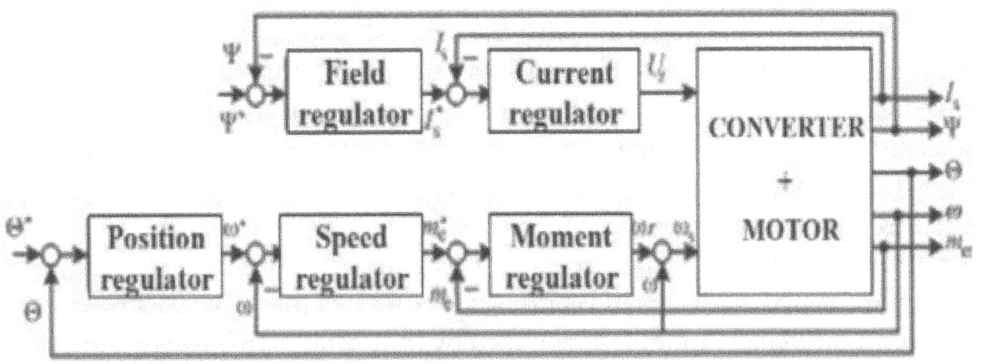

Fig. 10. Block diagram of AC motor regulator

Regulation (closed-loop control) involves control through negative feedback, whereby measured regulated parameters are compared to specified reference parameters, allowing for adjustments to command parameters to automatically achieve predetermined values of the controlled variables. A more complicated dynamic system is attained wherein inputs no longer exert control but serve as reference values.

6. CONCISE SUMMARY

Electric drive vehicles are one of the most modern solutions about environmental degradation. Recently, there has been a heightened interest globally in hybrid vehicles that exhibit reduced fuel usage and much lower emissions of pollutants. Hybrid vehicles can be defined as vehicles that include both energy production and storage systems. Two categories of cars are examined: parallel hybrids and serial hybrids. In parallel hybrids, a mechanical linkage exists between the power generator and the driving wheels, but in serial hybrids, such a connection is absent. Serial hybrids possess distinct advantages to parallel hybrids, including mechanical simplicity, design flexibility, and ease of integrating new technologies. An essential element in every hybrid or fully electric car is energy storage. Potential solutions include accumulators, supercapacitors, flywheels, hydraulic systems, and innovative materials for hydrogen storage. It has been previously noted that accumulators possess a distinct power issue. Flying wheels and hydrogen energy storage are still under research, necessitating significant technological advancements before implementation. Supercapacitors are the sole technology currently available that offers high power (exceeding 1kW/kg) and substantial cycle longevity at a reasonable cost. Supercapacitors possess additional attributes that render them intriguing for hybrid vehicles, including their capability for complete energy regeneration during braking (termed regenerative braking), which enhances energy efficiency, requires no specialized maintenance, optimizes electric energy utilization, exhibits low toxicity, and facilitates straightforward post-use storage. The most stringent specifications are established for capacitors utilized in electric drives or future cars. Batteries with substantial capacitance, reaching several hundred Farads, and operational voltages of several hundred volts are currently in production. In addition to substantial capacitance and comparatively elevated operating voltage, these capacitors must also possess high specific energy and power due to spatial constraints in vehicles. They possess a significant advantage regarding specific power relative to accumulator batteries; nonetheless, they are vastly inferior in terms of specific energy. The optimal configuration is a parallel connection of accumulator and capacitor batteries. In a steady state (normal operation), the vehicle's motor is powered by the accumulator battery, but during abrupt acceleration, it is provided by the supercapacitor. A crucial aspect is that during sudden braking, all mechanical energy can be converted back into the system as electrical energy just in the presence of supercapacitors with high specific power. The internal resistance of the utilized supercapacitors must be quite low for the aforementioned reasons. Leakage current is insignificant. Vehicles featuring this type of drive remain underutilized, primarily due to economic considerations. This chapter presents the theoretical foundation, practical implementation, and viability of utilizing supercapacitors in conjunction with accumulator batteries or fuel cells within the power supply system of electric vehicles. It also offers regulatory solutions and other critical power solid-state assemblies in optimized electric vehicles.

REFERENCES

Arbizzani, C.; Mastragostino, M. & Soavi, F. (2001). New trends in electrochemical supercapacitors. *Journal of Power Sources,* Vol.100, No1-2, (November 2001), pp. 164- 170 ISSN 0378-7753

Ardizzone, S.; Fregonara, G. & Trasatti, S. (1990). »Inner« and »outer« active surface of RuO2 electrodes. *Electrochimica Acta,* Vol.35, No1, (January 1990), pp. 263-267, ISSN 0013- 4686.

Bugarinović, S.; Rajčić-Vujasinović, M. & Stević, Z. Construction of double layer capacitors, *Proceedings of 39th International October Conference on Mining and Metallurgy*, pp. 470- 476, ISBN 987-86-80987-52-1, Sokobanja, Serbia, October 7-10, 2007.

Bugarinović, S.; Rajčić Vujasinović, M. & Stević, Z. (2008). Supercapacitors based on activated carbon, *Proceedings on 40th International October Conference on Mining and Metallurgy*, pp. 417-422, ISBN 978-86-80978-60-6, Sokobanja, Serbia, October 5-8, 2008.

Conway, B.E. (1999). Fundamentals of Electrochemical Capacitor Design and Operation. In: Electrochemical Supercapacitors: Scientific Fundamentals and Technological Applications, Kluwer Academic/Plenum Publishers, *ISBN 0-306-45736-9, New York, USA.*

Guerrero, M.A.; Romero, E.; Barrero, F.; Milanés, M. I. & González E. Supercapacitors: Alternative Energy Storage Systems. *Power Electronics & Electric Systems (PE&ES),* School of Industrial Engineering (University of Extremadura). Available from: andes.unex.es/archives%5CP126.pdf

Hadartz, M. & Julander M. (June 6, 2008). *Battery-Supercapacitor Energy Storage,* Master of Science Thesis in Electrical Engineering, Chalmers University of Technology, Göteborg, Sweden.

Kotz, R. & Carlen, M. (2000). Principles and applications of electrochemical capacitors. *Electrochimica Acta*, Vol.45, No15-16, (May 2000), pp. 2483-2498, ISSN 0013-4686.

Miller, J.M.; Dunn, B.; Tran, T.D. & Pekala, R.W. (1997). Deposition of ruthenium nanoparticles on Carbon Aerogels for High Energy Density Supercapacitor Electrodes. *J. Electrochem. Soc.*, Vol.144, No12, (December 1997), L309-311, ISSN 0013-4651.

Park, J. & Mackay, S. (2003). *Practical Data Acquisition for Instrumentation and Control Systems* (1st edition), Newnes, Oxford, Burlington, ISBN 07506 57960, Mumbai, India.

Stević, Z.; Andjelković , Z. & Antić, D. (2008). A New PC and LabVIEW Package Based System for Electrochemical Investigations. *Sensors*, Vol.8, No3, (March 2008), pp. 1819-1831, ISSN 1424-8220.

Rajčić-Vujasinović, M.; Stanković, Z. & Stević, Z. Consideration of the electrical circuit analogous to the copper or coppersulfide/electrolyte interfaces based on the time transient analysis. *Russian Journal of Electrochemistry*, Vol.35, No3, (March 1999), pp. 320- 327, ISSN 1023-1935.

Stević, Z. Supercapacitors based on copper sulfides, Ph.D. Thesis, University of Belgrade, 2001

Stević, Z. & Rajčić-Vujasinović, M. (2006). Chalcocite as a potential material for supercapacitors. *Journal of Power Sources*, Vol.160, No2, (October 2006), pp. 1511- 1517, ISSN 0378-7753.

Stević, Z.; Rajčić-Vujasinović, M.; Bugarinović, S. & Dekanski, A. (2010). Construction and Characterisation of Double Layer Capacitors. *Acta Physica Polonica A*, Vol.117, No1, (January, 2010), pp. 228-233, ISSN 0587-4246.

Stević, Z.; Rajčić-Vujasinović, M. & Dekanski, A. (2009). Estimation of Parameters Obtained by Electrochemical Impedance Spectroscopy on Systems Containing High Capacities. *Sensors*, Vol.9, No9, (September 2009), pp. 7365-7373, ISSN 1424-8220.

Stević, Z.; Rajčić-Vujasinović, M.; Nikolovski, D: & Antić, D. (2010). Hardware and software of a system for electrochemical and bioelectrochemical investigations, *Book of abstracts on CD ISIRR 2010, 11th International Symposium on Interdisciplinary Regional Research*, pp. 139, Szeged, Hungary, October 13-15, 2010.

Stević, Z.; Rajčić-Vujasinović, M. & Stanković Z. (2002). Achievments and perspectives in supercapacitors development and applying, *Proceedings of 34th International October Conference on Mining and Metallurgy*, pp. 435-440, ISBN 86-80987-17-4, Bor Lake, Serbia, September 30 – October 3, 2002.

Stević, Z.; Rajčić-Vujasinović, M. & Stanković, Z. (2002). Galvanostatic investigations of copper sulfides as a potential electrode material for supercapacitors, PO 320, *Book of Abstarcts 3rd International Conference of the Chemical Societies of the South-Eastern European Countries on Chemistry in the New Millennium - an Endless Frontier*, Vol.II, pp. 97, Bucharest, Romania, September 22-25, 2002.

Stević, Z.; Rajčić-Vujasinović, M. & Stanković, Z. (2003) Modelling of copper sulphide/ electrolyte systems as a potential material for supercapacitors *Proceedings of XXII International Mineral Processing Congress*, pp.463, ISBN 0-9584663-4-3, Cape Town, South Africa, September 28 – October 3, 2003.

Stević, Z.; Rajčić-Vujasinović, M. & Stojiljković, Z. (2004). Testing of system for electrochemical impedance spectroscopy developed at Technical faculty in Bor, *Proceedings of 36th International October Conference on Mining and Metallurgy*, pp. 415- 418, ISBN 86-80987-27-1 Bor Lake, Serbia, September 29 – October 2, 2004.

Van Voorden, A.M.; Ramirez Elizondo, L.M.; Paap, G.C.; Verboomen, J. & Van der Sluis, L. (2007). The Application of Super Capacitors to relieve Battery-storage systems in Autonomous Renewable Energy Systems. *Power Tech,* pp. 290-295.

Zheng, J.P.; Huang, J. & Jow, T.R. (1997). The limitations of energy density fro electrochemical capacitors. *J. Electrochem Soc.*, Vol.144, No6, (June 1997) pp. 2026- 2031, ISSN 0013-4651.

8 Chapter

INCORPORATION OF ELECTRIC VEHICLES INTO ELECTRIC UTILITY SYSTEMS

1. INTRODUCTION: AN OVERVIEW

In recent decades, energy consumption for electricity generation and the transportation sector has more than doubled (IEA - WEO, 2007) and currently encounters numerous issues pertaining to reliability, security, and environmental sustainability. Scientific data regarding climate change (IPCC, 2007) has emphasized the necessity for immediate cross-sector emission reductions, with electrified transportation being one of the technological solutions that could address the issue (IEA - ETP, 2008). In the majority of OECD countries, transportation and electric power systems account for the predominant share of CO2 emissions (IEA - WEO, 2008), and a significant portion of the fossil fuels (coal, natural gas, and oil) utilized for electricity generation and transportation is imported in many of these nations. Oil constitutes the predominant portion of primary energy imports, with over 60% utilized for transportation, primarily road transport, so contributing significantly to emissions within the transport sector. These factors compel decision-makers and manufacturers to take action in the road transportation industry by launching more efficient vehicles and diversifying energy sources. The technological advancement of Electric Drive Vehicles (EDV), including Hybrid Vehicles (HEV), Battery Electric Vehicles (BEV), and Fuel Cell Vehicles (FCV), will result in the gradual integration of EDVs into the transportation sector, supplanting Internal Combustion Engine Vehicles (ICEV). The forthcoming advancement in electric drive vehicle technology, as disclosed by several leading automakers, is the capability to connect to a normal electric power outlet for charging batteries using grid-supplied electric energy. Numerous firms, including significant and specialized global entities, are claimed to be developing models for the next years in the sectors of battery-powered electric vehicles, Plug-In Hybrid Electric Vehicles (PHEVs), and fuel cell electric vehicles (EV ReportLinker, 2007). Electric vehicles may significantly contribute to the integration of two essential components of the energy system—power generation and transportation—by transferring existing non-electric loads to the grid. In a situation where a pledge is established to diminish emissions from power generation, the integration of new intermittent power capacity poses challenges for the operation of electric networks (Skea, J, et al., 2008) and typically necessitates substantial expenditures in energy storage. The incorporation of additional demand from electric vehicles into the electricity grid can pose challenges, even if both systems operate more efficiently and contribute to a reduction in overall emissions. Moreover, future energy systems characterized by significant transportation electrification may leverage Vehicle to Grid (V2G) concepts, which provide potential storage capacity and utilize battery-stored energy to assist the grid during shortages. Individually, each car exerts a minimal influence on the power system; however, a substantial aggregation of vehicles could substantially affect the system, either as an added load or as a source of distributed generation capacity (Kempton and Tomic, 2005a; Kempton and Tomic, 2005b). This chapter examines the potential effects of electric vehicles on electricity systems, emphasizing the increased power demand, emissions from power generation linked to EVs, the role of demand side management (DSM) strategies in facilitating their integration, and the economic implications of EVs for electric utilities. The examination of the effects of widespread adoption of plug-in electric vehicles on electric utilities, considering electricity demand, CO2 and other greenhouse gas emissions, and energy costs, can be conducted within two distinct electric utility scenarios. A large electric grid synchronized with analogous systems on the same continent, and a smaller, isolated electric system on an island. Each situation possesses distinct characteristics; the most significant include the systems' robustness, the isolated system's requirement for additional backup power, and its typically limited diversity in production methods. A significant distinction is that in a tiny island, due to its size and isolation, there is insufficient capacity to operate an electricity market, resulting in the whole electricity supply being managed by a regulated monopoly. These disparities impact the ultimate determination of electricity pricing. Numerous studies on battery electric vehicles and plug-in hybrids are being conducted in various nations. In the United States, the capacity of the electric power infrastructure across various regions was analyzed concerning the additional load from PHEV penetration (Kintner-Meyer et al., 2007) and the economic implications of PHEV adoption for vehicle owners and electric utilities (Scott et al., 2007). Subsequent research (Hadley, 2006) examined the implications of introducing one million PHEVs in a US sub-region, assessing potential alterations in demand, effects on generation adequacy, and impacts on transmission and distribution. This analysis was subsequently expanded to encompass 13 US regions, incorporating GHG estimations for each of the seven scenarios evaluated for each region (Hadley, 2008). The capacity to arrange both charging and restricted discharging of PHEVs could markedly enhance power system utilization. The assessment of optimal PHEV charging effects, predicated on the assumption that utilities will manage charging times—either directly or indirectly—thereby offering consumers the lowest possible driving energy costs through the utilization of inexpensive off-peak electricity, was also examined (Denholm and Short, 2006). This study analyzed existing electricity demand and driving patterns across six geographic regions in the United States, concluding that when plug-in hybrid electric vehicles (PHEVs) utilize 40% of their mileage from electricity, no additional electric generation capacity is necessary under optimal dispatch conditions for a 50% PHEV market penetration. A comparable study was conducted by NREL (National Renewable Energy Laboratory), concentrating exclusively on a single location and evaluating four charging scenarios regarding their influence on the grid and greenhouse gas emissions (Parks et al., 2007). The findings indicated that off-peak charging would enhance efficiency regarding grid stress and energy expenses, while a notable decrease in CO2 emissions was anticipated; however, an increase in SO2 emissions was also projected due to the substantial reliance on coal generation during off-peak charging. The findings derived from one location on Earth cannot be applied to other places; only the approaches may be utilized. In addition to factors associated with driving behavior and road topology, the electrical generation source mix varies by location, with some areas incurring higher costs and others utilizing a greater proportion of renewable resources. This chapter will also concentrate on these variations and their impact on the fuel/energy costs of electric vehicles and the emissions equilibrium between the power generating and road transportation sectors in relation to electric mobility.

2. ELECTRIC UTILITY SYSTEMS

This section provides a description of electric power systems demand, highlighting its daily evolution and the potential role of electric vehicles in stabilizing the power consumption profile. Illustrations of standard load profiles incorporating various technologies (renewable sources, large hydro, and thermal units) are provided, along with the potential share of renewables in energy generation. Subsequently, the emissions related to the recharging of electric vehicles are considered. This analysis examines the economic consequences of the two case studies by delineating the distinct regulations governing technology dispatch within a market context and inside a conventional integrated electric system. This section will elucidate the formation of consumer prices, including electric cars, using examples from a market context and a vertically integrated corporation on an isolated island.

2.1 Electricity demand

Contemporary electric power systems are engineered to accommodate immediate customer demand. A primary characteristic of power consumption is the variation in demand across the hours of the day, the days of the week, and the seasons. Figure 1 illustrates the hourly demand characteristics of the Portuguese electric system. Each curve reflects a week's worth of data from four distinct seasons in 2008 and depicts the fluctuations in electricity usage. The yearly peak demand in this country occurs throughout the winter months of December or January, namely in the evening.

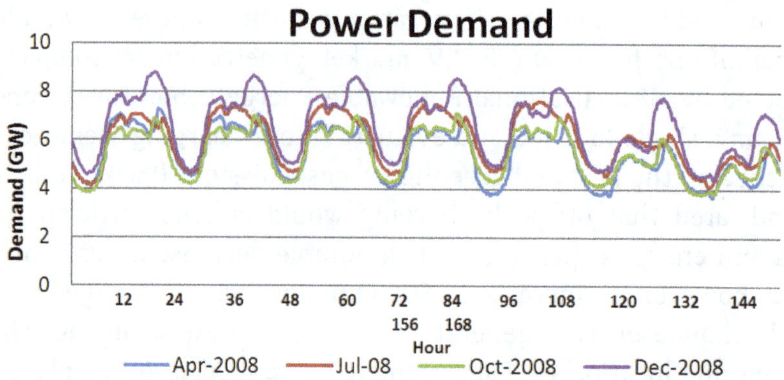

Figure 1. Power demand profiles in Portugal for different seasons

The fluctuation in daily and seasonal demand may indicate a persistent underutilization of capacity that might be leveraged during off-peak hours. Figure 2 illustrates the progression of hourly average power usage in Portugal during the 24 hours of each day during the entire year of 2008. This daily evolution, however, experiences a trough at night, accounting for around 60% of peak consumption, resulting in significant financial implications due to the necessity of maintaining multiple idle power plants and an underutilized network during nighttime hours. This condition presents an opportunity for electric vehicles to contribute to the stabilization of the power consumption profile.

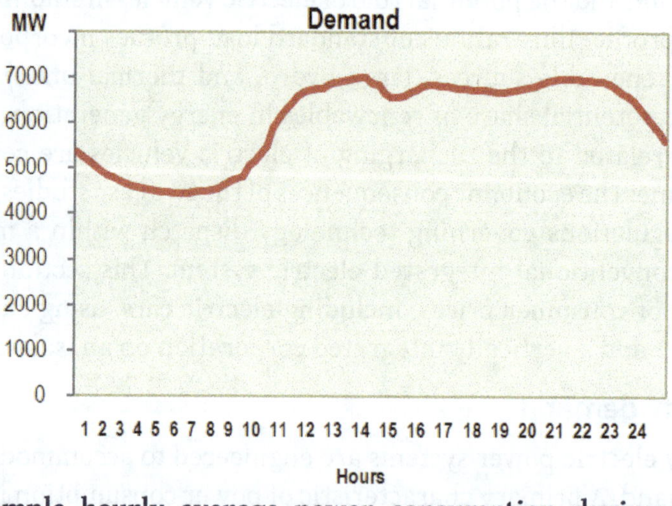

Figure 2. Example hourly average power consumption during (weekdays in Portugal mainland year 2008)

As an example, Fig. 3 shows the estimated contribution for the power consumption diagram levelling when considering different levels of the electric vehicles penetration. Portugal mainland was used as an example and it was considered that 85% of the electric vehicle charging happens uniformly during the valley hours (from 11pm to 8am) with the rest charge happening uniformly during the other 14 hours of the day. The extra energy that each electric vehicle should charge from the grid in average was considered about 2.5MWh per year, more or less 7kWh per day plus a 10% in transmission losses.

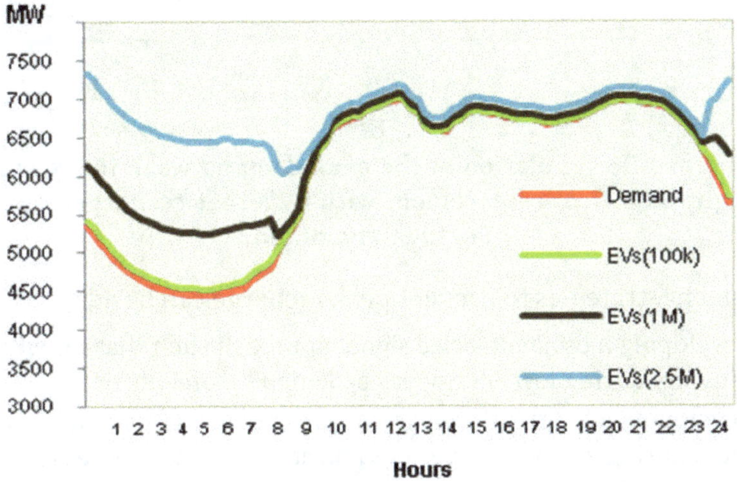

Figure 3. Electric vehicles contribution to the consumption diagram leveling

As can be observed in Fig.3 only with a high penetration of vehicles like ca. 1million units (almost ¼ of the country's light duty fleet), the impacts are visible in the consumption profile. The main question is how to incentivise off-peak charging?

It is easy to foresee major congestion problems in already heavily loaded grids and voltage profile problems in predominantly radial networks, particularly if the peak load periods coincide with EV charging periods. Hence, if no load management strategies are defined, significant technical problems will occur and their drawbacks might even be larger than the economic/environmental benefits arising from electric vehicles usage (Fig.4).

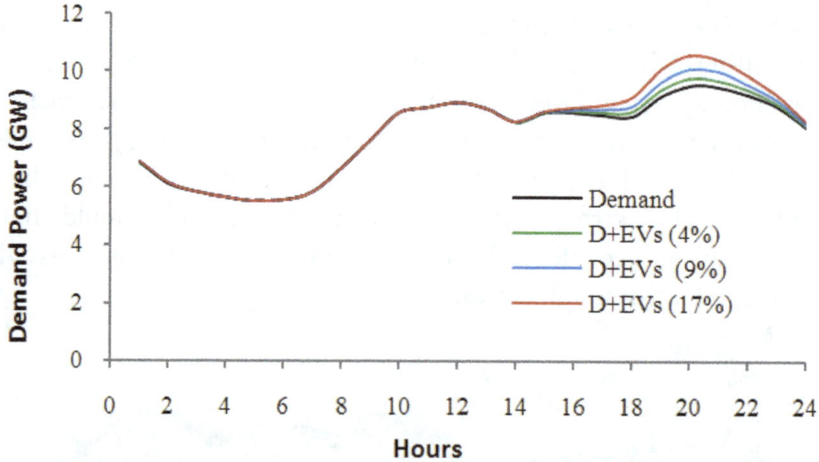

Figure 4. Load profile simulation of the peak demand weak for year 2015 with 4, 9 and 17% light duty fleet replacement with EV's that recharge at peak hours in Portugal mainland.

The management strategies to prevent peak recharge can be adopted in two ways:

1. By developing a dynamic price signal approach such that EV will charge predominantly during low energy price/demand moments or
2. By developing a technical management system such that charging can be distributed during valley hour periods and at times when there is large renewable power generation.

In a scenario of mass EVs penetration, the recharging profiles are important for the electric utility system's sake and their effects on peak consumption, electricity prices and grid congestion should be carefully foreseen and studied.

2.2 Electric power consumption and power plants technologies

The hourly average power consumption during a day presented in Fig. 2 has an equivalent diagram when analyzing the distribution during the 24 hours of the day of the annual energy consumed. Fig. 5 presents this distribution and also the different power technologies that generated this energy demand.

The great difference between the consumption at valley and at off-valley hours represents a barrier to the penetration of generation technologies that need to work more hours per year in order to became economically viable (usually with low unit variable costs but with large fixed costs) or non dispatchable units (renewable energies).

The effective use of the country´s power plant fleet can be illustrated by the load duration curve (LDC), where the hourly average power over the whole year is sorted by decreasing order constituting a curve that begins with the year peak at hour 1 and the smallest demand at hour 8760 as illustrated in Fig. 6. In this example, the unutilized capacity could be annually 4000 GWh. Nowadays some of this capacity is used in pumping during off-peak hours. In 2008, about 700GWh were used for pumping to provide energy storage in reservoirs available for the peak hours. Without increasing capacity it would be possible to account with around 3000 GWh a year for extra off-peak loads that could be used in plug-in electric vehicles' recharging. Considering a 3MWh/year of extra energy needed to charge an EV, it is possible to have one million EVs charging their batteries during valley hours, without any more investments in power installed and transmission lines.

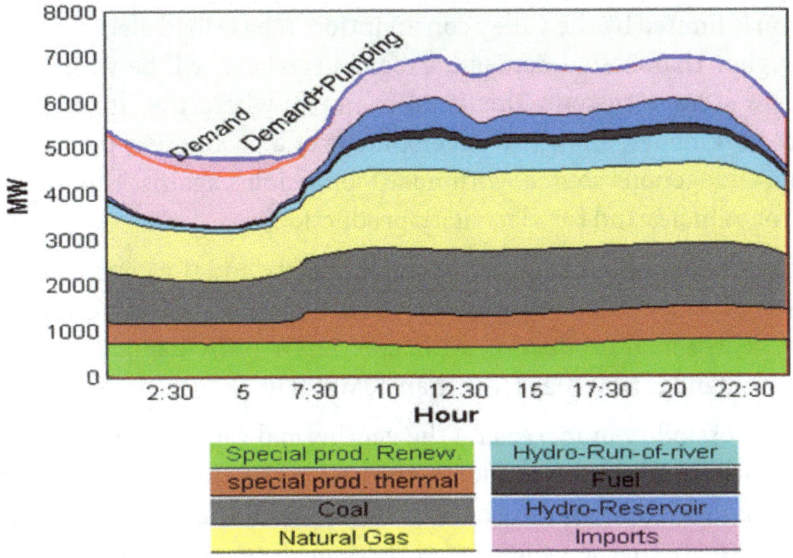

Figure 5. Example of hourly energy consumption with used power plants technologies

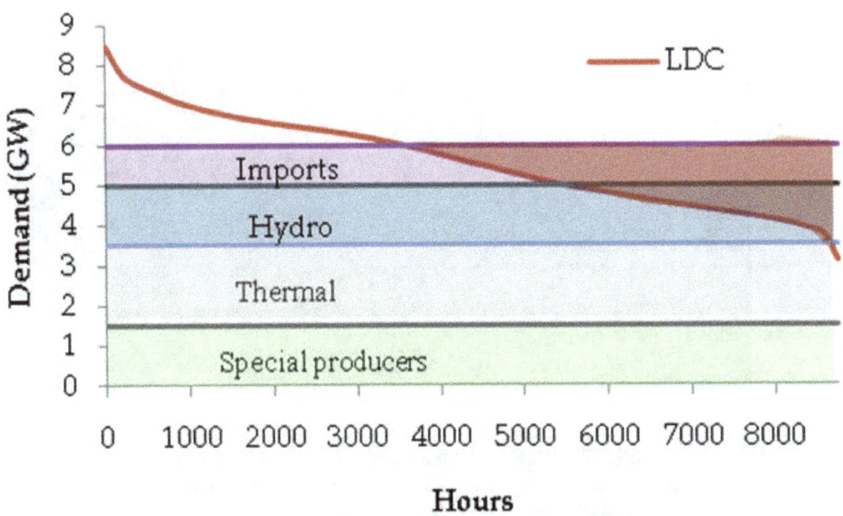

Figure 6. Load duration curve for Portugal in year 2007

In addition to the fixed costs associated with underutilized capacity, the significant cycling that occurs on a daily basis creates additional costs for plants due to the constantly varying loads that could require generators to operate well below the "design point" of optimum efficiency. Power plant cycling also increases operation and maintenance requirements.

This last situation is more visible for instance at St. Miguel Island in Azores, where the geothermal energy production (renewable energy without CO_2 emissions that should be used as base load due to its impossibility of production variation) penetration is limited by the valley consumption. If base load electricity generation is much higher than actual demand, excess electricity will be wasted unless it is coupled to a storage system. This is an example, where the use of EVs, charged during off-peak hours, allow the development of a production technology from a renewable, endogenous source, with no CO_2 emissions, against the systematic fuel imports (for vehicles and for electricity production).

S. Miguel has 27MW of existing geothermal capacity and the government wants to expand this capacity to meet future demand. Expanding existing capacity would mean that geothermal production will meet 40MW (EDA 2009). Current base load electricity demand is nowadays less than 40MW (Fig.7).

There is the possibility of increasing the geothermal capacity in 3MW by 2011 and 10MW in 2013, but these investments are limited by the off-peak demand. Even considering a demand increase of 4% for years 2010 to 2013, there is not enough off-peak demand to fill the valley and at the same time assure at least a 10% of fuel power to prevent any shortage of the renewable energy sources.

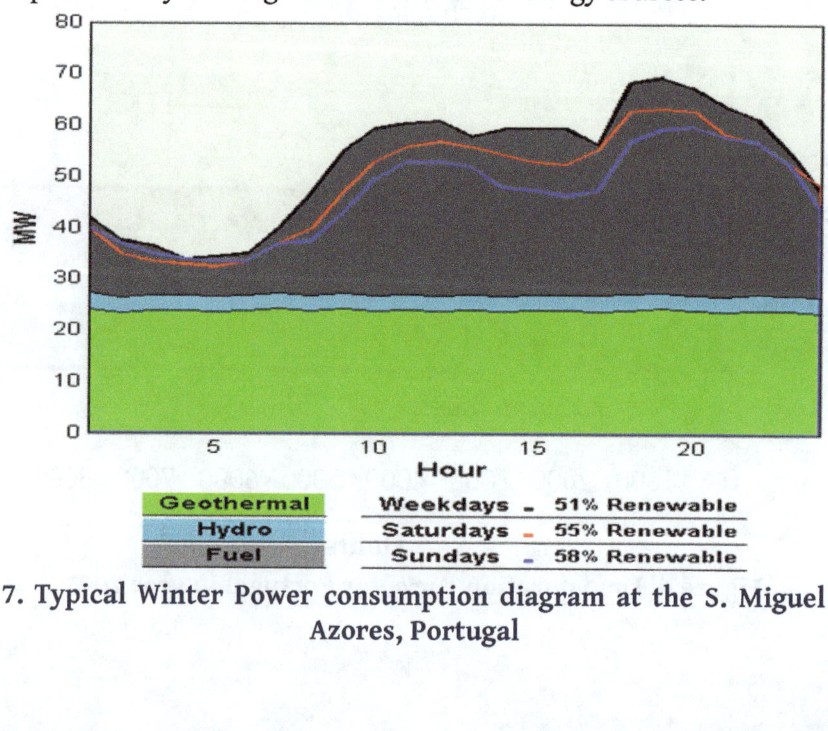

Figure 7. Typical Winter Power consumption diagram at the S. Miguel Island, Azores, Portugal

As stated in last section, the fact that the valley demand, in this case, is about half the value of peak demand, gives the opportunity for electric vehicles contribution for levelling the power consumption diagram. The extra demand for charging the vehicles at each hour of the day is computed using equation 1.

$$P_i = \frac{E_{Vavg}\left(1 + p_{loss}\right) p_{charge}}{h_{charge}} N \tag{1}$$

Where E_{Vavg} is the daily average energy needed for charging a vehicle, p_{loss} is the percentage of energy lost in the transmission lines, p_{charge} and h_{charge} are respectively the percentage that is charged in each period (valley and off-valley) and the length of the considered period (in hours) and N the number of vehicles.

Considering that 85% of the vehicles will uniformly recharge at night, the introduction of these vehicles will create and additional electricity demand that, in the context of today's electricity generation mix, will be fulfilled by the running fuel oil power plants (Fig. 8).

As expected, the percentage of renewable energy decreases and in terms of marginal emissions, they are simply being transferred from the tail pipes to the fuel power plants chimneys.

Considering an emissions rate of 830gCO_2/kWh for the fuel power plants, the additional CO_2 emissions to daily charge EVs (an extra energy of 86MWh), should be 71.4 tons of CO_2.

The TTW daily emissions avoided if these same vehicles were conventional was about 120 tons.

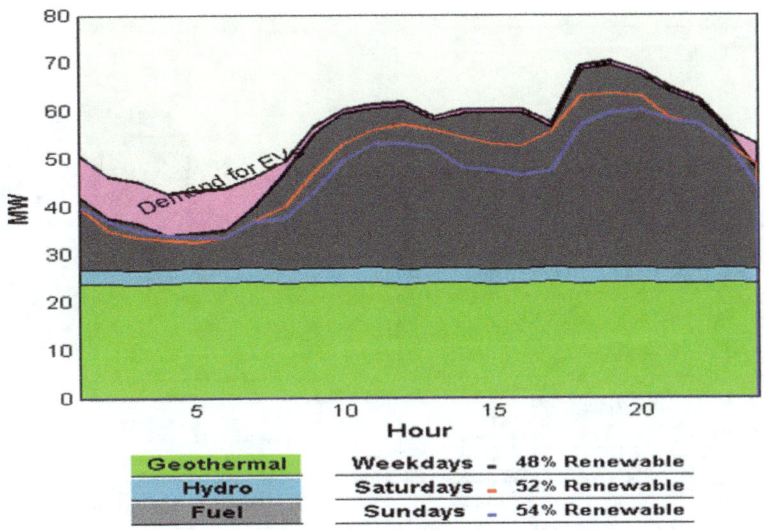

Figure 8. Load profile in S. Miguel for Winter season with EVs extra demand (30% of Light Duty fleet replaced for EVs)

However, if additional renewable electricity installed capacity is deployed, then the vehicles will also be running on renewable electricity. It can be considered that these EVs charged during off-peak hours are charged 100% with renewable energy. In fact there are three possible approaches:

It is considered the EVs as marginal load and as stated in Fig. 8, the energy and emissions associated with EVs charging would be mainly on fuel.

It is considered the EVs as a base load because they justify the increase in renewable base load generation and the energy and emissions associated with EVs are from renewable sources (zero emissions).

It is considered the EVs equal to the other load and the energy and emissions associated to EVs charging would be an average of the production mix. In this case the emissions associated with EVs recharging is as stated by equation 2

$$Emissions_{EV} = \frac{Demand_{EV}}{Demand_{TOTAL}} Emissions_{TOTAL} \tag{2}$$

These different approaches will lead to different CO_2 emissions associated to EVs' recharge (in this example, 8% of total daily electricity consumption).

In the first case, a daily 71.4 tons of CO_2 emissions for EVs recharge under the considered scenario of EV penetration, in the second case, a daily zero emissions for EVs recharge and in the third case a 37 tons.

It is possible to see in Fig. 9, for instance for the typical Spring load profile that the new geothermal plus wind power penetration are not possible unless they increase valley consumption.

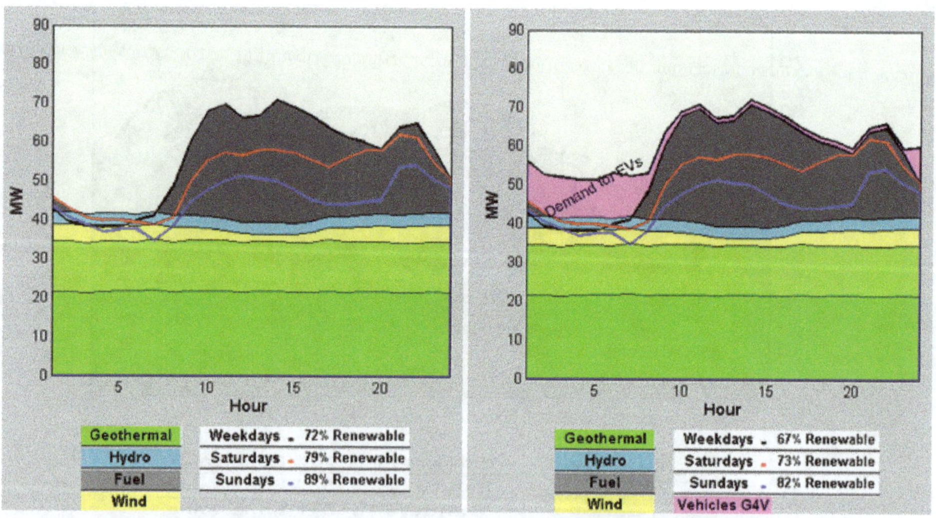

Figure 9. Expected load profile in S. Miguel for Spring (2013), 4% demand increase, 9 MW of wind, 13MW of new geothermal power and 30% LD fleet replacement by EVs.

The CO_2 emissions factors, unit consumption and daily usage of vehicles considered in this analysis are resumed in Table 1.

Table 1. Emission factors considered for the LD fleet and fuel power plants and average unit energy consumption.

	CO2 Emissions		Unit Consumption			Daily
	kg/litre	kg/kWh	litres/100km	kWh/100km	kg/kWh	km
ICEV Gasoline	2.31		7			32
ICEV Diesel	2.68		6			32
Fuel Plants		0.83			0.25	
EV				123		32

It is easy to see that even considering that EVs are recharged by fuel, there are advantages in the CO_2 emissions balance, as the daily 120 tons from the car tail pipes, are replaced by 72 tons in the fuel power plants chimneys. Although the correct mode is to attribute emissions for the percentage of electricity consumption to EVs' recharge, and that is only 37 tons.

2.3 Electricity costs

Electricity supply is considered a service of general economic interest. In the past, the electricity industry was organized as vertically integrated state-owned monopolies and consequently not subjected to the normal rules of competition. The liberalization process was put into force in many countries including in the European community (EC) in 1996 by Directive 96/92/EC, and led to the unbundling of activities. This directive defined common rules for the gradual liberalization of the electricity industry with the objective of establishing one common European market. Vertically integrated utilities have been vertically separated or unbundled and barriers to entry in generation and retailing were being removed to create competition.

In the two case studies presented in this chapter, two situations are analyzed: In the case of a small isolated system, a vertically integrated company is considered and the average unit production costs are used to define the merit order for technology dispatch. In the case of a continental European country, the marginal production costs are used to provide a supply curve and schedule the technology dispatch and find the wholesale price by market simulation.

2.3.1 Average unit production costs and rules for technology dispatch

The unit average cost for each technology is an average value that contains many simplifications. The cost functions are not linear and the average costs depend highly on the performance points in which the plants usually work. There are also the start-up costs (higher if there is a cold start) that should increase unit average cost if the plants don't work in a continuous state. In a monopoly context, the unit commitment service is centrally provided by the system operator as all the units belonged to the same company. In a market context this decisions are private and individual companies will predict the price of energy for the time under consideration and will then incur the cost of starting up if they expect to make a profit.

In this case, simple empirical rules are applied to distribute the energy source technologies to provide for the different scenarios of daily load profile. The rules used to order the power technologies that support expected demand plus EVs consider:

1. The special producers either renewable or thermal (special producers like co-generation) are the first in order;
2. The cheaper technologies follow afterwards. For each case, the cheaper technologies enter in increasing cost order until a certain percentage of the power installed is used and only then follows the next technology;
3. Production from reservoirs are used only in peak hours;

In any case the technical limits of each group committed to electricity production should be taken into account. In this case the imposed limitation is made by the power installed in each technology.

In the S. Miguel (Azores) case study, as a volcanic Island, the renewable source is mainly geothermal. Geothermal power plants are characterized by high capital investments for exploration, drilling wells and plant installation but low cost for operation and maintenance, so that they should be considered base load plants.

The unit costs per technology, considered in this example are described in Table 2 and the unit average production cost depends on the electricity mix. In table 2 there are two examples of electricity production mix and the associated average production costs. It is easy to see that the use of more fuel power plants instead of geothermal ones makes the unit variable costs higher due to fuel cost. Fuel prices tend to increase because of world oil demand increase and it must also not be forgotten that, associated to fuel plants use, there are the CO_2 emission costs that also tend to increase due to increasing environmental concerns.

It is also easy to see that the available technologies for electricity production make electricity costs expensive in S. Miguel Island. This has to be added to the fact that, an isolated system needs more redundancy and power installed in thermal plants to assure a high level of security of supply. This also increases fixed costs (the cost of power assurance).

Table 2. Unit average costs per technology, percentage of production scenarios and unit average production costs.

Production Technology	Unit costs €/MWh	Production scenario 1	Production scenario 2
Geothermal	33	37%	51%
Wind	82	4%	4%
Hydro	80	3%	3%
Fuel	100	56%	42%
Average unit prod. cost (€/MWh)		74	65

Added to the production costs there are also the costs associated to transmission lines use. Customers have to pay for the net's investments and maintenance and at the end for the service of distribution, accounting and billing. As all these services are supplied for the same vertically integrated company in a monopoly context, the final prices that end customers have to pay for electricity supply are regulated.

The impact EVs have on overall costs, depend on the additional variable cost associated with generating more electricity to serve the Plug-in vehicles' charging. Average variable costs may rise with increased electricity sales because higher cost generation plants are brought to service. Assuming though, no new infrastructure investment, average fixed costs will decrease, because the existing debt-service obligation is spread over more energy sold.

The time of vehicles charging is crucial for this analysis. An off-peak recharge can, in this example increase the penetration of geothermal power, decrease variable unit costs and fixed unit costs while at the same time reduces emissions and fuel oil consumption and imports.

2.3.2 Marginal production costs and rules for technology dispatch

To use the marginal production costs to schedule technologies to the expected load, a market simulation can be done with the demand and supply curves interception. A supply curve per technology should be generated. These curves should be added to generate a whole supply curve. Supply curves per technology are generated based on the expected power installed per technology and the supply curves verified in the past. As an example it can be observed the supply curves for the Portuguese units for the different technologies that play in the Iberian Electricity (MIBEL) spot market. In Fig.10 is an example of the supply curves per technology that occurred in two winter days, (2009/01/12) at 12h and (2010/01/19) at 22h.

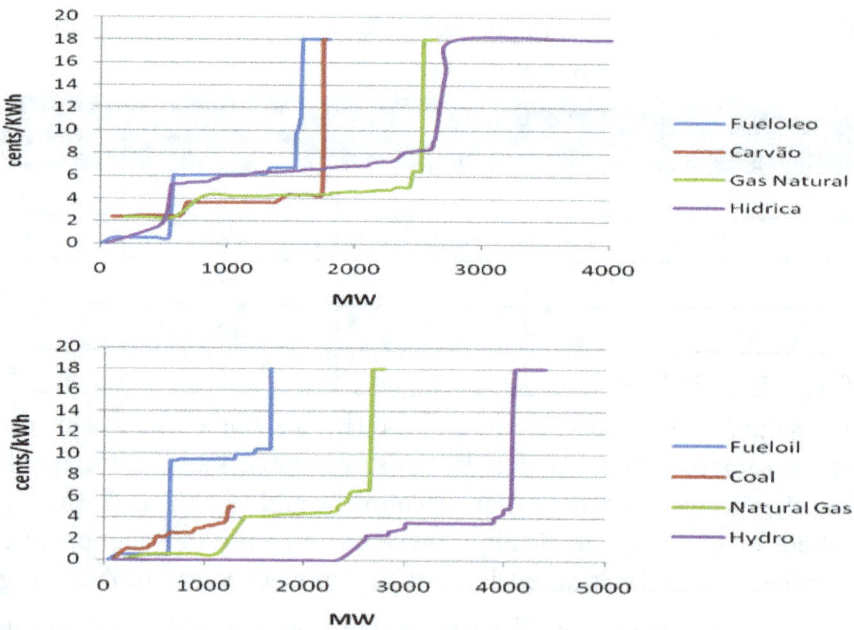

Figure 10. Supply curves per technology for day 22 Jan 2009 at 12h and 19 Jan 2010 at 22h in the Portuguese market

The supply curves in 2010 (the right hand side graphic in Fig. 10) show very low priced bids in hydro due to climate influence and also some low priced bids in natural gas due to more available supply (860MW were installed in the production system during the year 2009).

The supply curves to be generated would represent different scenarios according to the expected power installed per technology, a dry or wet year scenario and the season. There are in the majority of the cases, 2 steps in the supply curves for each technology. The amplitude of the slops depends on the hydrologic conditions of the hydro resources and the natural gas availability. If there is too much of these resources, low priced platforms are expected at the beginning of the supply curves as observed for year 2010 in Fig. 10. The power where the steps occur and its amplitude depends on the power installed in the technology, the climate conditions scenarios, season and hour of the day. The rules for supply curves generation are the following:

1. There are four cases of supply curves per technology associated with each season (Spring, Summer, Autumn or Winter);
2. The expected power installed in each technology in the year in study is an input that limits power generation;
3. The scenarios to be studied (a dry year with high priced hydro resources or a wet year with low priced hydro and natural gas offers) are inputs that lead to different supply curves shape;

4. Two or three slops are generated for each technology with more or less the same shape of the past supply curves observed as seen in Fig. 10.
5. According to the scenario studied, the season and power installed, the slops of the supply curves are generated with the same shapes of past technology's supply curves observed in similar conditions. A small randomness that represents the observed variations is also programmed.
6. The supply curves generated for each technology are added in ascending order of price to form a global supply curve that will intersect demand.

An example of supply curves for year 2020 in Portugal mainland can be seen in Fig. 11.

1. In this case, the rules for technologies dispatch are:
2. The special producers either renewable or thermal (special producers like co-generation) are the first in order;
3. The technologies with lower marginal costs follow afterwards. For each hour, according to demand level, the intersection gives market price and quantity per technology that will fulfill the daily load profile.

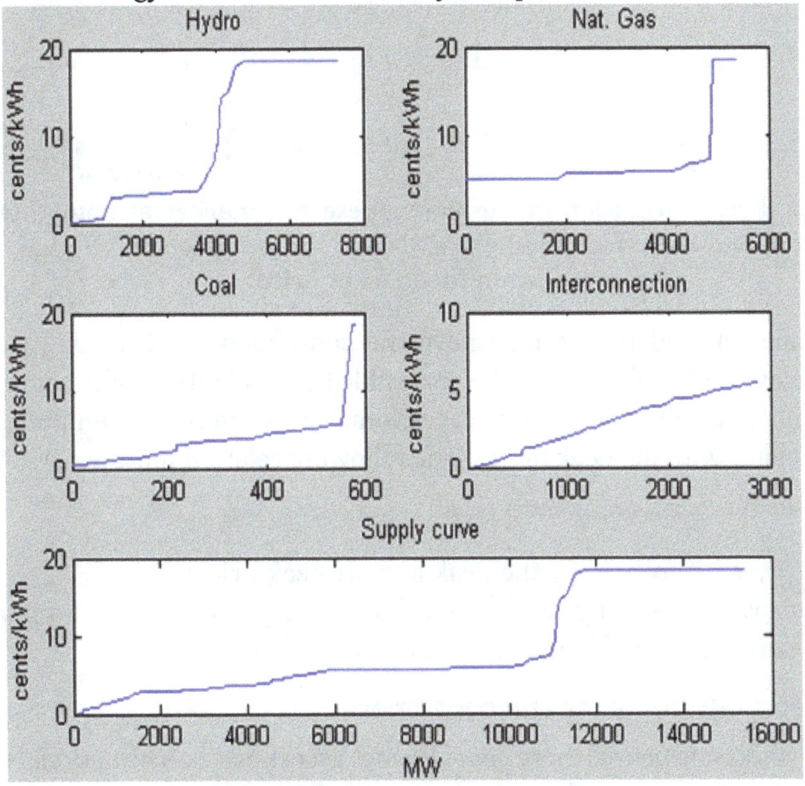

Figure 11. Scenario of supply curves per technology for year 2020 in the Portuguese market

In this case, the wholesale electricity price is formed by supply and demand curves intersection. An increase in demand makes demand curve shift to the right and so electricity prices rise as the curves intersection occurs at a higher point in the slop. As an example, it is possible to see price formation for instance in hour 5, in the Portuguese sub-market, for the 23th Jan 2008 with the real data available at Market operator (OMEL 2009) site. It is possible to see the effects of an increase in off-peak load to plug-in vehicles recharging, in the market prices. As an example, considering that a huge amount of EVs were charging at this hour (5 a.m.) with an extra 800 kW power demand. The demand curve shifts to the right (Fig. 12). The price rises for almost three cents and some of the previous low cost clients (the pumping charges) were not supplied because they had lower priced buying bids.

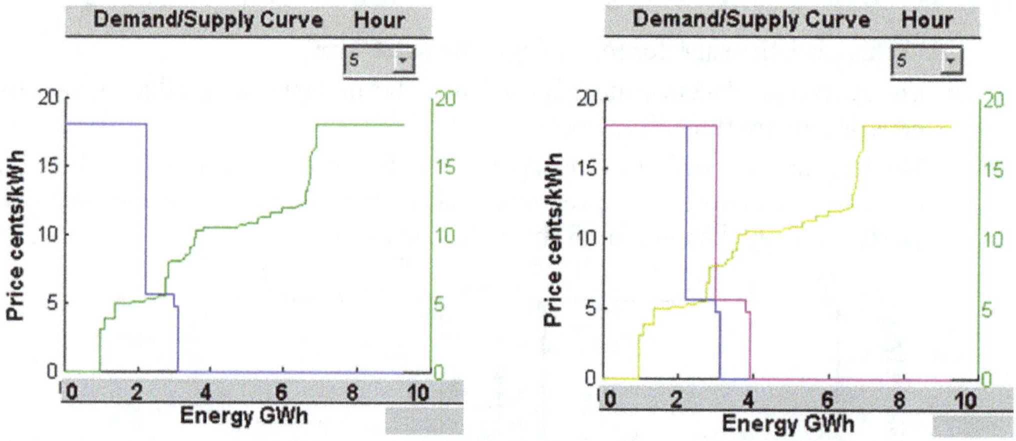

Figure 12. Price formation in the Portuguese sub-market at hour 5 (economic simulation) and the effect of an extra 800MW of power for EV off peak recharge has in the market price.

The leveling of load profile may have the consequences of making the hydro-pumping less profitable. In fact, the reversible hydro plants' profits come from the discharging and selling at peak prices revenues less the costs from pumping with energy bought with off-peak prices. The following relation must occur:

$$\pi_{peak}E_{peak} > \pi_{off-peak}E_{off-peak} / \eta_p \quad (3)$$

Where π_{peak}, $\pi_{off\text{-}peak}$ represent the peak and off-peak prices, E_{peak}, $E_{off\text{-}peak}$ the peak energy sold when discharging and the off-peak energy bought for pumping and η_p the total performance of the hydro reversible system.

2.3.3 Electricity costs for end consumers

Over the wholesale prices, there come the net access prices. The national regulator establishes the access and pricing rules of the transmission and distribution activities. The revenues for transmission and distribution network's operators are assured by the payment of a use tariff. The transmission and distribution network use tariffs are established by the regulator. In Table 3 there is the peak and off peak low voltage net access tariffs imposed in Portugal in the last years.

Table 3. Evolution of low voltage net access regulated tariffs in Portugal

Low voltage net access tariffs	2008	2009	2010	2011
Peak [cents/kWh] Off peak [cents/kWh]	7.45 3.28	5.39 0.68	9.09 3.76	8.92 3.86

Apart from the networks access regulated tariffs and wholesale prices, consumers should have to pay also retailing electricity costs. These activities work in competition as the established model imposed the existence of electricity wholesale and retail markets and the possibility for any customer to choose between different retailers. Depending on the retailer that each end consumer chooses, the cost of electricity to be paid by consumers should include the cost that the retailer paid for the energy at the wholesale electricity market plus the value of the regulated net access tariffs plus the revenue for the retailer. Considering a 4% on the wholesale electricity cost for the retailer revenue and the values presented in Table 3, Fig. 13 depicts the evolution of the final cost of energy that a low voltage domestic consumer should pay along the day.

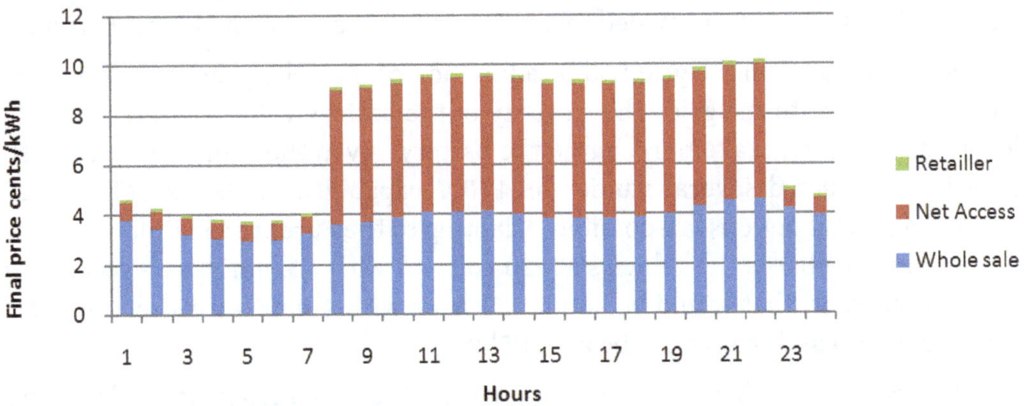

Figure 13. Average daily evolution for final costs of energy for a domestic consumer in 2009

This difference in costs that is reflected more or less in the regulated prices between peak and off-peak consumption should influence EVs' recharging profiles.

3. MODELING OF ELECTRIC VEHICLES

Electric drive vehicles equipped with plug-in functionality are just starting to become commercially accessible. An understanding of the mass manufacturing capabilities and market penetration of these vehicles is essential, contingent upon the development phases of its important components, particularly the battery. The mobility offered by plug-in automobiles is a crucial factor since it affects the vehicle's charging requirements. The demand aspect of the transportation sector requires examination, as consumer and passenger behavior will significantly influence future mobility. The fleet profile of each country is significantly influenced by economic and political/taxation factors; for example, Portugal possesses a reasonably modern fleet, although assessing the fleet's renewal rate is challenging due to a lack of incentives for disposal. These will be critical factors, as they will pertain to the market penetration of new vehicle technology. The energy demands of the electric vehicle and the anticipated power capacity of the charging circuit, along with the battery capacity, are critical factors that determine the charging duration and the car's all-electric range. The anticipated development of grid capacity for on-demand recharging and energy production technologies is crucial for calculating the energy consumption and emissions related to electric vehicles. In remote areas, such as islands, the use of intermittent renewable energy sources presents more challenges due to the lack of connection to a solid continental grid. How can one leverage off-peak capacity and the potential for charging to align with renewable energy sources?

3.1 Energy needs

How much power can be provided to an EV or a PHEV and during how much time?

The power demand on the grid will be a function of the voltage and current of the connection point to the grid. The capacity of the battery will then determine the length of time it will take to recharge the battery, given the connection strength. EPRI has conducted several studies on PHEV capabilities (EPRI, 2006), (Duvall, 2006) and several options are possible. Adapting to the European mainland reality, it is possible to charge the vehicles at 230 volts AC with 2,5 mm^2 wire's section (16 A circuit current) that would be about 3,5kW or with 4 mm^2 wire's section (25 A circuit current) and that would be about 5kW.

It is also possible to charge fast in stations, and this at a higher power level that may go till 50kW. While a slow charge can last 6-8 hours at 100% state of charge (SOC), a fast charge can last 15-30 minutes but only allows 80% SOC and with less battery efficiency (while a slow charge can have an efficiency of 80% a fast charge's is reduced to 55%).

The average daily energy needs to charge an EV depends on the daily mileage expected for the vehicle. The time and power of the recharge will define the EV's charging profile and the aggregation of all EVs, the extra load the electric system has to supply. A few EVs have no effect in the power grids but a mass penetration of plug-in vehicles should be simulated so that the electric utilities are prepared with the best strategies for a synergetic combination of the transportation and electricity production systems.

3.2 EVs market penetration

There are many factors regarding user preference for electric vehicle technology. The most pragmatic and objective causes that may a rational potential vehicle buyer decide not to choose this technology are:

- Fuel availability (the availability of charging infrastructures)
- Range limitations for the BEV configurations
- Acquisition costs

3.2.1 Fuel availability/charging network

PHEVs and BEVs are engineered to be charged in a house garage overnight. Garages are often absent in densely populated urban areas, which are precisely the locations where electric vehicles serve as an optimal alternative for personal mobility. In Lisbon, just one in six vehicles is parked in private garages, which encompass apartments and condominiums with designated parking spaces requiring electric charging (CML 2002).

In addition to the lack of garage access, the limited electricity-only range of plug-in vehicles will prompt the desire for drivers to "top off" their batteries when away from their normal overnight charging location. Electric vehicles must be plugged into the grid to refuel, but a public infrastructure to provide this service does not yet exist. Prospective plug-in car owners want the assurance that they can charge their vehicles at home, while at work, or parked anywhere for extended periods. There is a need for parking/charging points for slow charge and fast charging stations.

On average, cars are parked roughly 23 hours per day in home garages, apartments, condominiums and hotel garages, employee parking locations, public lots and curbside. To meet driver demand for convenient charging, these are the locations in which charging points should be installed.

3.2.2 Range limitations for BEV configurations

The limited driving range of pure EVs creates what is known as "range anxiety", which affect drivers as soon as the battery charge falls below 50%. Fast charging could alleviate "range anxiety" by supplementing home slow charging with convenient on-road charging at opportunistic charging points. In one 10-minute charge cycle, fast charge technology can provide enough energy to allow an EV to operate for another 60 km (Szczepanek A., Botsford C., 2009). With a network of fast chargers, consumers could charge anytime, anywhere - practical infrastructure akin to the gasoline fill-up model. This fast charge capability can help to enable rapid growth of the EV market by minimizing vehicle downtime. Fleets can fast charge during opportunistic breaks to maximize productive drive.

Battery manufacturers believe that they could develop a battery that could cope with fast charging although the priority is to mass produce batteries at low costs, while maintaining high quality standards, safety requirements and guaranteeing a life time of more than ten years rather than introducing batteries with super fast charging capabilities.

The successful launch of Li-Ion batteries for electronic goods such as laptops and mobile phones opened the door to further developments and it may be assumed that Li-Ion batteries will be the key technology for PHEVs and EVs in the coming decade. Regardless of the future development in the area of battery based energy storage it can be concluded that the current level of performance is now so good that the automotive industry has decided to include partly or fully electrical drivelines and traction batteries in many of their near future products.

When fast charge with safe batteries become available with increased capacity at competitive prices, the pure BEV can cover a wide range of the transport needs, especially in urban areas, and be a strong alternative to conventional transport powered by liquid fuels.

The challenge lies in setting up a commercially viable, convenient system for end customers. The difficulty is how to change drivers' mobility behavior, so that instead of going to a gas station just before the tank is empty, drivers need to charge their cars every other time they park.

There is also a third alternative for recharging EVs, the battery swapping service. This can be done in recharging stations and the change can be done in about 10 minutes for a full charged battery. This would relieve "range anxiety" giving the customer three different modes for recharge:

- Slow charge (7/8 hours for 100%);
- Fast charge (30 minutes for 80%);
- Batteries change (5 minutes for 100%).

The cost of the recharge service should escalate as time diminishes. A just pricing for each of these recharging models should be determined, ensuring it is sufficiently high to recover infrastructure investments while being low enough to benefit the user. This service is expected to elevate the expense of electricity for recharging.

3.2.3 Acquisition costs

Despite growing environmental awareness in society, studies have repeatedly shown that customers are only willing to pay a limited price for being "green". This means that EVs and PHEVs must be attractively priced, not only in terms of initial purchase price, but also the ongoing costs each month. The costs of EVs are still much higher than for conventional cars due to a low production volume and expensive battery packs. This reflects directly the manufacturing costs. On the other hand, the expenses per travelled km are quite attractive comparatively to gasoline use, for EV technological solution it is around 1/4. With present prices pure electric vehicles pay off in terms of cost only if long distances are driven (higher than 200 000 km) (Baptista, 2009). This fact is important when calculating eventual tax incentives to purchase these kinds of technologies, having in mind that the final consumer is extremely sensitive to the "km for breakthrough". These limitations force Governments incentives to push the penetration of electric vehicles. There are predictions on the evolution of EVs' batteries capacity, density and price for the coming years. According to some consultancy analysis (Roland Berger, 2008) battery costs are expected to decrease to half the price from 2010 to 2020.

3.3 EVs recharging profiles

There are many possibilities of vehicles charging profiles. Some could be more likely to happen but the uncertainty is high as it depends on vehicles charging requirements and drivers mobility patterns. Many studies performed included different charging scenarios.

3.3.1 Uncontrolled charging

Under this charging profile it is considered that each EV begins charging as soon as it is plugged in, and stops when the battery is fully charged in case of normal time charge or 80% SOC in case of fast charge. This can be considered a reference case without any intelligent control of how and when charging occurs, or incentives (such as time-of-use rates) to influence individual consumer behavior. The majority of charging is home charging though a little services charging is also considered. The uncontrollable charge could be as depicted in Fig. 14.

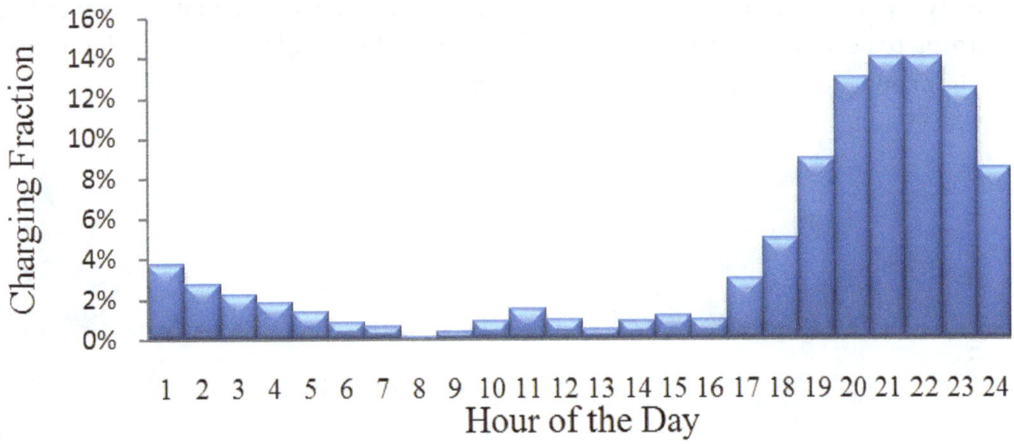

Figure 14. Expected charging profile for uncontrolled charging scenario

3.3.2 Off-peak charging

This charging scenario assumes that almost all charging occurs at home in the overnight hours. Given existing incentives for off-peak energy use it attempts to better optimize the use of low cost off peak energy by delaying initiation of household charging until 10 pm (Fig. 15) or 11pm according to the utility policy to promote off-peak demand.

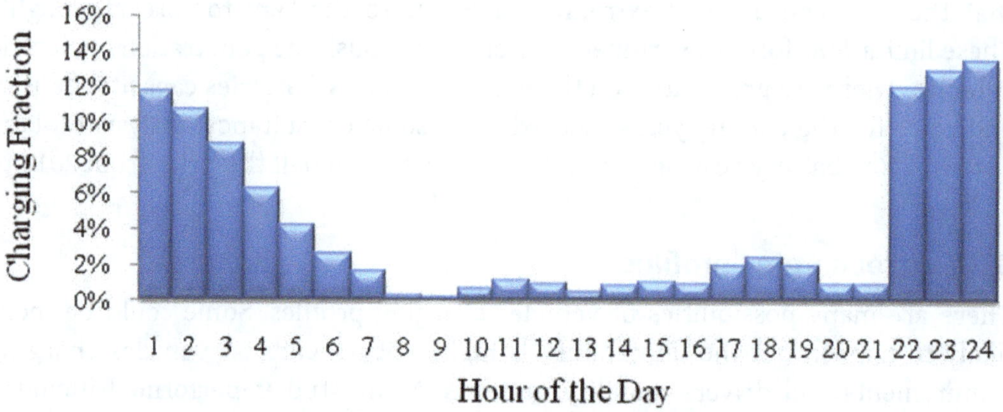

Figure 15. Expected charging profile for off-peak charging scenario

Under this recharging profile, a peak at 10 pm should be expected in a high EV penetration scenario. This peak could be smoothed with scheduled strategies for off-peak recharge (Fig. 16).

3.3.3 Optimal charging

This charging scenario also assumes that almost all charging occurs at home during the off-peak hours, however it attempts to provide the most optimal low cost charging electricity by assuming that the vehicle charging can be controlled directly by the local utility. This allows the utility to precisely match the vehicle charging to periods of minimum demand, allowing the use of lowest cost electricity, and improving overall utility system performance.

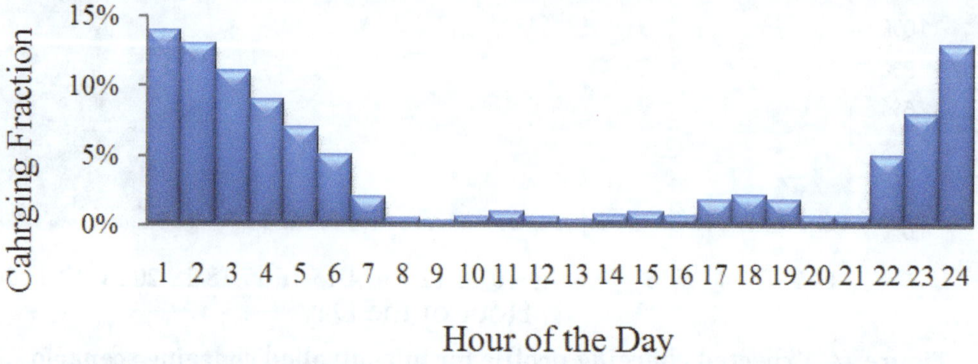

Figure 16. Expected charging profile for a smothing off-peak charging scenario

4. EFFECTS ON PRIMARY ENERGY USE, FOSSIL FUEL UTILIZATION, GREENHOUSE GAS EMISSIONS, AND ELECTRICITY EXPENSES

The impacts of EVs and PHEVs charge on the electric utilities depend on the first place of the combination of vehicles' penetration and charging profile scenarios. The case studies considered in this chapter arise from the combination of different scenarios of electric vehicle penetration and recharging operation. For each scenario, the following variables should be analyzed:

- Primary energy needs for the electric and transportation sector working as separate systems (BAU) and working together (EVs scenarios);
- Effects in electricity load profiles.
- Impacts on overall energy and CO_2 emissions
- Impacts on electricity costs.
- Impacts on fossil fuels use and imports.

4.1 Primary energy consumption and fossil fuels use and GHG emissions

Energy consumption, fossil fuels use and CO_2 emissions from electricity production and transportation (the light duty fleet segment) can suffer great reductions with the integration of these two sectors by transport electrification. The replacement of a great amount of ICEVs by EVs in a country in which power generation accounts with more than 50% of renewable sources has great impacts in fossil fuels use and CO_2 emissions reductions.

In terms of electricity production mix, with plenty renewable sources for generation, two extreme cases can be considered:

1. A dry year, with high prices offers for hydro and natural gas production, the supply curves per technology could have the following forms considered a certain power installed per technology (Fig. 17a).
2. A wet year, with low prices offers for hydro and also natural gas production, the supply curves per technology could have the forms depicted in Fig. 17b.

The expected load profiles for 2020 in a dry year hypothesis for a 2 million EVs scenario (about 1/3 of LD fleet replacement) should be as depicted in Fig. 18. A peak effect in load diagram in an uncontrollable recharge profile can be observed. By transport electrification, even in the worst situation of uncontrollable recharge, a 2% decrease in primary energy use, 9% in fossil fuels use (due to great reductions in the transportation sector) and 8% in CO_2 emissions can be achieved by transport electrification when compared with the case without EVs. An off-peak recharge profile with mass EVs penetration has more effects in energy, fossil fuels use and emissions reductions. In fact it was verified a 4%, 10% and 9% reduction respectively.

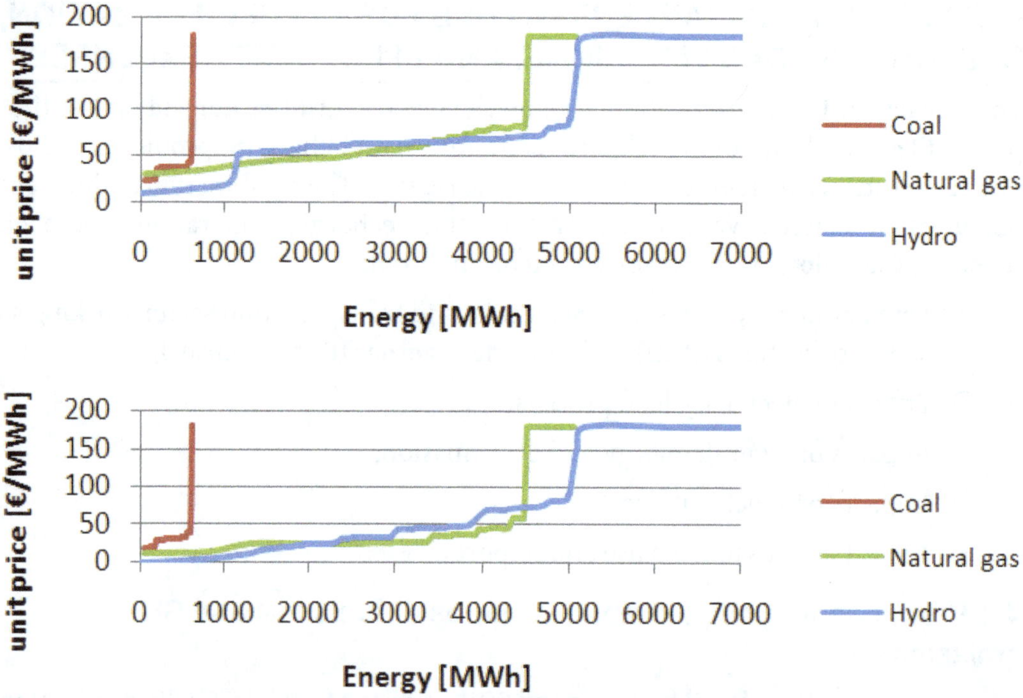

Figure 17. Example of supply curves per technology with a) high prices for hydro and natural gas production; b) low prices for hydro and natural gas production

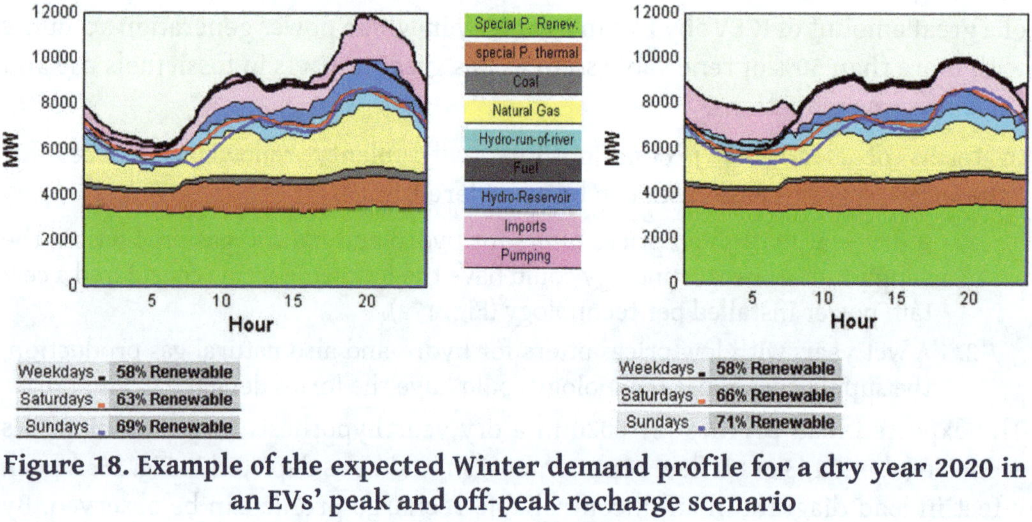

Figure 18. Example of the expected Winter demand profile for a dry year 2020 in an EVs' peak and off-peak recharge scenario

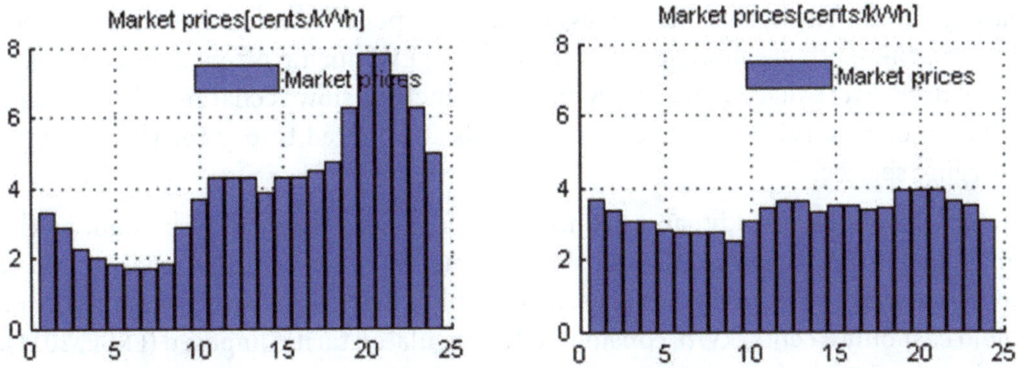

Figure 19. Expected Winter spot prices in a peak and off-peak recharge scenarios for 2M EVs in Portugal mainland

In any situation the integration of the electricity generation and transportation sectors has energy and environmental advantages. The decrease of fossil fuel use brings also economic advantages in the trading balance as all fossil fuels are imported in this example country.

In terms of the electricity wholesale prices, in a high EV penetration scenario, the leveling of the spot prices follow the leveling of the load profile (Fig. 19).

It is expectable that the peak effect of uncontrollable charging would cause price sparks at the peak hours. A time of use price for EVs recharge must be done to avoid this situation.

In the Island example as stated in sections 2.2 and 2.3.1, uncontrollable EVs recharge has worse consequences in renewable penetration and production prices.

4.2 Electricity costs, the fuel costs for EVs

As stated in section 2.3., electricity costs for EVs' recharge, either in a market context or as a vertically owned company is a sum of different costs related to production, net use and retailing of the product to the final client. To illustrate the differences, we have two case studies, the mainland, where there is competition in the production and retailing markets and S. Miguel Island where the same company monopolises all activities.

Adding all the costs associated to electricity supply service, for the mainland case, in a scenario of low hydro production the price could reach 17 cents/kWh, for 2 Million EVs charging mainly at peak hours, this includes the whole sale price (8c€/kWh) plus the net access tariff (the 2010 regulated tariff was used in this example, 9.09c€/kWh) plus the retailer revenue (a 4% of the wholesale, 0.32c€/kWh). In a high hydro production and low wholesale prices, an off peak recharge could reach the 5.6 cents/kWh. In these extreme conditions EV energy prices could be between 0.9€ to 2.8€ per 100 km. Compared with the most efficient ICEV cars and the present gasoline and Diesel prices in Portugal that can spend between 4.5€ and 8.2€ per 100 km, there are great advantages in charging EVs during off-peak hours. These simulations were made considering for the vehicles a slow recharge at low voltage at home or at service and no extra charge is considered to pay for the parking/charging service.

For the Island case study, an uncontrollable recharge must be forbidden, as the final electricity costs would reach the 20 cents/kWh, with all the environmental consequences as EVs would be recharged mainly with fuel oil. An off-peak recharge could cost only 8 cents/kWh, considered the regulated tariffs imposed (ERSE, 2011). In the Island EVs energy costs can be between 1.3€ and 3€ per 100 km.

5. CONCISE SUMMARY

Electric vehicles (EVs) and plug-in hybrid electric vehicles (PHEVs), which derive their energy from the grid through battery charging, are poised for mass market introduction and anticipated to aid in the decrease of oil use. PHEVs and EVs present a significant opportunity to mitigate CO2 emissions from transportation if the electricity utilized for charging their batteries is produced using low-carbon technology. Besides environmental concerns, electric vehicles (EVs) present techno-economic challenges for utilities due to their significant load flexibility, as they remain parked 93% of their lifespan. This allows for charging at home, work, or parking facilities, indicating that their charging times can vary considerably throughout the day. The substitution of a significant number of ICEVs with EVs in a nation where over 50% of power output derives from renewable sources substantially affects fossil fuel use and CO2 emissions. The simulation results indicate that a significant adoption of electric vehicles in Portugal reduces energy consumption, diminishes the reliance on fossil fuels (primarily oil), and lowers CO2 emissions from the two sectors—electricity production and transportation—that currently contribute most to emissions from fossil fuel combustion. The imperative to produce electricity from indigenous low-carbon resources in most nations renders natural transport electrification a viable strategy for reducing emissions and fossil fuel use in the transportation sector. The cost and range of the car continue to be the primary obstacles to the penetration of electric vehicles. The expense of charging the electric vehicle is primarily influenced by the power producing technologies, the structure of the electricity market (whether competitive or not), and the timing of the recharge (peak or off-peak periods), which collectively affect the ultimate expenses. This chapter presents scenarios of electric vehicle (EV) penetration and recharge profiles, alongside extreme cases of anticipated electricity production, resulting in varying wholesale prices and hourly price profiles, as well as differing fossil fuel consumption and emissions related to EV charging, which ultimately affects the cost of EV fuel per kilometer.

In light of rising oil prices, the energy cost per kilometer for electric vehicles (EVs) has become significantly more favorable compared to the fuel expenses of internal combustion engine vehicles (ICEVs), sufficiently offsetting the disparity in initial purchase costs.

REFERENCES

1. Baptista P. Camus C. Silva C. Farias T. 2009Impact Of The Introduction Of Electric Based Vehicles In São Miguel Island, 10th Conference on Energy for a Clean Environment, July 2009
2. CML Lisbon City Hall, Lisbon 2002The mobility challenge, 2002, available at http://ulisses.cm-lisboa.pt/data/002/002/pdf/mobilidade.pdflast accessed 09/05/25)
3. Denholm P. Short W. 2006An Evaluation of Utility System Impacts and Benefits of Optimally Dispatched Plug-In Hybrid Electric Vehicles, Technical Report NREL/TP-620 40293October 2006.
4. Duvall, Dr. Mark S. (2006), "Plug-in hybrid electric Vehicles Technology Challenges", EPRI Sep 2006
5. EDA- Electricity of Azores, available at www.eda.pt
6. EPRI 2006Comparing the Benefits and Impacts of Hybrid Electric Vehicle Options for compact sedan and sport utility vehicles, EPRI,1006892. July 2002
7. ERSE 2011http://www.erse.pt/pt/electricidade/tarifaseprecos/Paginas/default.aspx.last accessed 11/01/15)
8. Hadley Stanton. W. 2006Impact of Plug-in Hybrid Vehicles on the Electric Grid, October 2006
9. Hadley Stanton. W. 2008Potential Impacts of Plug-in Hybrid Vehicles on Regional Power Generation, ORNL/TM-2008/150, January 2008.
10. IEA 2007International energy agency- World Energy Outlook 2007, (93assessed at 09/05/19) at http://www.iea.org/textbase/nppdf/free/2007/weo_2007.pdf
11. IEA 2008International energy agency- World Energy Outlook 2008. (508assessed at10/10/19) available online at http://www.iea.org/Textbase/about/copyright.asp
12. IEA 2008Energy Tecnology Perpectives 2008, Chapter 15 Transport
13. IPCC 2007Fourth Assessment Report. Climate Change 2007: Synthesis Report. Summary for Policy Makers", 2007.

14. EVWorld 2009http://www.evworld.com/index.cfmlast assessed at 20/05/09)

15. EVreportlinker 2009http://www.reportlinker.com/059605 Electric-Vehicles.htmllast assessed 20/05/09)

16. Kempton W. Tomic J. 2005aVehicle-to-grid power fundamentals: calculating capacity and net revenue, J. Power Sources 144 2005

17. Kempton W. Tomic J. 2005bVehicle-to-grid power implementations: from stabilizing the grid to supporting large-scale renewable energy, J. Power Sources 144 2005

18. Kintner-Meyer Michael, Kevin Schneider, Robert Pratt, 2007Impacts assessment of plug-in hybrid vehicles on electric utilities and regional u.s. power grids", Part 1: technical analysis, Pacific Northwest National Laboratory (a), November, 2007.

19. OMEL 2009Electricity Market Operator, available at www.omel.es,assessed at 09/10/10

20. Parks K. Denholm P. Markel T. 2007Costs and Emissions Associated with Plug-in Hybrid Electric Vehicle Charging in the Xcel Energy Colorado Service Territory, Technical Report NREL/TP-640 41410May 2007

21. Roland Berger Strategy Consultants 2008Value for EDP in Electric Vehicles, Lisbon, July 29th, 2008

22. Scott Michael. J. Kintner-Meyer Michael. Elliott Douglas. B. Warwick William. M. 2007Impacts assessment of plug-in hybrid vehicles on electric utilities and regional u.s. power grids: part 2: economic assessment. Pacific Northwest National Laboratory (a), November, 2007.

23. Skea J. Anderson D. Green T. Gross R. Heptonstall P. Leach M. 2008Intermittent renewable generation and maintaining power system reliability", Generation, Transmission & Distribution, IET 2 1January 2008 Page(s):82- 89

24. Szczepanek A. Botsford C. 2009Electric Vehicle Infrastructure Development: An Enabler for Electric Vehicle Adoption, EVS24,Stavanger, Norway, May 13-16, 2009

9 Chapter COMMUNICATION PERTAINING TO ELECTRIC VEHICLES

1. INTRODUCTION: AN OVERVIEW

Although electric vehicles (EV) are prevalent in specific applications, such as forklifts and baggage transport (cf. Rand et al., 1998), their use as personal automobiles remains restricted. For instance, about 2,300 electric vehicles were registered in Germany on January 1, 2011 (Federal Motor Transport Authority, 2011). Numerous factors are propelling the global shift towards the electrification of personal vehicle traffic: advancements in battery technology, elevated oil costs in 2008 and 2011, the recent automotive crisis in 2009, and the anticipated ecological benefits of electric vehicle utilization.Moreover, the combination of two major energy conversion systems, namely the electric utility system and the light vehicle fleet (e.g. individual motor cars), could create considerable synergies (Kempton & Letendre, 1997; Kempton & Tomic, 2005a; Tomic & Kempton, 2007). One proposed concept is Vehicle-to-Grid realized through "... vehicles with an electric drive motor powered by batteries, a fuel cell or a hybrid drive train, [that] can generate or store electricity when parked, and with appropriate connections can feed power to the grid ..." (Kempton & Tomic, 2005b).

EV can be subdivided into hybrid EV (HEV) and battery EV (BEV). The subset of HEV combines the (parallel or serial) electric drive motor with a combustion engine, a fuel cell or human power (e.g. Pedelec). The BEV rely only on a – mainly electro-chemical – energy storage. Although the energy to fulfill an HEV's mobility function could be provided alternatively (e.g. by fuel or a battery exchange), all EV are assumed to be plug-in vehicles in this article. This means they charge from and possibly discharge into the power grid. All in all, the term EV refers from now on to (at least partly) electrically propelled cars used for individual motor traffic which can be conductively or inductively connected to the power grid. A lot of research questions have been raised with respect to EV. Most of them approach the subject from a technical (e.g. battery performance), economic (e.g. total cost of ownership) or a user (e.g. driving behaviour) perspective, others investigate social, political and cultural barriers for broader EV usage (Sovacool & Hirsh, 2009). Information and communication technology (ICT) (definition cf. Krcmar 2006), is affected in all of these perspectives, since it has to ...

- be mastered technically,
- be used economically, and
- increase usability.

Furthermore, study inquiries can be classified according to an electric vehicle's life cycle, which, from a broad perspective, aligns with the sequential stages of manufacture, utilization, and recycling. Questions about production focus on efficiently manufacturing electric vehicles to meet market demand, while the recycling phase pertains to the economic and ecological disposal of electric vehicles. During the usage phase, inquiries regarding the electric vehicle's interaction with users and infrastructure must be addressed. The utilization phase will be the primary focus of the subsequent considerations.The subsequent sections of this chapter are organized as follows. The initial focus is on the fixed and intersecting points of energy transmission (section 2). Subsequently, essential parameters for energy transmission use cases are provided and analyzed (section 3). The informational dimension is elaborated upon in a distinct part (section 4) pertaining to an information system. The e-mobility roaming example demonstrates the necessity and rationale for a comprehensive overview of such an information system (section 5). Ultimately, the final remarks encapsulate the current state and provide a perspective on forthcoming study (section 6).

2. FIXED AND INTERSECTING POINT: ENERGY TRANSFER

The utilization of electric vehicles highlights a significant issue: energy.Similar to conventional vehicles, reliable and secure movement in an electric vehicle (EV) can only be ensured by an adequate supply of energy at all times.Consequently, "sufficient" is contingent upon the efficiency of the electric vehicle and the context of its utilization.A compact city car is highly efficient, typically requiring energy for no more than 100 km daily, however an SUV utilized by a traveling salesperson is considerably energy-intensive, maybe necessitating energy for 1000 km or more. Three primary methods of energy transmission for electric vehicles are presently under consideration: conductive charging, inductive charging, and battery exchange. One electric vehicle may utilize a combination of different transmission modes, such as typically recharging conductively at home while employing a battery exchange for infrequent long-distance travel to visit family. However, this amalgamation of several energy transmission systems undoubtedly incurs elevated expenses.

Due to high costs and a low energy density of its storage, especially BEV still have the disadvantage of a low energy storage capacity (cf. Figure 1, left). Neglecting the technically more complex battery exchange as well as inductive charging when driving, lifetime considerations of the (most often) electro-chemical storage requires slow charging and leads to a rather time-consuming replenishment. In contrast to these disadvantages in comparison with conventional vehicles, BEV profit from the fact that electricity is already quasi- omnipresent allowing for many potential suppliers. Conventional vehicles and (to a certain extent) fuel-based HEV rely on petrol stations operated by only a few companies at well distributed, but at a limited number of places (compare Figure 1, right).

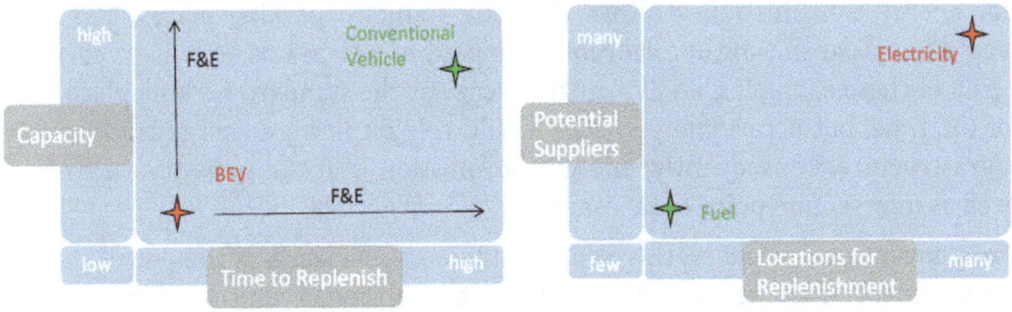

Fig. 1. Qualitative comparison of energy storage and availability between BEV and conventional vehicles

To sum up, the discussion at this point reveals that although BEV are unprivileged with respect to the energy storage, the potential availability of energy is significantly higher. Therefore, the energy transmission between BEV and the power grid is an important factor of all BEV applications. Although HEV are not fully reliant on the power grid, the idea of sustainable mobility by charging renewable energy is realizable only for the charged energy. Henceforth, all EV business models are forced to define how and when the EV are to be served with energy. This is why the energy transmission between the power grid and an EV can be seen as a fixed point of EV usage (cf. Fig. 2), in case of BEV a particular important one.

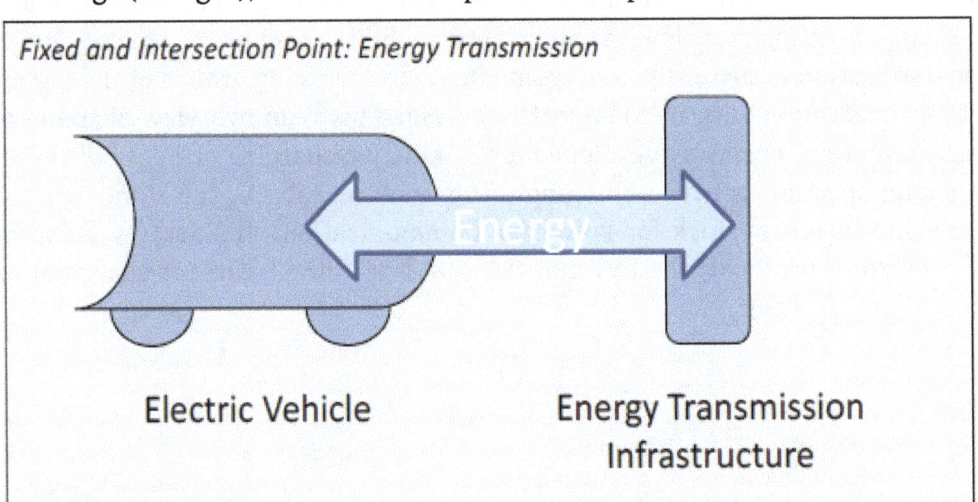

Fig. 2. Fixed and intersection point of EV business models

The energy transmission is not only a fixed point of EV usage, but also a point of intersection of different EV business models. For example, an EV manufacturer could assume the EV buyer to normally charge via a standard power outlet at a standard parking place (e.g. at home). However, mobility of EV implies that the EV could be anywhere else other than this standard parking place when the battery runs out of energy. Henceforth, the EV manufacturer needs to offer the possibility for energy transmission at other than the standard parking places. The same applies for a car sharing or a car rental company as well as taxi or delivery services. In all of these examples, an EV can get energy at the standard parking place most of the time, but occasionally, the EV would rely on foreign energy transmission infrastructure. Consequently, energy transmission is at the same time a fixed as well as intersection point of EV usage.

For conventional vehicles, this point of intersection clearly is the petrol station. Independent of the exact vehicle usage, a conventional vehicle can be refueled through the standardized filler necks. The petrol stations are not available everywhere, but they are well distributed over high frequented places. However, this concept works well only for vehicles with a huge energy storage capacity as well as a fast refueling process. Henceforth, it is improper for all the EV that need to charge regularly, but can not store huge amounts of energy.

3. ESSENTIAL DIMENSIONS FOR ENERGY TRANSMISSION APPLICATIONS

By recognizing the energy transmission as a fixed as well as intersection point of business models, it becomes apparent that isolated infrastructure is inefficient and renders all business models unprofitable. Therefore, significant efforts are necessary to standardize the energy transmission – best on an international level to keep production costs low and allow for e- mobility roaming (cf. chapter 5). This standardization is currently undergoing in several working groups of the ISO and IEC in cooperation with national institutes. Fig. 3 gives an overview of standards that exist or are currently developed for conductive charging of EV. With respect to communication between charging infrastructure and EV, the standards ISO/IEC 15118 (protocol stack for Powerline Communication), IEC 61850-x (adaption of substation protocols for EV) and IEC 61851-24 (direct current charging) are particularly relevant.

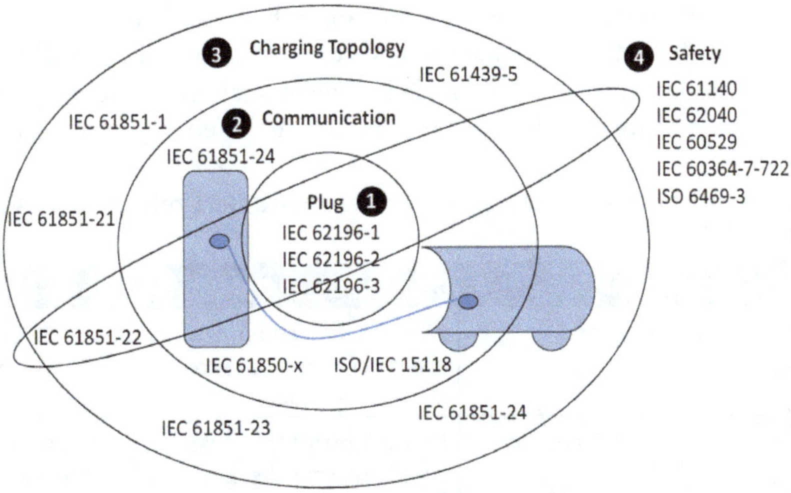

IEC 62196-1: Plugs, socket-outlets, vehicle couplers and vehicle inlets Conductive charging of electric vehicles, Charging of electric vehicles up to 250 A a.c. and 400 A d.c.
IEC 62196-2: Plugs, socket-outlets, vehicle connectors and vehicle inlets - Conductive charging of electric vehicles, Dimensional compatibility and interchangeability requirements for a.c. pin and contact-tube accessories
IEC 62196-3: Plugs, socket-outlets, and vehicle couplers - conductive charging of electric vehicles, Dimensional interchangeability requirements for pin and contact-tube coupler with rated operating voltage up to 1 000 V d.c. and rated current up to 400 A for dedicated d.c. charging
IEC 61850-x: Communication networks and systems in substations
ISO/IEC 15118: Vehicle to grid communication interface
IEC 61439-5: Low-voltage switchgear and controlgear assemblies, Assemblies for power distribution in public networks
IEC 61851-1: Electric vehicle conductive charging system – General requirements
IEC 61851-21: Electric vehicle conductive charging system – Electric vehicle requirements for conductive connection to an a.c./d.c. supply
IEC 61851-22: Electric vehicle conductive charging system – a.c. electric vehicle charging station
IEC 61851-23: Electric vehicle conductive charging system – D.C electric vehicle charging station
IEC 61851-24: Electric vehicle conductive charging system – Control communication protocol between off-board d.c. charger and electric vehicle
IEC 61140: Protection against electric shock - Common aspects for installation and equipment
IEC 62040: Uninterruptible power systems (UPS)
IEC 60529: Degrees of protection provided by enclosures (IP Code)
IEC 60364-7-722: Low voltage electrical installations, Requirements for special installations or locations - Supply of Electric vehicle
ISO 6469-3: Electrically propelled road vehicles, Safety specification, Protection of persons against electric shock

Fig. 3. Standards for conductive charging of EV (German National Platform for E-Mobility, 2010)

The fixed point of energy transmission can be characterized and designed with respect to various dimensions. Three fundamental dimensions shall be distinguished in the following:

- Electrical (e.g. charger or specification of plug)
- Organizational (e.g. parking place type or infrastructure owner)
- Informational (e.g. ICT hardware or authentication protocols)

The electrical dimension is the core of energy transmission. That is why international standardization efforts have focused on this dimension for a long time. Though, organizational and informational aspects become more and more important. Organizational questions appear with respect to relevant roles and business models (cf. Fig. 4). Informational aspects are to be investigated since ICT is a promising enabler for EV usage due to several reasons:

1. Process efficiency for energy supplier and infrastructure operator could be increased.

2. Consumer protection laws require instant, reliable and verifiable information about transferred energy quantity and corresponding prices for the EV user.
3. The possibility to offer value added services such as the routing to and the reservation of functioning and free places at the energy transmission infrastructure.
4. More sophisticated concepts such as Vehicle-to-Grid rely on sophisticated real-time information.

Role	Example(s)	Organizational Question(s)
User	• Social services • Delivery services	Where, when and how should one's EV be charged? How to ensure that the EV are always charged adequately?
Vehicle Provider	• Car sharing company • Car rental company	How to ensure that customers are able to charge where ever needed in a comfortable manner?
Mobility Provider	• Taxi	What is the charging status of each EV and which EV is therefore should take which customer?
Energy Provider	• Energy supplier • Distribution network operator	How can a customer be provided with energy in transit? How is it possible to avoid overcharge a certain grid string when lots of EV are charging?
Parking Area Operator	• Supermarket • Visitor parking area	How to best comfort customers by providing energy for EV?

Fig. 4. Organizational questions of energy transmission to EV

One could regard the informational dimension linking the electrical and organizational aspects in order to harmonize the process before, during and after the energy transmission.

The organizational dimension does not need to include the commercial (financial, contractual, etc.) aspects, in contrast to the business level identified elsewhere (Bolczek, 2010). Instead, it can be assumed that basic organizational aspects are describable without assuming certain roles or business models. For example, procedures to access infrastructure have to be consistent throughout roles and business models and therefore need to be described independently.

There are abstract use cases that can be characterized around the fixed and intersection point of energy transmission throughout the three fundamental dimensions. The use cases themselves can be part of one or several business models.

In each dimension, a number of attributes can be identified while an attribute can be instantiated by one or several values. This can be seen as a simple morphological box (cf. Zwicky & Wilson, 1967). The combination of values over attributes and dimensions leads to a particular use case. Many and manifold use cases are possible, however, only a few particular types will fit to the requirements of business models in the long run. An excerpt of such a box is given in Fig. 5.

<table>
<tr><th>Dimension</th><th>Attribute</th><th colspan="6">Value(s)</th></tr>
<tr><td rowspan="3">Electrical</td><td>Energy Transmission</td><td colspan="2">Conductive</td><td colspan="2">Inductive</td><td colspan="2">Battery Exchange</td></tr>
<tr><td>Plug Specification</td><td>CEE 7/7</td><td>CEE 3P+ N+PE, 6h</td><td>JARI Level 3 DC</td><td>IEC 62196-2</td><td>SAE J1772</td><td>SCAME</td></tr>
<tr><td>...</td><td colspan="3">...</td><td colspan="3">...</td></tr>
<tr><td rowspan="3">Organizational</td><td>Entitled to dispose of infrastructure</td><td colspan="2">Natural Person (private: individual)</td><td colspan="3">Legal Person (private: company)</td><td>Legal Person (public: authority)</td></tr>
<tr><td>Access Level</td><td colspan="2">EV parking place</td><td colspan="3">EV supply equipment</td><td>Electricity</td></tr>
<tr><td>...</td><td colspan="2">...</td><td colspan="3">...</td><td>...</td></tr>
<tr><td rowspan="3">Informational</td><td>Top Level Elements</td><td colspan="2">User</td><td>EV</td><td colspan="2">EV Supply Equipment</td><td>IT-Backend</td></tr>
<tr><td>Authentication Hardware</td><td colspan="2">RFID</td><td colspan="3">SmartCard</td><td>Mobile Phone (SimCard)</td></tr>
<tr><td>...</td><td colspan="2">...</td><td colspan="3">...</td><td>...</td></tr>
</table>

Fig. 5. Excerpt of attributes in the morphological box in three dimensions

With respect to the communication with and for EV, the informational dimension is pertinent and will be focused on in the following.

4. AN INFORMATION DEVICE FOR ELECTRIC CARS

The informational dimension can be described via an information system. An information system is a social and technical system that combines human and mechanical components to achieve the optimal allocation of information and communication (cf. Heinrich, 2005; Krcmar, 2007). The description of such an information system is the base to design an adequate information flow within and throughout the manifold business models. As seen before, the point of energy transmission is a fixed point for all EV. Henceforth, this fixed point is a useful starting point for the description of an information system for EV usage. Top level elements of such an information system are elements that play a prominent role when it comes to using or processing information. Those elements were identified to be the user of the EV, the EV itself, the point of energy transmission, as well as the supporting IT- Backend (cf. Fig. 6). Deliberately, the IT-Backend resembles a cloud, since - with some ups and downs - there is a continuous trend towards the flexible utilisation of IT-Infrastructure and services over (high speed) internet connections.

An information system can be detailed via an information system architecture. Several information system architectures have been proposed (The Open Group Architecture Framework, 2010; Scheer, 2000; Krcmar 2007). However, on the top level the architecture components are very similar and the architectures differ mainly with respect to their illustration. Although, the information system architecture concepts were originally developed for enterprises to describe and design their information system, it can be adapted to fit the needs for an information system around EV usage.

According to Krcmar (2007), an information system architecture consists of six components, namely

- Infrastructure
- Communication architecture
- Data architecture
- Application architecture
- Organizational architecture

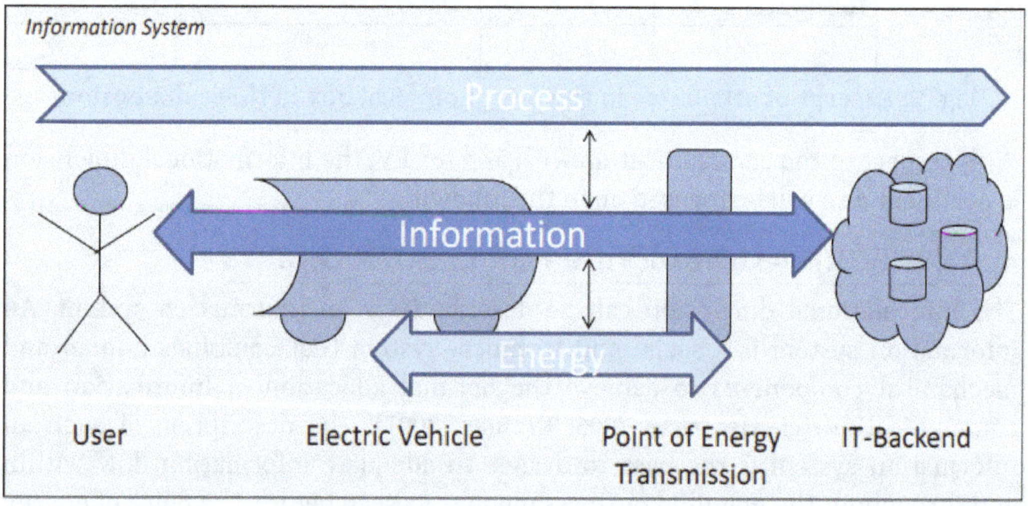

Fig. 6. Top level elements of the information system for EV

The components are structured hierarchically, from rather physical ones (infrastructure) to rather logical ones (strategy). In order to emphasize that these components must be well balanced to achieve efficient information flows, Krcmar used the spinner to illustrate the relations of the components (cf. Fig. 7).

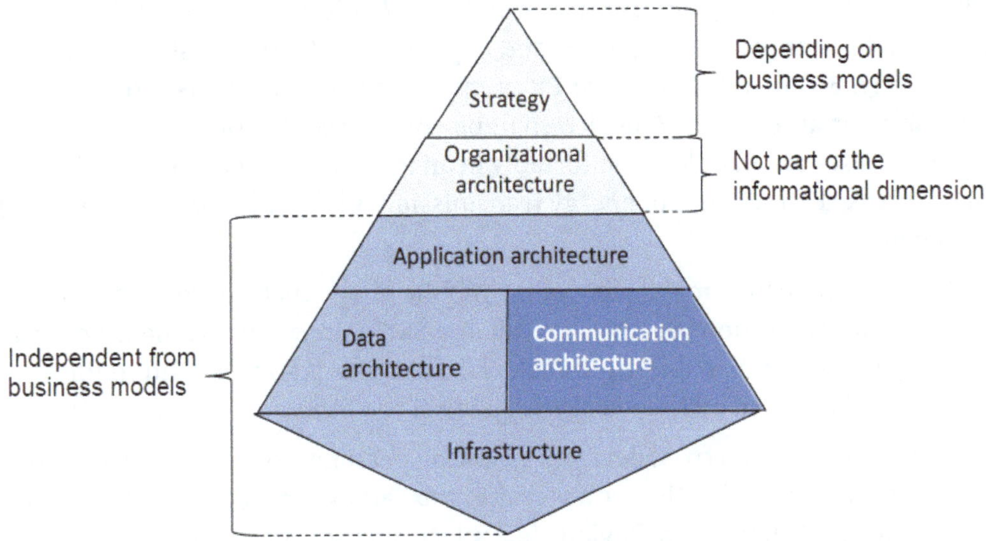

Fig. 7. Information system architecture as spinner (adapted from: Krcmar, 2006)

The architecture originally was designed for enterprises which realize at least one business model. Thus, the creation of use case types independent of business models requires an adaption of the general architecture. At first, due to the close relation to long term profit, the strategy has to be left out. Secondly, the organizational architecture that stands for a company's process and structural organization is not part of the information dimension (cf. section 3). Well adapted for a base for use case types are the infrastructure, the data architecture, the communication architecture as well as the application architecture. They can be characterized as follows:

- The infrastructure contains all design elements of ICT hardware, for example which kind of microcontroller, CPU, graphics card, etc. are used.
- The data architecture consists of data models and objects that are necessary to provide information to EV users and other stakeholders.
- The communication architecture describes the topology for communication as well as the used protocols.
- The application architecture is built up on the first three components in order to provide defined functions for users and systems.

However, these components of an information system can hardly be interpreted in an isolated manner, but are in close relation and interaction. Thus, despite the focus on the communication architecture, the remainder of this chapter will also contain aspects of infrastructure, data and functions.

With respect to the communication with and for EV, it is useful to adapt the general information system overview (Fig. 6) and explicitly model the most important IT-Systems (Fig. 8). In addition to the EV, a mobile device as well as a fixed device is available for an EV user. A device shall be considered "mobile" in this context when an EV user can handle it comfortably in an upright position. For the following examples, the abstract level of energy transmission is left and conductive charging is assumed:

- An example for a mobile device is a mobile phone that can be used to unlock a charging station via a hotline call or a SMS. Alternatively, the widespread smartphones are able to unlock the charging station via an application ("App") offered by the energy supplier of the EV user.
- An example for a fixed device is a desktop personal computer or a notebook. Via a client application or a website, they allow for an offline check of the individual charging behaviour (e.g. place, energy and cost).

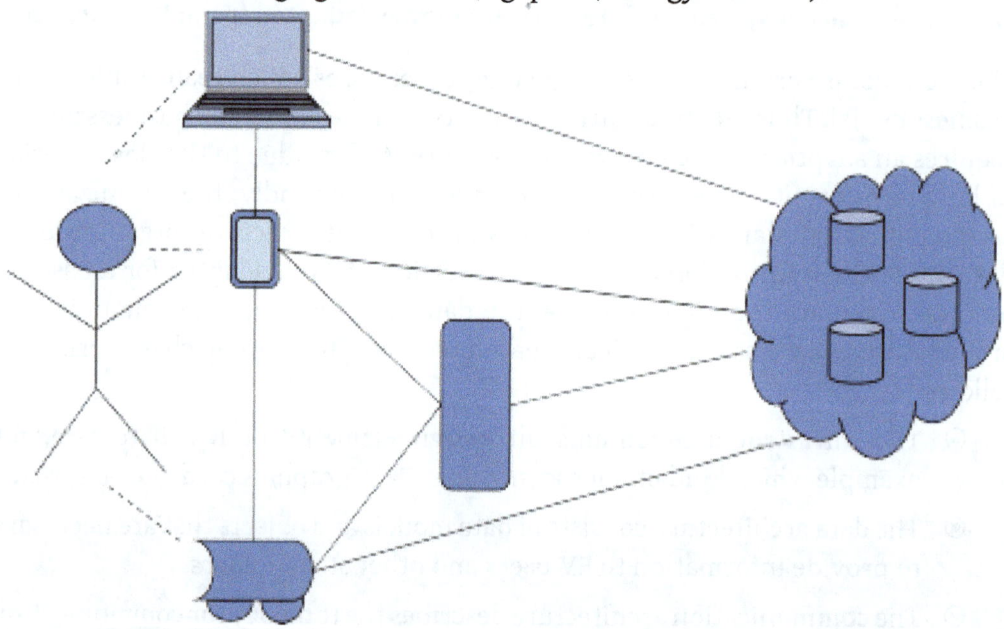

Fig. 8. Communication architecture from an EV user perspective

This adapted view of the information system is the infrastructural base of communication from the EV user perspective. In this chapter, the server landscape in the IT-Backend is not further detailed. Although, design and operation of applications and services in the IT- Backend is challenging, the questions that are to be asked and answered are not necessarily specific for e-mobility. In many cases, the integration of e-mobility in existing software solutions is the core problem. This is done via interfaces of which four already appear in the chosen illustration (Fig. 8).

After this introduction to the information system at the fixed point of energy transmission to and from an EV, an example shall illustrate how this information system view can help to master the current informational challenges of EV usage. Interesting examples would be:

- Access to own infrastructure under consideration of authentication, authorisation and accounting.
- Access to foreign infrastructure under consideration of authentication, authorization and accounting.
- Collection and transmission of metering data with respect to the conflicting goals of usage transparency, data reliability and data privacy.
- Value added user services such as routing to or reservation of infrastructure.
- Vehicle-to-Grid services such as frequency regulation.

Since Vehicle-to-Grid services and value added user services are not solving one core problem of e-mobility ("enough energy for mobility"), they might be less interesting at this point. For a discussion of the collection and transmission of metering data, the different national laws play an important role. However, access to infrastructure is a prior subject in all countries, and when using foreign infrastructure an informational interesting one. Therefore, the following chapters use this subject, later defined more precisely as e-mobility roaming, to illustrate the information system in detail. For the sake of convenience, again the conductive charging is assumed. For inductive charging or battery exchange, similar considerations can be made.

5. E-MOBILITY AND ROAMING

5.1 Motivation

Considering private car owners, they can be distinguished by the type of their standard parking place (e.g. at home). Either car owners have a dedicated parking space such as a garage, or they park curbside somewhere nearby. The former can charge the EV by simply plugging it into a fixed socket-outlet in their garage. In this case, no identification and authorization is needed – however possibly useful in the future for providing grid services (Vehicle-to-Grid). While charging at the standard parking space can cover most trips, recharging is necessary from time to time if the trip distance is higher than a typical EV- range or recharging at the standard parking space is not possible. Hence, in such cases, these EV owners rely on foreign charging infrastructure for which physical access, authentication, authorization and accounting (AAAA) is needed. The latter group, i.e. EV owners parking curbside, need foreign charging infrastructure every day.

Carsharing companies can have dedicated parking spaces for their cars. For EV usage, they can upgrade them with electric vehicle supply equipment (EVSE). Thus, whenever an EV is parked there, it can be recharged by using individual methods for AAAA. Since many carsharing offers still restrict the start and end of any rental to these dedicated parking spaces, regular recharging is guaranteed. However, for two reasons this is not enough: First of all, in some cases, the user might need to additionally recharge somewhere else. Secondly, the product "carsharing" mainly sells easy access to individual mobility. Therefore, the most successful carsharing companies will allow the start and end of a rental anywhere which endangers regular recharging at the dedicated parking spaces. A carsharing company can enable recharging not only at those own dedicated parking spaces, but also at foreign charging infrastructure by guaranteeing AAAA there.

Term	Definition
E-Mobility Provider	Contracts with EV Users in order to offer services (e.g. charging) and can be EVSE Operator at the same time.
EV User	Uses an EV within a contract with an E-Mobility Provider.
EVSE Operator	Operates at least one EVSE as a service for E-Mobility Providers, but has no continuous contractual relation to EV Users.

Fig. 9. Terms to define e-mobility roaming

All in all, these two simple and known business models already show that the usage of foreign charging infrastructure would occur regularly. This is often referred to as e-mobility roaming, in analogy to the mobile communications sector. More precisely, roaming with cellular phones means the uninterrupted availability of all services while moving out of range of the former carrier to another one (Schiller, 2003). On this base as well as with the definitions in Fig. 9, e-mobility roaming (or just roaming) in this paper refers to the situation in which an EV User is using an EVSE within a contract with an E-Mobility Provider that is not the EVSE Operator of the used EVSE (cf. Fig. 10).

Due to the still prevailing regional and local character of many power markets (e.g. in Germany), roaming is likely to happen not only international, but even within one town or street. Having a contract with every single provider is very uncomfortable. Hence, mechanisms to enable AAAA for roaming are inevitable. In order to guarantee a user- friendly e-mobility roaming experience, there are several challenges to cope with. Paying cash or via credit card is uncomfortable and requires more expensive infrastructure than identifying as a user through an adequate contract.

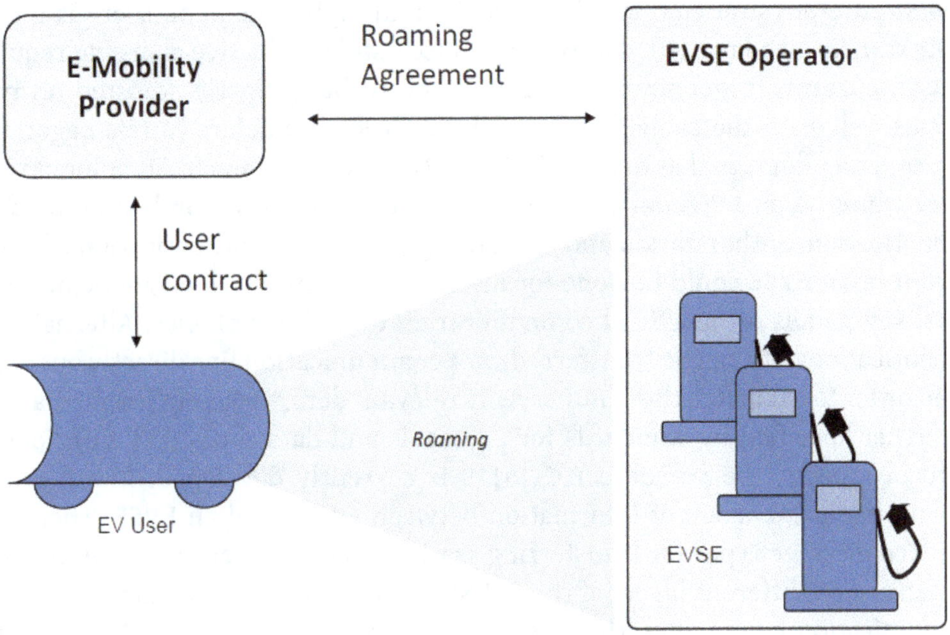

Fig. 10. E-mobility roaming

5.2 Major Challenges of roaming

Based on the aforementioned comprehension of e-mobility roaming and its commercial context, an examination of the prerequisites for roaming is conducted. Roaming, involving many parties, necessitates preconditions that are intricately linked to interoperability and the application of standards. The prerequisites for roaming can be categorized into electrical and commercial considerations, each pertaining to elements of the underlying medium or its utilization (Fig. 11).

	Electrical	Commercial
Medium	I	III
Use of Medium	II	IV

Fig. 11. Categories of requirements for roaming in e-mobility infrastructure

For example, a straight forward requirement for an electrical medium (I) is - assuming conductive charging - a standardized EV plug. Since the usage of adapters is very uncomfortable, an EV plug should fit into the outlet of all EVSE. The International Electrotechnical Commission (IEC) therefore currently revises the international standard IEC 62196. Considering other ways of getting power into an EV, such as induction or battery exchange, different requirements must be fulfilled. For inductive charging, a consistent form and position of the charger and the inductor is vital. For the battery exchange, especially the size and interface of the batteries as well as the security concept must be compatible.

Beyond pure physical characteristics of the underlying medium, there is a need for its standardized use (II). For example, successful conductive charging requires voltage, current, frequency and charge mode to be correctly adjusted on both sides as well as to the cable diameter. These basic parameters can be negotiated via a control pilot signal as defined in SAE J1772. From a commercial point of view, the charging of an EV requires a medium for containing or conducting data for authentication, authorization and accounting (AAA) (III). Authentication of a user in front of an EVSE could be done for instance via RFID cards, magnetic or smart cards, key panels or near field communication by cellular phones. Alternatively, authentication data can be transferred via a communication line directly out of the EV. In order to exchange the commercially relevant data, the use of the media must be further specified by standards for protocols and data types (IV). Considering protocol aspects, the standard IEC 15118 is currently developed. It will enable the automatic exchange of information between an EV and an EVSE. Therefore, standard message types for transferring session, status, metering and billing data are defined on different layers of the OSI Model. In addition to protocols using the communication connection, there is a clear commercial need for the definition of basic identifiers (IDs) that can be used throughout the information systems of involved companies. The remainder of this paper focuses on identification issues and discusses possible and necessary IDs for roaming with EV.

5.3 Identifiers for roaming

Every Identifier (ID) has a certain scope in which it is valid. For roaming, the distinction of intra-company and inter-company IDs (henceforth called uniform IDs) is essential. While intra-company IDs such as customer numbers are sufficient for many commercial applications, roaming requires uniform IDs for involved objects to allow for inter-company data exchange. Since uniform IDs require significant standardization efforts, it is worth to investigate which IDs should to be uniform in which cases. The cases clearly depend on the underlying business model(s) and technical choices. However, two abstract scenarios can cover many of them. Both scenarios differ from each other only with respect to the sequence of communication steps (Fig. 12).

In scenario A, the EV User (or its EV on behalf of him) passes all information needed for authentication through the EVSE (A1) to the EVSE Operator (A2). The EVSE Operator forwards the information to the E-Mobility Provider and requests AAA for the EV User (A3). If the response (A4) is positive, the EVSE Operator unlocks the EVSE for charging (A5). In scenario B, the EV User directly connects to the E-Mobility Provider (B1) for AAA. If authorization is successful, the E-Mobility Provider requests the EVSE Operator (B2) to unlock the particular EVSE for charging (B3).

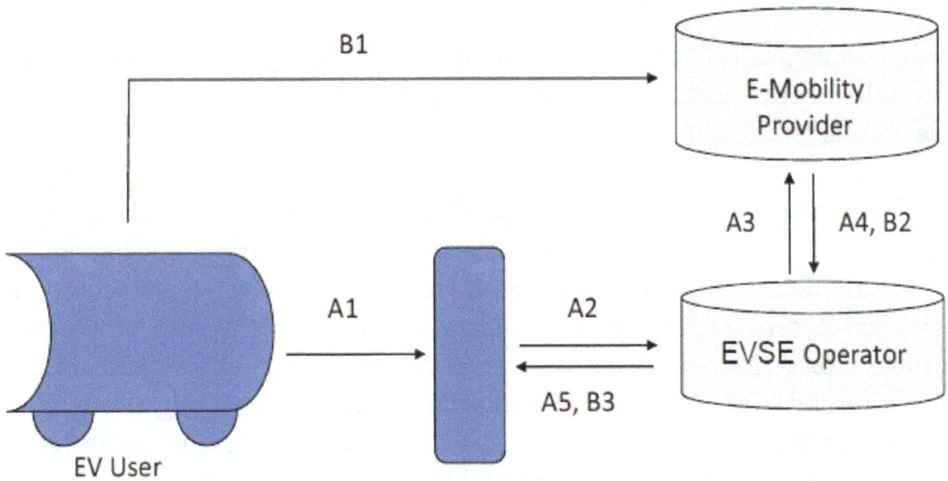

Fig. 12. Two scenarios for the sequence of communication steps

Identifiers	Scenario A	Scenario B
E-Mobility Provider	Required *operator need to know which provider to contact*	Optional *provider known by operator*
EV User	Optional *user known by provider*	Optional *user known by provider*
EVSE	Optional *EVSE known by operator*	Optional *EVSE known by operator*
EVSE Operator	Optional *operator known by provider*	Required *provider need to know which operator to contact*

Fig. 13. Requirement of uniformity depending on scenario

Analysis of four pertinent roaming IDs indicates that, in terms of the necessity for uniformity, each situation mandates a minimum of one consistent ID (Fig. 13). Nonetheless, even when uniform IDs are not mandatory, the standardization of these IDs is beneficial. In scenario B, where an EV user is authenticated using a cellular phone, the user must transmit the IDs of the EVSE and the EVSE Operator to the E-Mobility provider. The usability significantly diminishes if the EV User must manually input these numbers into their mobile device, given that various EVSE Operators employ disparate formats for these IDs. An ideal application would enable users to capture an image of a code (such as a barcode, QR code, or a standardized numerical format) and retrieve the EVSE ID on their smartphone.

6. CONCISE SUMMARY

This chapter begins by elucidating the significance of energy transmission to electric vehicles (EVs). The characteristics of a fixed and intersecting point were elucidated. Three fundamental dimensions—electrical, organizational, and informational—were presented and addressed for this fixed and intersecting point. Subsequently, the informative dimension was elaborated upon utilizing an information system. The significance and use of the information system were ultimately demonstrated by the example of e-mobility roaming. In summary, as electric vehicles achieve wider market penetration, the integration of these vehicles into infrastructure and their interaction with user services becomes increasingly significant. While information and communication technology is a beneficial enabling tool for electric vehicle usage, it is important to note that ICT itself requires resources to effectively meet the needs of electric vehicle stakeholders. Despite the initiation of numerous efforts (e.g., standardization), much more effort is required to utilize ICT for electric vehicles successfully and inexpensively. The suggested outline of an information system that explicitly integrates the user perspective with ICT components at a suitably selected fixed and intersection point ("energy transmission") may serve as a valuable foundation for the amalgamation of ongoing research activities and the formulation of additional research inquiries.

REFERENCES

Bolczek, M. (2010). Business Models for Electric Vehicles. *Proceedings of the 2nd European Conference on Smart Grids and E-Mobility,* p. 34, ISBN 978-394185144. Brussels, Belgium, October 20-21, 2010

Federal Motor Transport Authority (2011). *Emissionen, Kraftstoffe - Deutschland und seineLänder am 1. Januar 2011.* 17.3.2011. Available from http://www.kba.de/cln_015/nn_269000/DE/Statistik/Fahrzeuge/Bestand/EmissionenKraftstoffe/2011 b emi eckdaten absolut.html

German National Platform for E-Mobility, NPE (2010). *The German Standardization Roadmap - E-Mobility Version 1.* Figure 12, p. 28. 30.11.2010. Available from: http://www.elektromobilitaet.din.de/sixcms_upload/media/3310/Normung- Roadmap_Elektromobilit%E4t.pdf

Heinrich, L.J. & Lehner, F. (2005). *Informationsmanagement.* Oldenbourg.ISBN 978-3486577723, München, Germany

Kempton, W. & Letendre, S (1997). Electric vehicles as a new power source for electric utilities, In: *Transportation Research Part D - Transport and Environment 2 (3),* Elsevier Science Ltd., pp. 157-175, Great Britain

Kempton, W & Tomic, J. (2005a). Vehicle-to-grid power implementation: From stabilizing the grid to supporting large-scale renewable energy, In: *Journal of Power Sources 144 (1)*, Elsevier Science Ltd., pp. 280-294, Great Britain

Krcmar, H. (2006). Information Management, Springer, ISBN 978-3540206286, Berlin, Germany

Kempton, W & Tomic, J. (2005b). Vehicle-to-grid power fundamentals: Calculating capacity and net revenue, In: *Journal of Power Sources 144 (1)*, Elsevier Science Ltd., pp. 268- 279, Great Britain

Rand, D.; Woods, R. & Dell, R. (1998). *Batteries for electric vehicles*, Research Studies Press Ltd., John Wiley and Sons Inc., ISBN 978-0863802058, N.E. Bagshaw

Scheer, A.-W. (2000). *ARIS - Business Process Modeling*, Springer, ISBN 978-3540658351, Berlin, Germany

Schiller, J. (2003). *Mobile Communications*, Addison Wesley, ISBN 978-0321123817, p.113, München, Germany

Sovacool, B. & Hirsh, R. (2009). Beyond batteries: An examination of the benefits and barriers to plug-in hybrid electric vehicles (PHEVs) and a vehicle-to-grid (V2G) transition, In: *Energy Policy 37 (3)*, Elsevier Science Ltd., pp. 1095-1103, Great Britain

The Open Group (2009). The Open Group Architecture Framework - TOGAF Version 9. Van Haren Publishing, ISBN 978-9087532307, Great Britain

Tomic, J. & Kempton, W. (2007). Using fleets of electric-drive vehicles for grid support, In: *Journal of Power Sources 168 (2)*, Elsevier Science Ltd, pp. 459-468, Great Britain Zwicky, F. (1967). The morphological approach to discovery, invention, research and construction, pp. 273-297 In: *New Methods of Thought and Procedure. Contributions to the Symposium on Methodologies*, May 22-24, 1967, Pasadena. Springer-Verlag, New York. 17.3.2011. Available from: http://www.swemorph.com/pdf/new-methods.pdf

10 Chapter UTILIZATION OF SR DRIVE SYSTEMS IN ELECTRIC VEHICLES

1. INTRODUCTION: AN OVERVIEW

With the ongoing increase in global vehicle manufacturing and ownership, the issues associated with automobiles are becoming increasingly evident each day. The severity of these issues is far greater in China. Consequently, the development of zero-emission electric vehicles has emerged as a primary scientific research initiative in numerous countries globally in the 21st century. Energy-efficient motor drive technology has emerged as a critical factor in the commercialization of electric vehicles. Current electric vehicles utilize multiple primary drive systems, including the chopper system for DC motors, the variable frequency drive system for induction motors (IM), the drive system for permanent-magnet motors (PM), and the switching reluctance drive system (SRM), among others. The DC machine has increasingly become obsolete in electric car drive systems due to its high starter current, substantial size, low efficiency, and inadequate reliability. Moreover, the carbon body and the commutator, which are unsuitable for high-speed operation, require frequent replacement. The variable frequency drive system of the induction motor exhibits minimal torque variation; nonetheless, it demonstrates low efficiency, particularly at low speeds. During gradient climbing, the electric vehicle exhibits low torque output and excessive current consumption. Despite the high efficiency of permanent-magnet motors, their manufacturing process is intricate, and the machines may experience a decline in performance due to demagnetization at elevated temperatures. Thus, it is not the optimal method. The SR Motor's structure is robust and stable. The SRD system exhibits excellent reliability, extensive speed control range, elevated efficiency, little starter current, and substantial torque output, making it particularly well-suited for the operational conditions of electric vehicles. The implementation of SRD in electric vehicles is influenced by torque fluctuations and significant noise. The performance comparisons of the three motors are presented in Table 1 below.

Because of its own characteristics,electric vehicles motor drive system should meet the following demands:

1. Output a large torque under base speed to meet the requirement of starting, accelerating, climbing and some other complicated working conditions.
2. Output constant power above the base speed in order to adapt max speed, overtaking and so on.
3. Maximize motor efficiency over the whole speed range to extend endurance mileage.

From the table, the SRM has more advantage than the other motors.

Many control different strategies have been proposed for the torque fluctuation task. Full rotor pitched insulating non-magnetic colloid techniques of SRM and SRM fuzzy logic adaptive torque control system based on instantaneous torque sum are proposed in this chapter.

Table 1. Performance comparisons of IM, PM and SRM

Motors Items	IM	PM	SRM
System efficiency	lower	higher	higher
Starting torque	lower	higher	highest
Power density	lower	highest	higher
Workmanship	simple	complicated	simplest
Reliability	higher	lower	highest
Life	longer	short	longest
Manufacturing cost	lower	highest	lowest

2. DESIGN OF SR MOTOR ON EV

Noise sources can be classified into four primary categories: magnetic, mechanical, aerodynamic, and electronic. Consequently, the magnetic flux within the machine traversing the air gap in a nearly radial orientation generates radial forces on both the stator and rotor, leading to magnetic noise and vibrations. Thus, a 12/8 configuration is employed in the SRM design, and a novel rotor structure is introduced in this section.

2.1. Electric vehicle power demand

The selection of driving motor on electric vehicles mainly depends on rated power and rated speed. The more power grade is choosed the more reserve-power is got and the better vehicle driving feature is. But the volume and weight of the motor will increase rapidly by the same time and lead to the decline of the motor efficiency. So, the motor power should not too large. The calculation of power matching of EV motor is as follows:

A simplified model of the road vehicle dynamics can be used to estimate the tractive requirement of the vehicle drive-train, from which the individual component specifications can be rated with-regard-to their peak and continuous duties. The vehicle model accounts for the resultant forces acting against the vehicle when starting and when in motion, as illustrated in following figure 1. These forces can generally be considered as comprising of four main components, viz.:

$$F_d = F_w + F_r + F_a + F_j \tag{1}$$

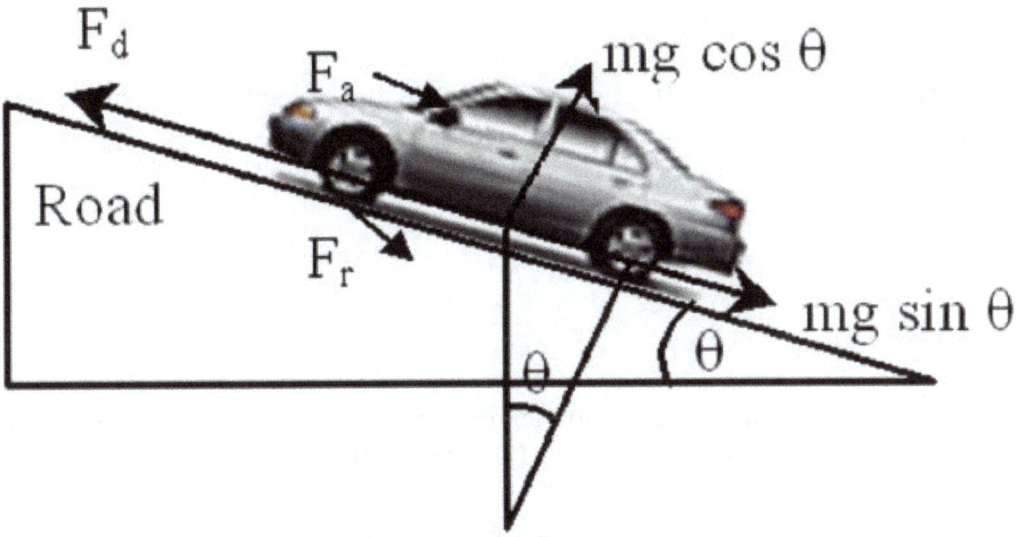

Figure 1. Vehicles dynamics analysis

Where the force to overcome the tyre to road power loss, or rolling resistance, $Fr = kr \cdot mg \cdot \cos\theta$, a resistive force related to the road gradient, $F_w = mg \cdot \sin\theta$, an aerodynamic resistance or drag force, $Fa = 1/2 \cdot \rho C_d A_f v^2$, and the transient force required to accelerate or retard the vehicle, $F_j = m \cdot dv/dt$. Where:-

kr is the rolling resistance coefficient which includes tyre loss and is approximated to be independent of speed and proportional to the vehicle normal reaction force; *m* is the vehicle and payload mass; θ is the road gradient; *g* is the gravitational constant; ρ is the density of air; *Cd* is the drag force coefficient; A_f is the vehicle frontal area, and *v* is the vehicle linear velocity. Having determined the forces acting upon the vehicle, the road wheel torque can be calculated from the equation of motion, viz.:

$$T_w = J_w \cdot \frac{d\omega_w}{dt} + d_f r_w F_d \tag{2}$$

Where J_w, ω_w, r_w, are the wheel inertia, angular velocity and mean radius, respectively, and d_f is a factor proportioning torque distribution on the rear axle. By way of example, for a direct rear wheel drive scenario, it is assumed that there is an equal share of the required tractive force between each wheel drive machine (i.e. d_f= 0.5). For an on-board drive machine option, a gear stage is included in the drive-train, thus the output torque of the traction machine is related to the road wheel torque by the total transmission gear ratio, n_t, transmission efficiency, η_t, and the machine rotor inertia, J_m. Incorporating these into the equation of motion yields a general expression for traction machine torque:

$$T_m = J_m \cdot \frac{d\omega_m}{dt} + \frac{1}{n_t\eta_t} T_w \tag{3}$$

Expressing the wheel and traction machine angular velocities in terms of the vehicle linear velocity yields:

$$\omega_w = \frac{v}{r_w} \tag{4}$$

$$\omega_m = n_t \frac{v}{r_w} \tag{5}$$

From which the machine torque equation can be expressed in terms of the vehicle linear velocity by substituting eqns.(1), (2), (4) and (5) into eqn.3:

$$T_m = \left(\frac{n_t \cdot J_m}{r_w} + \frac{J_w}{n_t\eta_t r_w} + \frac{d_f r_w m}{n_t\eta_t} \right)\frac{dv}{dt} + \frac{d_f r_w}{n_t\eta_t}\left[(k_r \cos\theta + \sin\theta) mg + \frac{1}{2}\rho C_d A_f v^2 \right] \tag{6}$$

Mechanical power is torque multiplied by mechanical speed:

$$P_m = T_m \omega_m \tag{7}$$

Acording to the confirguration of the PEUGEOT 505 SW8 showed in table 2, the drive motor power can be calculated using above equations. P_m = 26.36kW .Considering batteries (weight 650 kg) will be added on the vehicle, the motor should be enlarged to 30kW.

Table 2. The parameters involved are listed in table 2.

the weight with full load	2000kg
rolling resistance coefficient	0.0267
air resistance coefficient	0.3
windward area	1.75m^2
the average efficiency of motor	0.94
the density of air	1.205Kg/m^3
the road gradient	15º
the max speed of EV	180km/h

Assume there are two motors with same rated power, the one with higher rated speed is smaller and lighter. In the view of vehicle performance, there will be less mechanical loss if the rated speed is higher. Meanwhile, it can provide large speed range to the drive system. Although the higher rated speed is favorable, the drive gear will be much more and more complicated. So, the above mentioned factors should be all considered in the selection of motor rated speed.

2.2. Designed SRM for PEUGOT 505 SW8

The parameters design of SR drive motor are contained the preliminary selection of frame size, the number of stator poles Ns and the number of rotor poles Nr, the stator and rotor pole angle selection, the bore diameter and the stack length, the selection of the conductor and the winding design, the calculation of the minimum and maximum inductance according to specifications for the SRM selected above, viz. 30kW, 4000r/min, 300v SRM. Then the motor verification is designed by Finite Element Analysis.

Following are the parameters of SR drive motor (Table.3), we choose 300V lead acid storage battery as power supply system. The mortor's performance curves are showed in figure 2, 3,4. The fig.2 gives the profile of flux linkage vs. current of unaliagned and aliagned stator and rotor tooth. The fig. 3 decribes when the rotor's angel changing, the stator's winding current changs. The fig.4 represents the composed torque of the motor changes with the angel. It shows big variety occurs commutation between the winding phase A and B or Band C et. Thus measurement should be taken to avoide the torque ripple.

Table 3. The parameters of SR drive motor

Rated power (kW)	30	Rated voltage (V)	300	Rated speed (r/min)	4000
Poles and phase	12/8, 3	Maximum value of phase inductance (mH)	3.98401	Minimum value of phase inductance (mH)	0.379135
Effective rated value of phase current(A)	88.6784	Motor average efficiency	0.94	Rotary inertia($kg*m^2$)	0.0486939

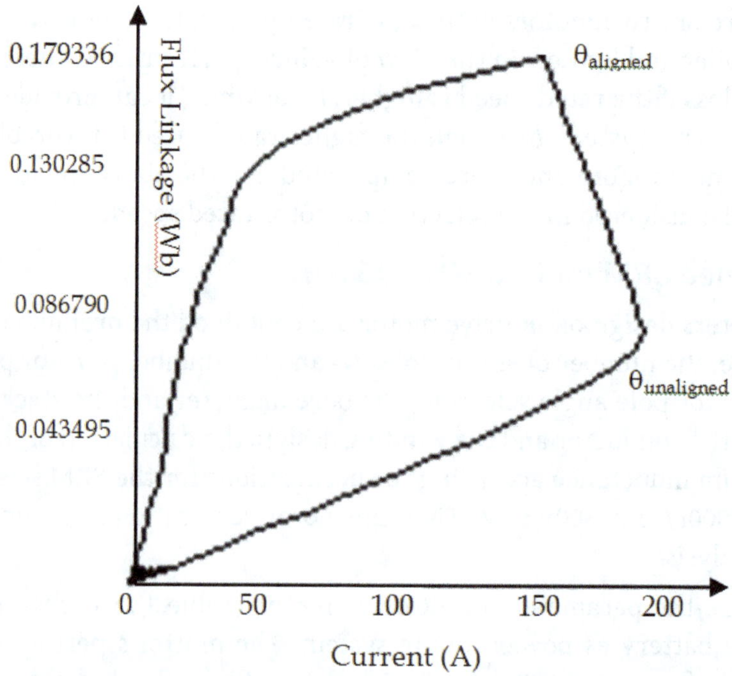

Figure 2. Profile of designed SRM flux-linkage-current

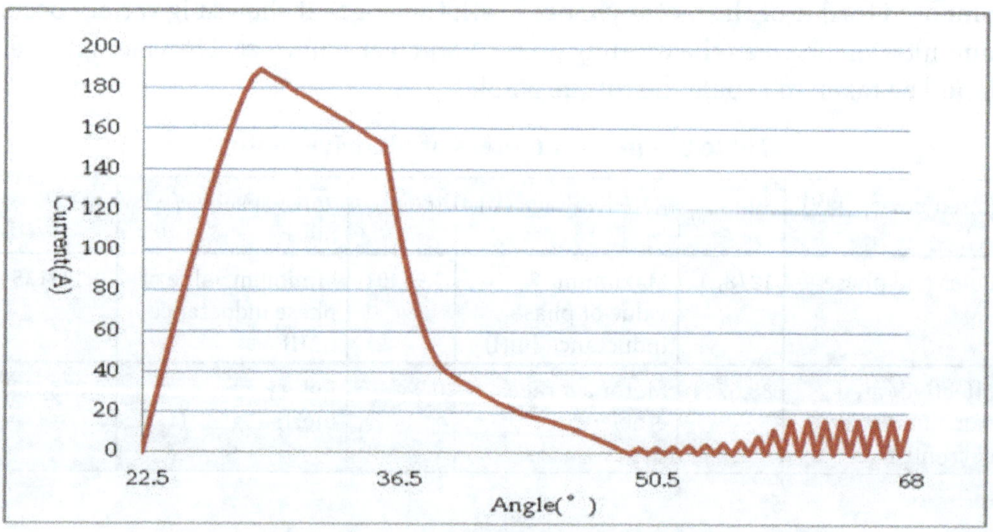

Figure 3. Profile of designed SRM current vs. angle

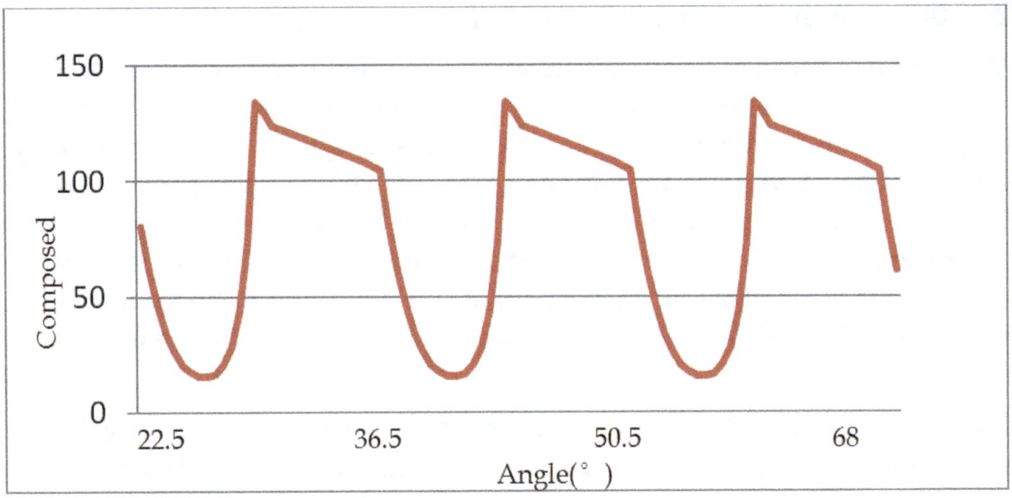

Figure 4. Profile of designed SRM composed torque vs. angle

2.3. New rotor structure

There are large spaces between the present SRM rotor teeth, which will cause strong noise when the rotor rotating. A new type of rotor structure is proposed in the chapter. Figure 5 (a) shows the originally one, (b) (c) is the new structure diagrammatic sketch.The structure include 1 shaft, 2 rotor tooth, 3 yoke part, 4 screw bolt and nut, 5 insulating non-magnetic colloid, 6 copper collar, 7 steel ring. The insulating non-magnetic colloid is filled in the yoke part between rotor teeth. The two copper collars which are used to fix insulating non-magnetic colloid by screw bolt and nut are connected through the rotor shaft.

The expansion factor of insulating non-magnetic colloid is similar to rotor silicon-steel sheet, which can avoid the fissure between insulating non-magnetic colloid and rotor teeth. There are small amount of heat and noise when the new SRM rotor structure is applied during high speed rotating. It is obvious that the working efficiency is higher than the existing one.

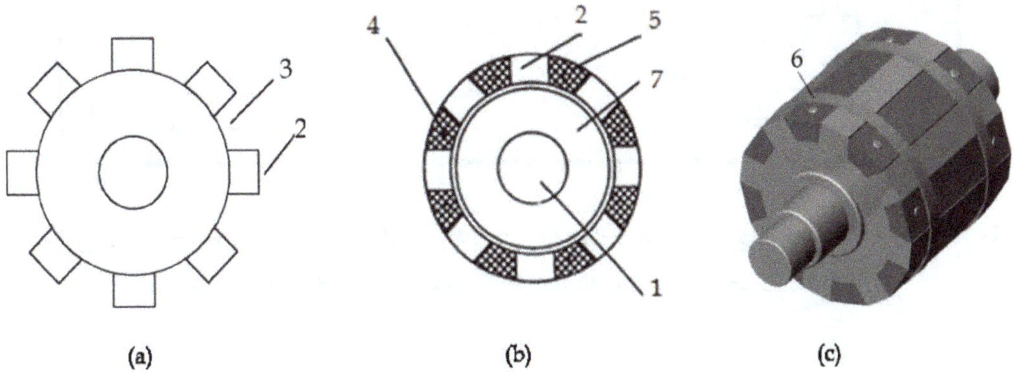

Figure 5. A novel rotor structure for SRM

2.4. Drive mode for PEUGEOT 505 SW8

At beginning development of electric vehicles, inorder to concentrate to develope battery cell and drive motor system, electric vehicles conversion design is usually adopted. The most defferent between the electric and regular fuel vehicles is energy system. Dynamic system of the EV is composed of battery system and drive motor system. The battery provides direct current supply. The supply passing through electric apparatus which is made of controller and power main electical circuit is changed into electical power which can be used by the motor. Then the motor runs and the wheels are drived. The energy power transmission is showed in following figure 6.

The PEUGEOT 505 SW8 is a kind of wagon which is drived by rear-wheel. So the conversion EV can be designed as the motor connecting to gear-box via clutch, transaxle driving the wheels through transmission shaft. The drive mode is shown in figure 7.

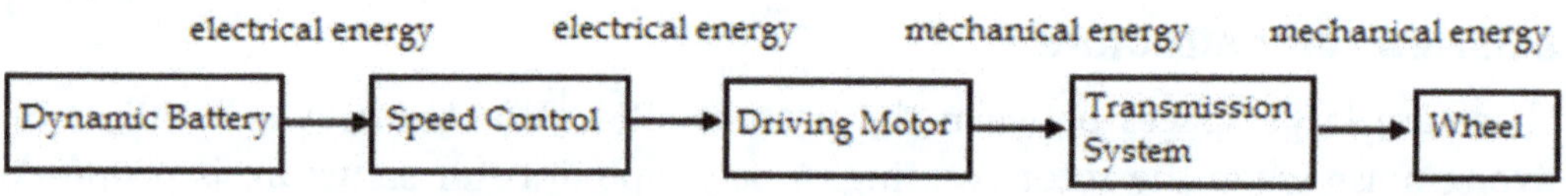

Figure 6. The power transport process of EV

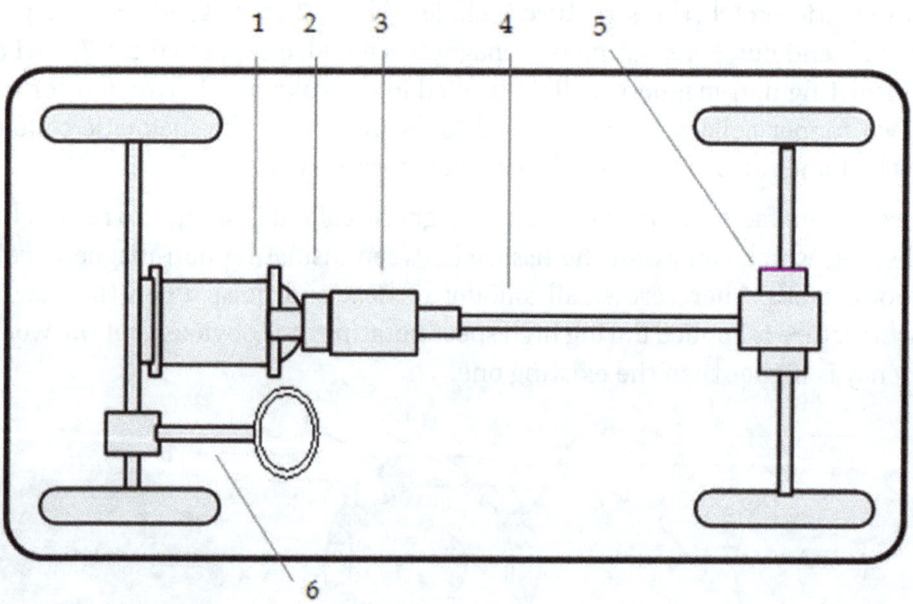

1-motor, 2-clutch, 3-gear-box, 4-transmission shaft, 5-transaxle, 6-steering gear

Figure 7. 1- Drive mode for PEUGEOT 505 SW8 conversion

3. SR DRIVES POWER CONVERTER

The cost and performance of the SRM drive have been determined by many or converter topologies invented, and unlike the conventional inverts-fed induction machine, the SRM drives are highly dependent on the convert topology used to dive the SRM. An asymmetric half bridge converter topology is adopted in the 3-Phase SRM drive for EV, which is showed in figure 8. The main electrical circuit composes of power electrical lead acid batteries U_s, support capacitor $C_{s,}$ asymmetric half bridge IGBTs from V11 to V62 and freewheel diodes from VD11 to VD62. It permits control of the individual phases fully independent of each other and thus permits the widest freedom of control. During normal operation, the electromagnetic flux in an SR motor is not constant and must be built for every stroke. In the motoring period, these strokes correspond to the rotor position when the rotor poles are approaching the corresponding stator pole of the excited phase. In the case of Phase A, shown in figure 8, the stroke can be established by activating the switches V11 and V42. At low-speed operation the Pulse Width Modulation (PWM), applied to the corresponding switches, modulates the voltage level. When the switches V11 and V42 are turned off in the same time, producing transformer electromotive force in Phase A break-over forward freewheel diodes VD11 and VD42, the Phase A current after flows through VD11, VD42 and Cs. Each of V11 with its anti-parallel diode VD11 and V12 with its anti-parallel diode VD12 is the asymmetric half bridge structure in the one IGBT module. KDC is a DC contactor which powers up to the drive system, and is controlled by key of the vehicle which starts the engine originally.

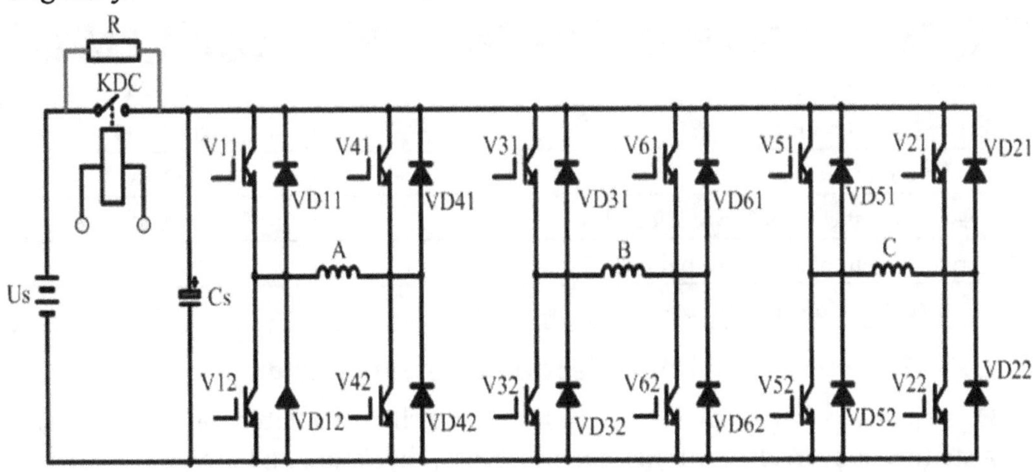

Figure 8. Schematic Diagram of Power converter in Electrical Vehicle

3.1. Parameter selection of the main circuit

The driving power U_s is composed of 25 series lead acid batteries which rated voltage is 12V, capacity is 85Ah. Thus the whole volume of U_s is 300V. Because safety working voltage arrange of the single battery is from 10.5V to 14.4V, the actual group of batteries working voltage arrange is from 262.5V to 360V. Selecting maximum DC bus voltage 360V, in the asymmetric half bridge main circuit the main switch IGBT will bear the maximum DC power voltage from the bus. Considering twice surplus capacity, the IGBT withdraw voltage is in equation 8.

$$U_r = 2U_{smax} = 2 \times 360V = 720V \tag{8}$$

The effective rated value of the phase current for the designed SRM is 88.6784A, considering the overloade current of the drive system two multiple of the phase current, the effective rated value of the IGBT current is in equation (9).

$$I_{srms} = 2I_N = 2 \times 88.6784A = 177.3568A \tag{9}$$

Having determined the effective rated value of the IGBT current, the peak current of the IGBT can be calculated in equation (10), viz

$$\hat{I}s = \sqrt{2} \times 2 \times \sqrt{q} I_N = 434.3553A \tag{10}$$

In the equation (9), *m* is the number of the phase of the SRM which eaquals 3 in the system. Actually power converter parameters involved in the designed SR drive motor are listed in table 4. To sum up, the main circuit switch IGBT can be selected as BSM300GA120DLC of EUPEC which has 1200V bear voltage (T_C = 80^{0}C), 300A effective current value, 600A peak current value (t_p = 1ms).

Table 4. Designed Parameters of SR Drive Motor

Electrical parameters	Values
Average current of main switch	46.1426 A
Effective current of main switch	84.7971A
Peak current of main switch	188 A
Average current of after flow bidiode	11.878A
Effective current of after flow bidiode	27.0406 A
Peak current of after flow bidiode	138A

3.2. IGBT drive and snubbed electronic circuits design

IGBT is simply voltage driven switches, because their insulated gate behaves like a capacitor. Conversely, switches such as triacs, thyristors and bipolar transistors are "current" controlled, in the same way as a PN diode. Because IGBT gate drive condition is closely related to its static and dynamic performance. The turn-on voltage, switching time, switching loss, short-circuit withstand capability etc. are ordinately effected by positive and negative voltage (V_{GE} and - V_{GE}) between gate and emitter, gate electrode resistance (R_G). Being reasonably designed drive and protection circuit is quite important.

3.2.1. IGBT drive conditions

The concrete requirement of IGBT drive circuit is listed flowering.

(1)Because IGBT is a type of voltage drive module, it has a threshold voltage value of 2.5 V to 5.0V and has a capacitive input resistance. IGBT is very subtle to electric charge accumulation of the gate electrode, so the drive circuit must be very reliable to grantee there is a discharge loop with a low resistance itself, viz. there is as short as possible wire between the circuit and IGBT.

(2)Charging and discharging the gate oxide capacity with small interresistance drive supply inorder to gurantee that the gate electrode controlling voltage has much enough steep forward and reverse edge. Because IGBT has nonlinear capacitive character, the driver must have enough instantaneous absorbing current capability. Thus the gate electrode voltage can be set up or disappear soon inorder to decrease the switch loss lowerly. On the other hand, after the IGBT turns on, the gate electrode driver should supply enough power to avoide to be destroyed because of exit saturation. The drive interresistance can not be lower than commanded R_G, otherwise there will have underdamped harmonic motion between the stray inductance and the gate electrode capacity in the drive loop. Meanwhile, the surge current increases during short switching time. This induces booth unsafty in the main circuit and disturbance in the control circuit.

(3)Although the higher $+V_{GE}$, the higher current limit becomes, $+V_{GE}$ should be remained under maximum rated G-E voltage, $+V_{GES}$ =+20V. The reason is once there is a over current or short current, the higher $+V_{GE}$, the higher current, the bigger probability of the IGBT destroy becaues the time of enduring short current capability decreases. Usually, the value of $+V_{GE}$ is considered as between 12V and 15V.

(4)Set the enough gate reverse bias voltage value($-V_{GE}$) to IGBT. While the IGBT is in off-state, there are some high frenquency osillation signals in the gate electrode circuit because the other part circuits are stiill working. These signals may let IGBT to be in micro on-state, it results in the power loss of IGBT increase. Therefore a recommended $-V_{GE}$ to IGBT is -5V to -10V (the maximum gate reverse bias voltage value is $-V_{GES}$ =-20V), so that the IGBT can be cut off reliably even if there are switch noises in the gate electrode of IGBT.

(5) The gate electrode resistance effects R_G on the switch loss, switch speed, and even involves in whether the drive circuit appears oscillation and the collector electrode generates surge current. Usually the value of R_G can not be selected to be over 10 times than the producer recommended value. In actual application, R_G should be ajusted carefully. R_G should be disposed closely to the gate electrode.

(6) There should be have a gate electrode amplitude limiting circuit to avoid to

breakthrough the gate electrode.

(7) The drive circuit should be isolated from the control circuit so that, when the IGBT is destroyed the other components can not be damaged. When the IGBT burns, the collector electrode high voltage usually pwties the drive circuit through destroyed the gate electrode, and then disrupts some components in the drive circuit. The gate electode drive circuit as far as possible should be simple and actual. It is much better that the IGBT itself has protection and very strong the ability of disturbance. When the IGBT is under the state of load short circuit or over current, the IGBT can prevent fault current automatically through decreasing the gate electrode voltage gradually to be switched the IGBT off softly. That is to prevent very high di/dt causing by the fault current because of IGBT turn-off swiftly. The high value di/dt can produce high spiking voltage under the stray inductance function, results in the IGBT to be unbearable and damaged. By the same rule, the soft turn-off process of the drive circuit should not be effected by the input signal disappear. The drive circuit should have a time logical for thegate voltage control function. When there has over current, no matter whether there is the input signal, the drive circuit should be turned off unconditionally.

3.2.2. IGBT drive electrinic circuits design

The IGBT-driving hydrid IC EXB841 produced by Fuji Electric is used in the EV power convter. It can drive 300A/1200V or 400A/600V IGBT moduel and the drive signal delay reaches 1.5μs, the maximum switching speed is 40KHz. The drive circuit is designed in figure 9 arround EXB841, in which C1 and C2 are two electrolytic capacitors to absorb the chang of supply voltage. MC1413,LED,R1,R2 are composed of the drive line from the control circuit to EXB841. V3 and V4 are two voltage-stabilizing diodes to limit amplitude of EXB841 output drive voltage, so that to avoide the drive voltage higher and destroy the IGBT. V3 and V4 are two voltage-stabilizing diodes which are in inverted series and connected in parallel with collector and emitter electrode. The over current protective circuit for IGBT is composed of R3, E1,R4, V1 and V2, in which V1 and V2 connecting to the sixth pin of EXB841 completes to monitor the collector electrode voltage. V1 is the fast recover diode which is ERA 34-10 made from Fuji Electric. V2 is the voltage-stabilizing diode which can change the controlled point of the current protection by adjusting the voltage-stabilizing value of the diode. Theoretically, if the over current protection of EXB841 takes effect is that the pin of EXB841 outputs a low electrical level when the collector electrode voltage monitoring the six pin is greater than 7.5V. Then the opto-coupler in the figure outputs a fault signal to the control board which is high level OC signal. The signal produces interruption on the board and controls CPU to block trigger pulse for IGBT. The theoretical protective value 7.5V of EXB841 is the operating point which the supply voltage is strictlly controlled at 20V. When the supply voltage has ripple or error, the protective value will change. When the supply voltage is greater than 20V, the protective value increases 1V as the supply voltage does. When the supply voltage of EXB841 is 20V, the drive voltage suppllied IGBT turning on is 15V. According to the profile of switching losses of IGBT-inverter of BSM300GA120DLC, the curve of collector electrode current vs. IGBT switching on voltage drop V_{CE} can be abttain. The voltage drop between the gate and the emitter electrode can be calculated in the following equation and figure 10.

$$V_{CE} = RI_C + V_0 \tag{11}$$

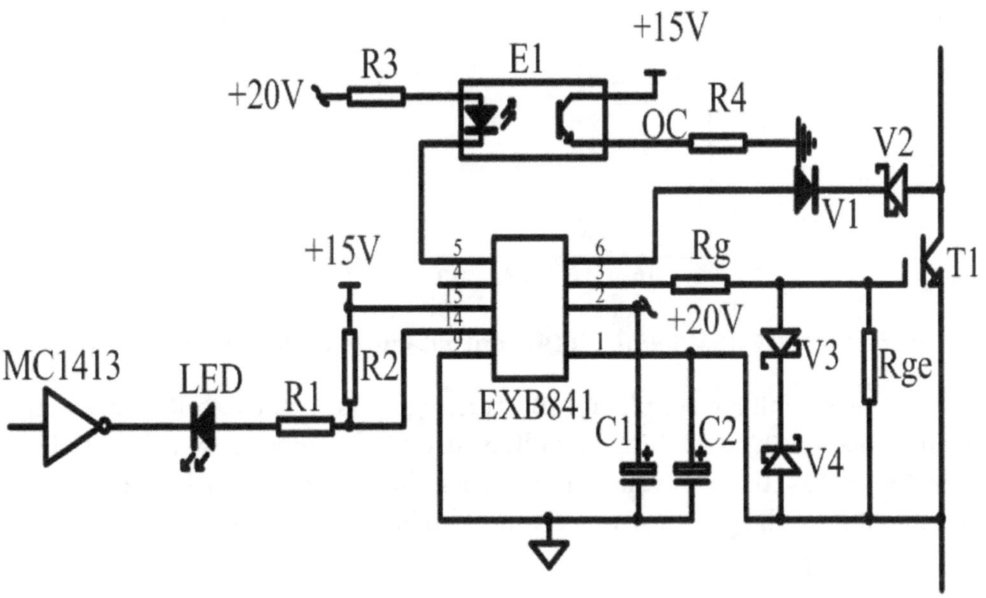

Figure 9. IGBT drive circuit

According to the figure 10 and 11, when T_{vj} = 125°C, IGBT output curve can be linearized by line *y=Ax+B*. Known the three points on the curve, respectively they are (2.0, 208), (2.5, 325), (3.0, 465). Using the points (2.0, 208) and (2.5, 325), the equation is listed in 12.

$$V_{CE} = 0.0042735I_C + 1.01282 \tag{12}$$

Taking I_C = 480 A, V_{CE} = 3.0641V, the error of V_{CE} and the percent rate of error are listed in following equations.

$$\Delta V_{CE} = 3.0641 - 3.06 = 0.0041 \tag{13}$$

$$\eta = \frac{0.0041}{3.06} \times 100\% = 0.133\% \tag{14}$$

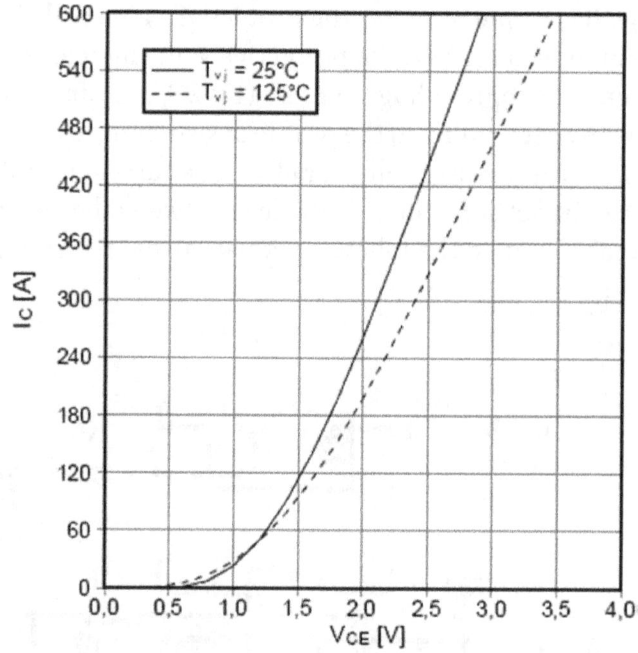

Figure 10. IGBT BSM300GA120DLC output curve under different voltage

When the drive voltage suppllied IGBT turning on is 15V, the collector electrode current is 600A, IGBT switching on voltage drop V_{CE} is about 3.5V calculated by 12 and in figure 10. Then the value of voltage-stabilizing diode V2 is selected as 9V. From the table 5, the gate resistance value can be chossed 3.3Ω.

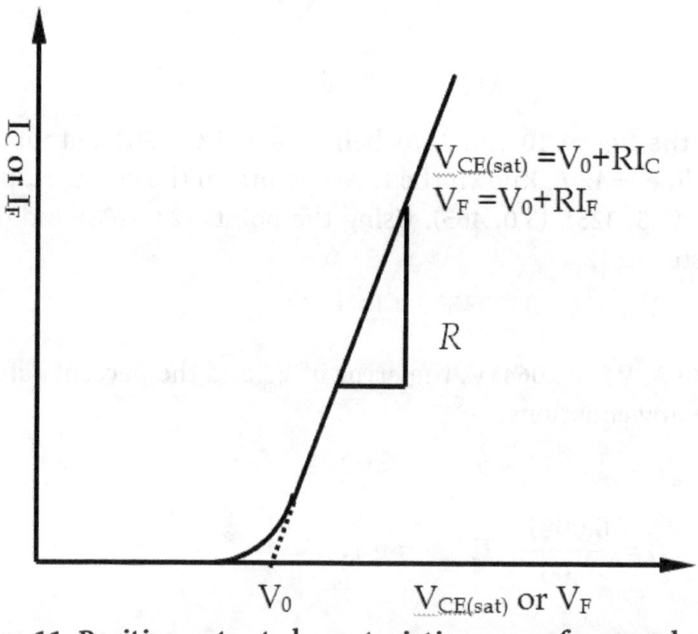

Figure 11. Positive output characteristic curve of power devices

Table 5. Recommended the gate resistance R_G and the current loss

IGBT Rating	600V	200A	300A	400A	-
	1200V	200A	150A	200A	300A
R_G		12Ω	8.2Ω	5Ω	3.3Ω
Icc	5kHz	20mA	22 mA	23 mA	27 mA
	10kHz	24 mA	27 mA	30 mA	37 mA
	15kHz	27mA	32 mA	37 mA	47 mA

3.2.3. IGBT snubbed electronic circuits design

When the power electronic device is used, buffer electronic circuit should be designed to inhibit respectively *di/dt* and *dv/dt* when the device switched on and off. The aim is to change switch locus of the device in order to avoid the maximum value of V_{CE} and i_c appear at same time. Thus the switching loss is decreased and reliability can be improved. The IGBT snubbed electronic circuit is put particular emphasis on the voltage absorbing and restraining in switch on state. That is because the IGBT working switch frequency is very high, tiny inductance in the electronic circuit can cause very big *L di/dt* and produce over voltage to endanger the IGBT security. RCD snubbed electronic circuit is often used which is designed in figure 12. The RCD bridges joint every IGBT module connecting with two ends of the power. Capacity and resistance value selection has much more relationship with snubbing voltage in the snubbed electronic circuit. If the selection is improper, it would affect the voltage absorb so far as to bring about osillation in the circuit.

Within the IGBT switching off process, the current of device drops fastly, the current in the snubbed circuit increases at same change rating. The magnetic energy stored in the parasitic inductance in the main electronic circuit will wholly transfere into the electrical energy in the absorbing capacity. Assumming the current flowing the main switch device drops linearly when switching off, the current flowing the snubbed circuit increases linearly because the switching off process is very short. These two current can be expressed in the following equations.

$$i_T = (1 - \frac{t}{t_f})I \tag{15}$$

$$i_C = I - i_T = \frac{t}{t_f}I \tag{16}$$

In the equations, i_T is the current in the main switch device; i_C is the charging current in the absorbing capacity; t_f is dropping time when switching off. *I* is the average current in the direct current bus line. Therefore, the voltage of capacity two ends is expressed in the equation 17.

$$U_{CS} = \frac{1}{C}\int_0^{t_f} i_C dt = \frac{1}{C}\int_0^{t_f} \frac{t}{t_f} I dt = \frac{It_f}{2C} \quad (17)$$

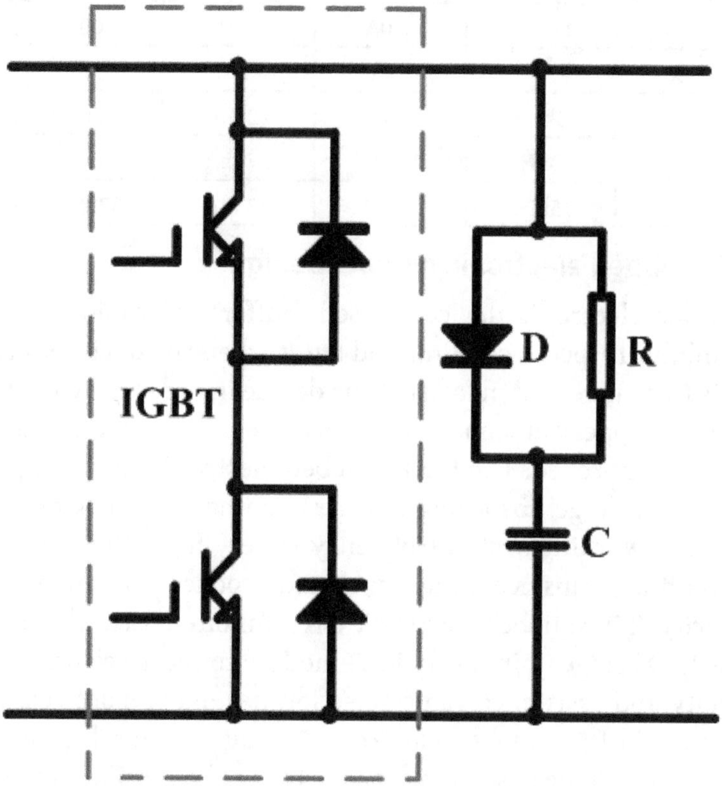

Figure 12. IGBT snubbed elctronic circuit.

The capacity voltage U_{CS} usually can be selected as 10 to 50 percent of the supply voltage U_{DC} at t_f. For example, selecting $U_{CS} = 0.5U_{DC}$, the capacity C is in the following equation.

$$C \geq \frac{It_f}{U_{DC}} \quad (18)$$

The value of the absorbing capacity can be selected as 0.47μF according to the the equation 18 and experience value. The selection rule of resistance R is to releasing electric charge thouroghly stored in the snubbed capacity before the IGBT switching off signal comes. If R is too large, it makes the discharging time of the capacity C lower. But if R is too small, the discharging current is too large and fast when the IGBT switching on, it can endanger the device security and cause the osillation. So period (T) of the switch device eaquals 1 or 2 times of $3RC$. So the value of the resistance R can be calculated in the equation 19.

$$R \leq \frac{1}{6 \times C \times f} \tag{19}$$

R calculated by the equation 19 can not be greater than 41.72 ohm, 39ohm is selcted in the RCD snubbed circuit. The maximum power loss on the absorbing resistance can be calcultaed as in the equation 20.

$$P_R = \frac{1}{2} C \Delta V^2 f = 20W \tag{20}$$

The fast recovery diode is selected as FR607. Beside the IGBT drive and snubbed electronic circuits design, the switched power supply circuit and control circuit should be designed. Because the length of the chapter, the content cannot be covered all the bases. As figure 13 shows, the control circuit structure of SRD system is given. With MICM2002 (Motor Intelligent Module) based on DSP and AT89C51 singlechip the folowing control strategy is realized. When the EV is powered, the controller goes into working state. The control signal from all kinds of fault signals and the driver operating system are coded and loaded the AT89C51 through prior coder. If the system is checked with no fault and no operation, the system is in standby mode. When the driver gives the start and throttle given signal, according to the SRM rotor position signal from the position sensor, the singlechip sends out the phase turn on/off signal and the MCIM produces the PWM signal, then the system integrates the protective and the current chopping signals to give the main circuit IGBT drive signal and control the power main circuit to supply the SRM windings electricity and move the motor. According to the positive and negative rotation setting and the position information of the motor, the singlechip controls the windings power-on sequences. When the motor rotates at low speed, current chopping control mode can be used. Firstly, the chip gives the upper limitation of the current chopping signal and puts it into the MICM through D/A converter; secondly, the MICM comparates the current limitation with the phase current detecting from the current sensor, then calculates, optimizes and sends out the chopping signal. When the motor speed arrives above basic speed, the control mode changes into angle control from the current chopping control. When the driver changes the operting signal, the control circuit changes the working logic and implements corresponding dive requirement through the power circuit. If there is fault in the motor running, the control circuit blocks the trigger pulse of IGBT and protects IGBT, and displays and alarms through display circuit. Meanwhile it communiates with CAN module and sends the fault signal to driver video facility.

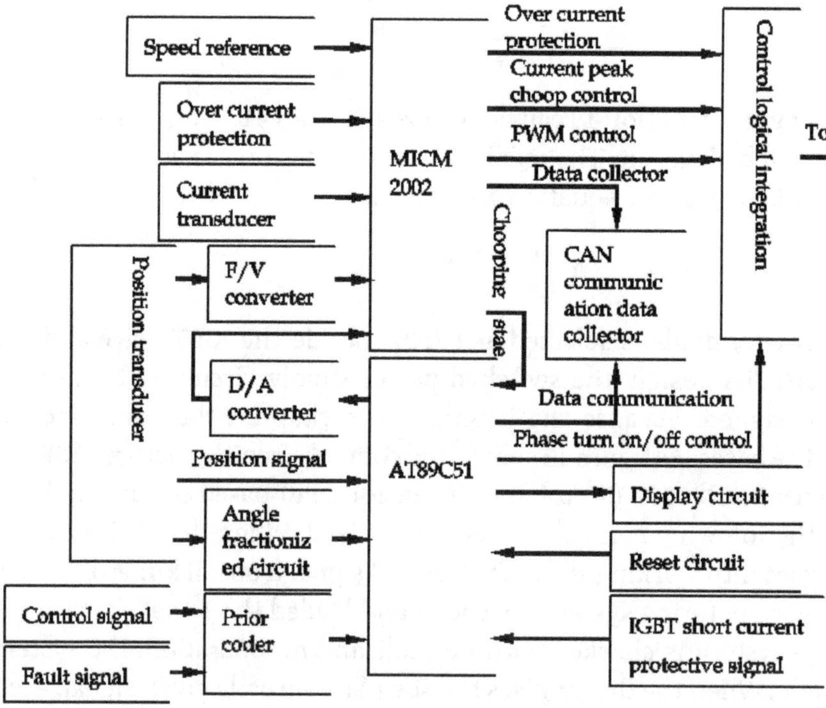

Figure 13. Control circuit structure of SRD systems

4. ADAPTIVETORQUECONTROLSYSTEMUTILIZINGFUZZYLOGIC FOR SRM

The non-linearity of the magnetic circuit renders the SRM challenging to examine and control. Moreover, constructing an appropriate mathematical model is challenging. Fuzzy control is primarily utilized in systems characterized by significant non-linearity or those for which obtaining a mathematical model is challenging. A fuzzy adaptive control for the SRD system is presented to enhance speed regulation precision and reduce torque fluctuations. This chapter introduces a novel SRM control system that integrates a PI regulator with fuzzy logic control, termed the SRM fuzzy logic adaptive torque control system based on instantaneous torque summation.

4.1. Speed PI regulator design

Figure 14 presents the principal frame of SRM fuzzy logic adaptive control. The conventional PI regulator is applied in the external loop and the fuzzy logic adaptive control based on instantaneous torque sum is used in the inner loop system. According to the voltage balance equation of SR motor:

$$V_i = R_s i_a + \frac{d}{dt}(La i_a) = R_s i_a + La\frac{di_a}{dt} + i_a\frac{dLa}{dt} = R_s i_a + La\frac{di_a}{dt} + i_a\frac{dLa}{d\theta}\bullet\frac{d\theta}{dt} = R_s i_a + La\frac{di_a}{dt} + i_a\omega\bullet\frac{dLa}{d\theta} \quad (21)$$

Suppose the inductance change rate is constant, which defined as g L .

$$g_L = \frac{dL_a}{d\theta} \quad (22)$$

So the equation 21 can be written as following equation.

$$V_i = (R_s + \omega g_L)i_a + La\frac{di_a}{dt} \quad (23)$$

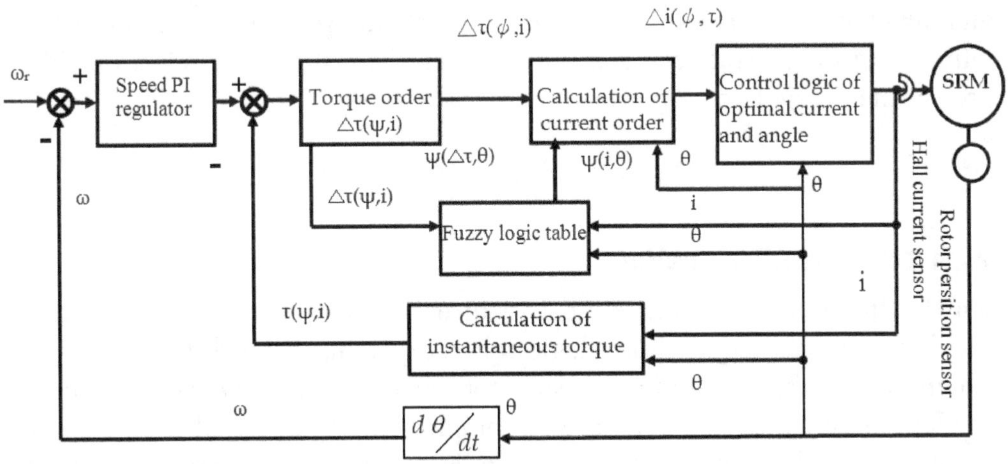

Figure 14. Principal frame of SRM fuzzy logic adaptive direct torque control system

The electromagnetic torque created by phase A is:

$$\tau_e(\theta,i) = \frac{1}{2}g_L i_a^2 \quad (24)$$

According to the torque balance equation 25:

$$\tau_e(\theta,i) = \sum_{n=1}^{3}\tau_{dn} = \tau_L(\omega) + k_\omega\omega + J\frac{d\omega}{dt} \quad (25)$$

When the system enters steady state, the torque is invariable:

$$\tau_e(\theta,i) = \sum_{n=1}^{3}\tau_{dn} = \tau_L(\omega) + k_\omega\omega \quad (26)$$

$\tau_L(\omega)$ is load torque and k_ω is coulomb friction coefficient.

Compare equation (7) with equation (8), we get:

$$\Delta\tau_e(\theta,i) = J\frac{d\omega}{dt} \qquad (27)$$

If the sample time T is small enough, load torque and k ω ω can be seen as constant in the sample interval. So $\Delta\tau_e(\omega)$ is proportional to $\Delta\omega$, equation (28) is:

$$\Delta\tau_e(\theta,i) \approx J \bullet \frac{\Delta\omega}{T} \qquad (28)$$

In the equation above, $\frac{d\omega}{dt} \approx \frac{\Delta\omega}{T}$. Thus we get the torque deviation signal from speed deviation signal through the PI regulator.

$$\Delta\tau_e(\theta,i) \approx K_p \bullet \Delta\omega \qquad (29)$$

The bandwidth of speed closed loop is small. It will be better when the proportional regulator used. But it is hard diminish to the steady-state error. So, in order to minimize the steady-state error and strengthen the disturbance rejection ability, we select the PI regulator. Equation 30:

$$\Delta\tau_e(\theta,i) \approx \left(K_p + {K_i}/{s}\right) \bullet \Delta\omega \qquad (30)$$

4.2. Fuzzy controller design

From the figure 14, the inner loop is a direct torque control loop which is a three dimension self-adjust fuzzy logic control system, in which the torque loop is composed of the instantaneous sum torque negative feedback control. The inner loop is completed by software to complish the feedback of the fuzzy logic control itself, so that the SRM can be controlled in an optimal state. The following detail is about the fuzzy logic controller design and the adaptive "soft feedback" complement.

4.2.1. { (θ , i) ⊠ ψ } Fuzzy logic tables

The SRM is multi input multi output controlled object. The fuzzy logic table describes the connection between input and output. In mathematics, this table can be seen as a two input single output non-linear functions. The steps to build the table are as follows:

First step: confirm input and output fuzzy domain and its membership function.

Input variable is the rotor position angle and winding current. Their corresponding variation range are 0—45^0 and 0—200A. In order to improve learning rate, we assign that the membership function of fuzzy system input is isosceles triangle and its vertexes are located in the centre of triangle bottom. Shown in figure15(a)(b). The membership function of output flux linkage is shown in figure15(c)and its corresponding range is 0-0.18Wb. The fuzzy subset of linguistic value which describes input and output value are: { S , M , B } , which S=Small,M=Medium,B=Big.

$$R_i : if \ \ \theta \ \ is \ \ A_i^{M\theta(n)}, \ \ i \ \ is \ \ B_i^{Mi(n)} then \ \ \psi \ \ is \ \ C_i^{M\psi(n)} \tag{31}$$

Second step: generate fuzzy rules from input data.

When every membership functions of input and output fuzzy domain is confirmed, we can get fuzzy rules from the measured data. Every input-output data pair is consist of current, rotor position and flux linkage which has specifically numerical relationship. In the first place, we get the membership degree from the corresponding fuzzy membership domain. Second, we assign the max membership degree to the variable in the domain. So, the value of nth data pair is θ*(n)*, *i(n)*, ψ*(n)*. The assigned value will point to the fuzzy domain with the max membership degree, which can be written as the following fuzzy rules. In this equation, $A_i^{M\theta(n)}, B_i^{Mi(n)}, C_i^{M\psi(n)}$ are respectively represent the fuzzy domain of discrete data to θ*(n)*, *i(n)*, ψ*(n)*. Table 2 is fuzzy rules. For example, if current is max and rotor position is max, the flux will be max.

Third step: confirm fuzzy rules membership degree.

When every new fuzzy rule is created from the input output data pair, a rule degree or fact is connected to this rule. The rule is defined as trust degree to the fuzzy rule. Actually the rule degree is related to the function, which describes the relationship between current, angle and flux linkage. The rule degree equals to the product of membership degree in each fuzzy domain. Like equation 31, can be depict as equation 32.

$$Degree(Rule) = \mu_{A_i^{M_{\theta(n)}}} \times \mu_{B_i^{M_{i(n)}}} \times \mu_{C_i^{M_{\psi(n)}}} \tag{32}$$

The instantaneous torque sum can be get from current, the rotor position angle and flux linkage in the following fuzzy logic table (Table 6).

4.2.2. Fuzzy model trainning

The phase current *I* is obtained by magnetic balance Hall current sensor and the angle θ by photoelectric position sensor. The flux linkage is calculated by the finite element analysis of current and the rotor position angle. The torque order is acquired from rotate speed order and then the current order is get from the torque order. Thus the current control can be realized. Figure 16 is the fuzzy controller based on MATLAB fuzzy toolbox. Figure 17 shows the finite element analysis ψ—*i*—θ graph. According to finite element data, the model is training in the offered Matlab fuzzy toolbox. Figure 18 presents the ψ—*i*—θ graph acquired by training the static data in fuzzy model. As we can see from the Figure 17, the established fuzzy rules are correct that we can get accurate flux linkage ψ from the input phase current and the rotor position.

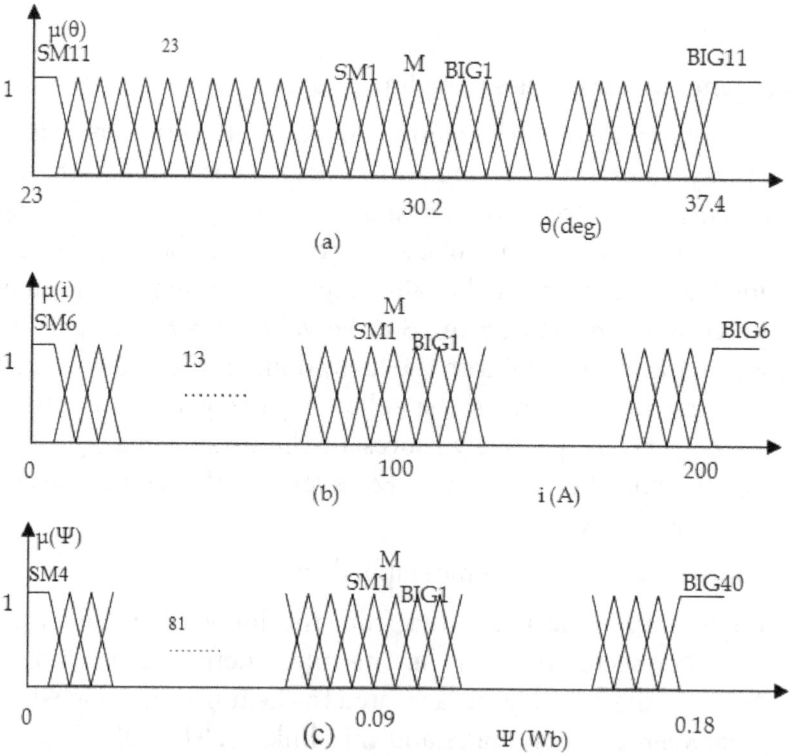

Figure 15. Fuzzy domain regions and membership for each variable (a) Rotor position, (b) Current, (c) Flux linkage

Table 6. Fuzzy logic table between ψ and i—θ

		Flux linkage		
Current	Medium (M)	S	M	B
	Big (B)	S	S	B
		Small (S)	Medium (M)	Big (B)
		Rotor position		

5. RESULT

5.1. Tests on the motor platform

Before the experiment on vehicle, we do the bench load test with the selected motor first. The experiment table includes three phase dynamometer, torque measurement oscilloscope, DC generator, resistance box and so on. The DC power needed by EV drive is supplied by 25 lead acid traction batteries. DC generator and resistance box make up the load of the drive motor, which is adjusted by excitation voltage.Figure 19 shows the two phase winding current waveform of SRM when the motor speed at 500r/min and load with rated torque. From this figure we can see that the effective current increases so as to output required torque. The out power is 3.2kW, the efficiency is 84% of the SR drive system. Figure 20 shows the two phase winding current waveform of SRM when the motor speed at 500r/min and load with peak torque.

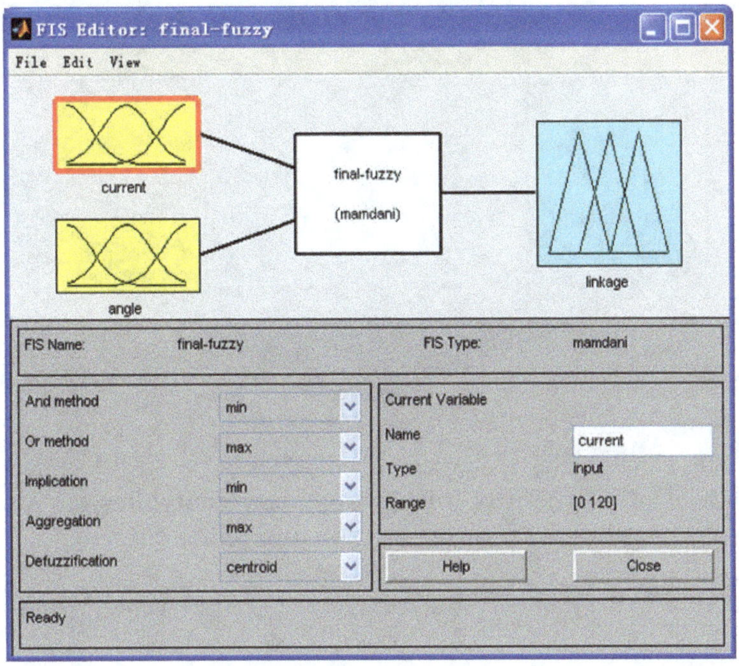

Figure 16. Variation of the flux linkages of FEM for a single phase winding with rotor position and phase current

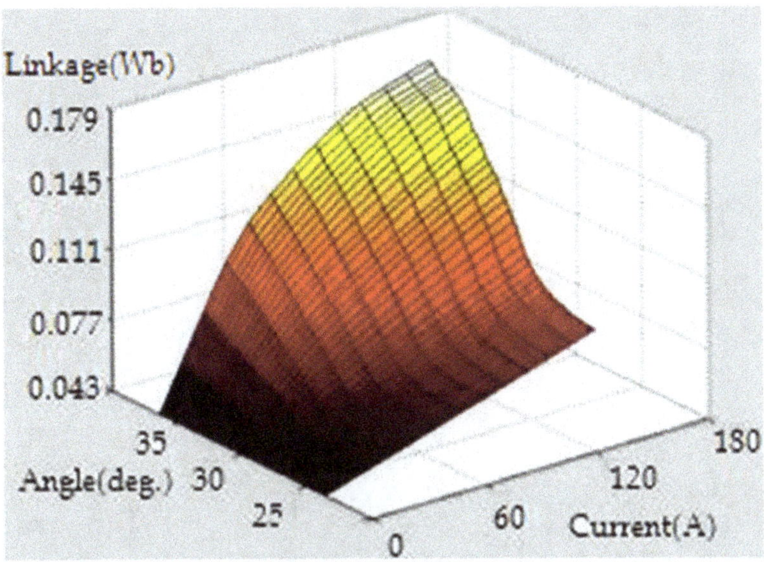

Figure 17. Variation of the flux linkages of FEM for a single phase winding with rotor position and phase current

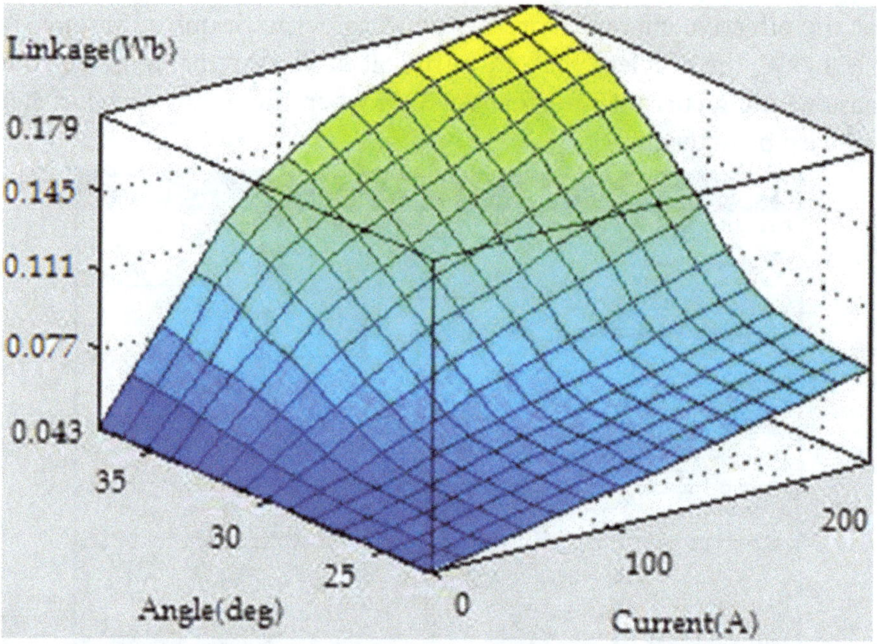

Figure 18. Variation of the flux linkages of fuzzy controller for a single phase winding with rotor position and phase current

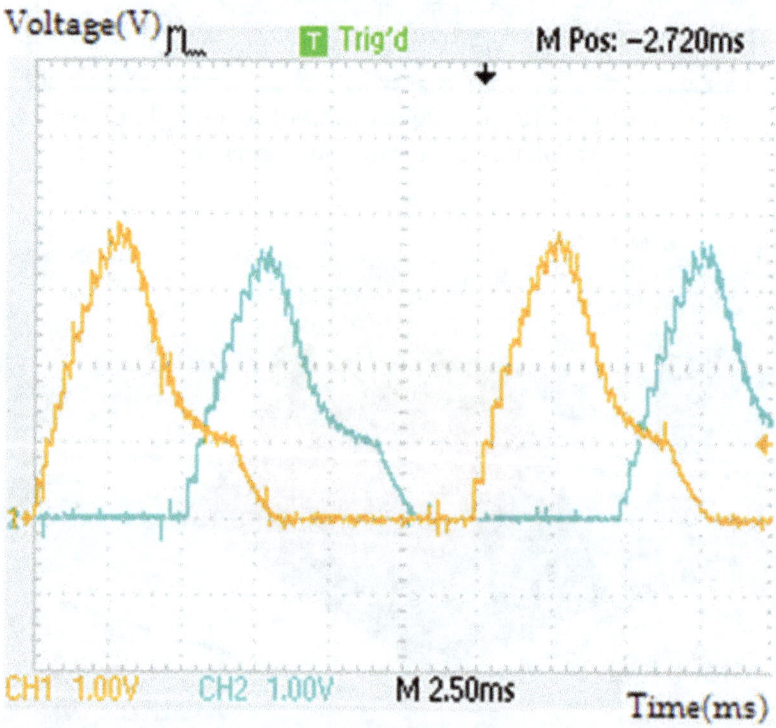

Figure 19. Winding current waveform of n=500r/min under loaded 72Nm

The output torque is 144N.m and output power is 6.4kw. It is obviously that winding current is controlled below the peak value (189A). The waveform of the current is flat top and the drive system is working with full load. This status is used to provide peak torque when EV startup or accelerate. In order to improve system reliability, it is allowed to work overload for one minute. After that, the control system will lock trigger pulse and give overload alert to prevent system damage. The figure 21 shows steady state torque profile at speed of 400r/min and output power is 4kW, it shows the torque ripple is only within less than 10 N m.

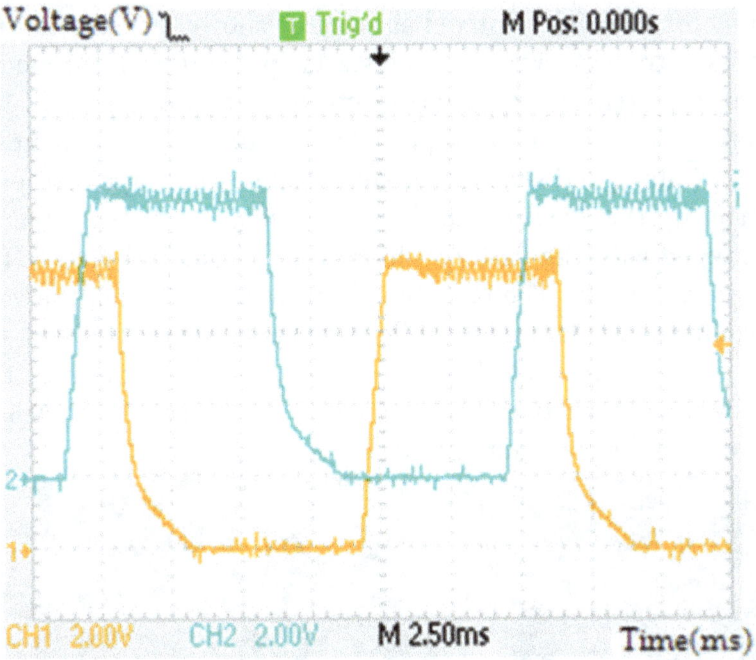

Figure 20. Winding current waveform of SRM when the motor speed at 500r/min

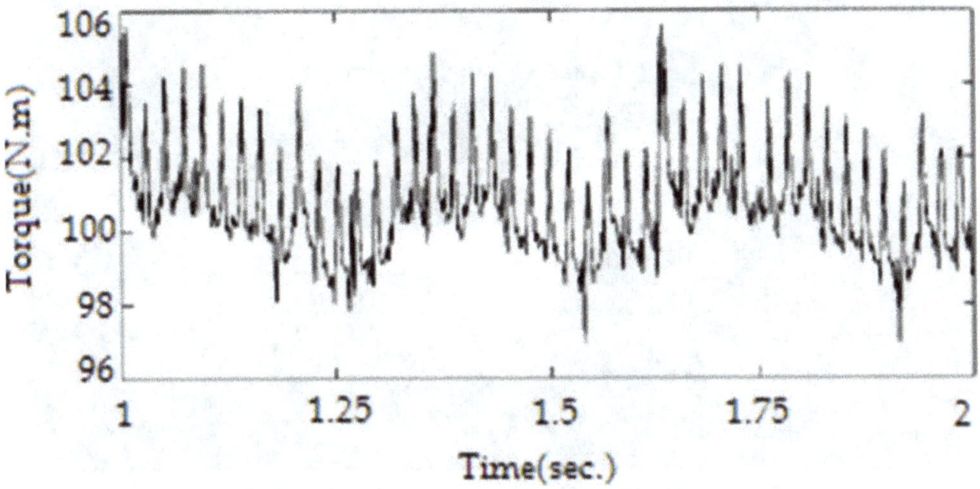

Figure 21. Steady state torque profile at speed of 400r/min

5.2. Tests on the PEUGOT 505 SW8

The SR drive system designed in this chapter was installed on the PEUGEOT 505 SW8 to do vehicle tests.The van preserves clutch, gearbox and other transmission mechanism. Thus we can reduce effect on vehicle traction performance. On the other hand, in doing so can improve startup torque. The installment of Lead-acid Battery mainly considers axis distribution and its structure. The battery is assembled by the space and axis load distribution rather than central installation to ensure the balance of front and rear bearing. The SR motor is in the position of engine and motor controller is fixed above it. It shows the excellent mechanical characteristics of the SRM when the van starts up. Pictures of the modificated EV and the SR drive system are showed respectively in figure 22 and 23. The starting torqueis almost twice the rated torque, which meet the requirement of starting, accelerating, climbing and some other complicated working conditions. The van starts up smoothly, the current of the bus is low which is less than 15A. The vehicle test was arranged with the battery which was charged full voltage (360V). The driving range was 205km. The battery voltage was 265V when the van stopped. Table 7 is running test data under different dears. Figure 24 shows the battery voltage and bus current when the EV climbed the hill which grade was greater than 25º. They were respectively 255V and 70A. The current was 120.5A when the EV accelerated and the maximum speed reached 165km/h.

Figure 22. Modification of PEUGEOT 505 SW8

Figure 23. SR drive system for EV

Table 7. Testing data of EV running parameter

Number	Gear	Speed(r/min)	Battery voltage(V)	Bus current(A)
1	3	40	360	38
2	4	50	360	43.3
3	5	75	355	57
4	5	95	355	77
5	5	85	355	60
6	5	80	335	50
7	4	80	335	60
8	5	80	335	52
9	5	90	330	57
10	5	65	330	33
11	5	73	330	45
12	5	80	320	60
13	5	70	320	49
14	5	80	312	55
15	5	80	313	67.4
16	5	75	290	72.5
17	5	70	280	62.7

Figure 24. Battery voltage and bus current climbing the hill

6. CONCISE SUMMARY

The refurbishment of the gasoline vehicle ensures that the engineered SR motor and drive system meet the requirements for dynamic performance, startup behavior, and acceleration attributes. During the startup phase, the SRM exhibits a current of 15A, with continuous torque and commendable acceleration characteristics. The highest speed attains 165 km/h, and the voyage extends to 205 kilometers or more. The novel rotor configuration reduces wind noise, with the SRM noise level measuring only 76dB. This chapter outlines the design of a 30kW SRD system utilized in the PEUGEOT 505 SW8. The system utilized fuzzy logic adaptive control to address significant torque fluctuations and substantial noise in the switched reluctance motor (SRM) by analyzing the instantaneous torque total. The automotive load experiment demonstrates that the implemented procedures are successful. The engineered SRD system has a low beginning current, minimal torque variation, and excellent efficiency, making it particularly well-suited for the dynamic characteristics of electric vehicles. It possesses extensive potential applications. When the batteries and power systems are planned concurrently, the engineered SRD system will demonstrate its superiority by optimizing and integrating the subsystems.

REFERENCES

1. J. C. Moreira, "Torque ripple minimization in switched reluctance motors via bicubic spline interpolation," IEEE Power Electronics Specialists Conference Record, Toledo, Spain, June1992 851 856

2. F. Filicori, C. G. L. Bianco, and A. Tonielli, "Modeling and control strategies for a variable reluctance direct drive motor," IEEE Trans. Industrial Electronics,40 1 105 115 1993

3. D. G. Taylor, "An experimental study on composite control of switched reluctance motors," IEEE Control System Magazine, vol. 11,2 31 36 1991

4. Nigel Schofield, Electric Vehicle Systems Notes, the University of Manchester, 2006

5. Yin Tianming A novelrotor structurefor S. R. M.China, Utility Model Matent. 03279782.6 2003

6. Technical information, IGBT-Module BSM300GA120DLC, EUPEC Power Electronics in Motion

11 Chapter

CATHODE MATERIAL: LIFEPO4

1. INTRODUCTION: AN OVERVIEW

Rechargeable batteries have predominantly supplanted primary cells, conserving resources and mitigating pollution. Recent surges in oil demand, coupled with related environmental sustainability concerns, are persistently exerting pressure on an already stressed global energy infrastructure. There is a demand for clean and efficient energy production from renewable sources in our environmentally aware society. Among secondary batteries, lead-acid and nickel-metal hydride batteries have diminished in market presence due to the emergence of a superior alternative, lithium-ion batteries. Lithium-ion batteries fulfill our requirements. The exceptional attributes of high capacity, elevated electrochemical potential, higher energy density, durability, and design flexibility expedite the replacement of traditional secondary batteries. Currently, they are predominantly utilized in portable electronic devices, with 57.4% of sales attributed to mobile phones, 31.5% to notebook computers, and 7.4% to cameras. Their applicability has also been expanded to other domains, including hybrid electric vehicles, aerospace, military vehicles, and others. Table 1 illustrates the distinctions among different batteries.

Table 1. The comparison between various batteries

Cathode	Li-ion	Pb-Acid	Ni-Cd	Ni-MH
lifetime/cycle	500~1000	200~500	500	500
Working potential/V	3.6	1.0	1.2	1.2
Specific energy/Wh kg^{-1}	100	30	60	70
Specific energy/Wh L^{-1}	240	100	155	190

Table 2. Electrochemical parameters of several cathode materials

Cathode	$LiFePO_4$	LiFePO4 +5%C	$LiMn_2O_4$	$LiCoO_2$	$LiNi_{0.8}Co_{0.2}O_2$
Density/g cm^{-3}	3.60	3.48	4.31	5.10	4.85
Potential/V	3.50	3.50	4.05	3.90	3.6
Specific capacity /mAh g^{-1}	169	159	148	274	274
Specific energy /Wh g^{-1}	0.59	0.56	0.56	0.98	0.98

$LiCoO_2$ is first chose to work as cathode materials when Li-ion batteries come out in 1990. Its long history supports $LiCoO_2$ a big progress. During that process, other cathode materials are discovered, $LiNiO_2$, $LiMn_2O_4$, $LiNi_{1/3}Co_{1/3}Mn_{1/3}O_2$, $LiFePO_4$ et al. Comparisons of electrochemical parameters of several cathode materials are listed in Tab.2.

Each of them has their own characteristics. For example, $LiCoO_2$ is costly and toxic, and its resource is no longer abundant (A. G. Ritchie, 2001). $LiMn_2O_4$ owns a much lower capacity and inferior cycle stability (Yuan Gao & Dahn J. R, 1996). Iron-based compounds look attractive as Fe is abundant, inexpensive, and less toxic than Co, Ni, or Mn. The phospho-olivine $LiFePO_4$ is currently under extensive studies due to its low cost, low toxicity, high thermal stability and high specific capacity of $170mAhg^{-1}$. Reduced reactivity with electrolytes results in the very flat potentials during charge-discharge processes.

The potential of material is partly decided by the Fermi level (A. K. Padhi et al, 1997). Much lower Fermi level is wanted to attain a higher working voltage. Among the iron-based compound, especially in $LiFePO_4$, $(PO_4)^{3-}$ lowers the Fe^{3+}/Fe^{2+} redox energy to useful levels. Strong covalent bonding within the polyanion $(PO_4)^{3-}$ reduces the covalent bonding to the iron ion, which lowers the redox energy of iron ion. The Fe^{3+}/Fe^{2+} redox energy is at 3.5 eV below the Fermi level of lithium in $LiFePO_4$. The lower is the Fe^{3+}/Fe^{2+} redox energy and the higher the V *vs.* lithium for that couple. In $LiFePO_4$, approximately 0.6 lithium atoms per formula unit can be extracted at a closed-circuit voltage of 3.5 V vs. lithium. The most prominent advantages of $LiFePO_4$ are (1). The structure of material hardly changes while Li ion intercalation and deintercalation; (2). It holds a long voltage platform.

The working principle of Li-ion battery is revealed in Fig.1. Lithium ions extract from anode to insert in cathode in the discharge process. The route is inversed as charge takes place. $FePO_4$ is the second phase that is present on electrochemical extraction of lithium from $LiFePO_4$. The extraction of lithium from $LiFePO_4$ to charge the cathode may be written as Formula (1) and the insertion of lithium into $FePO_4$ on discharge as formula (2).

$$LiFePO_4 - xLi^+ - xe^- \rightarrow xFePO_4 + (1 - x)LiFePO_4, \quad (1)$$

$$FePO_4 + xLi^+ + xe^- \rightarrow xLiFePO_4 + (1 - x)FePO_4, \quad (2)$$

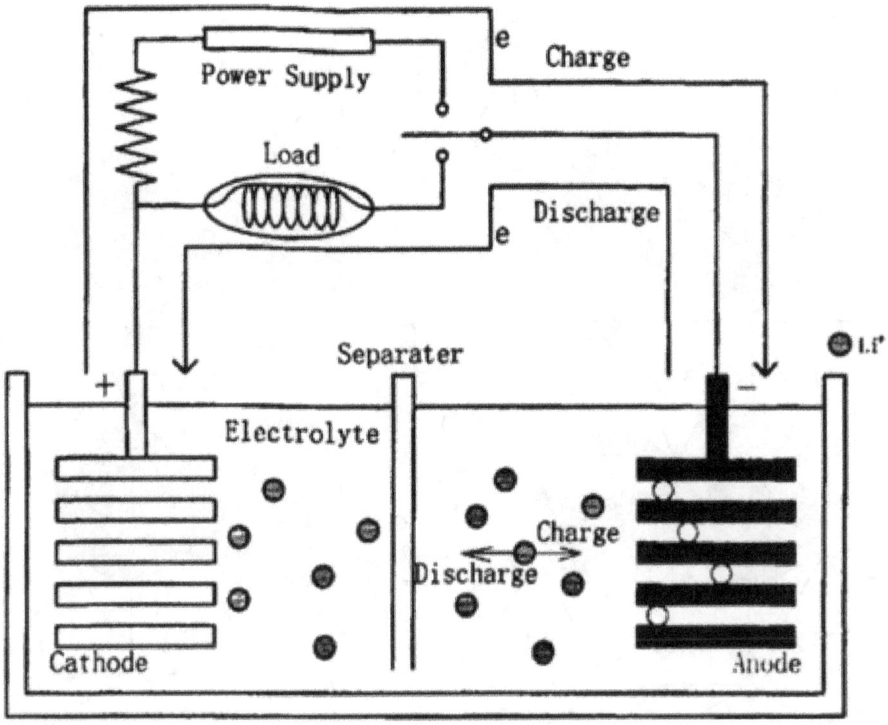

Figure 1. The schematic diagram of working principle for lithium battery

More efforts are conducted on the investigations of new electrode materials for lithium-ion batteries (Li-ion). The iron based olivine type cathodes (mainly lithium iron phosphate, $LiFePO_4$) are regarded as possible alternatives to cathodes based on rare metal composites (i.e. the transition metal oxides $LiCoO_2$, $LiNiO_2$).

2. FEATURE OF LIFEPO$_4$

2.1. Crystal structure of LiFePO$_4$

$LiFePO_4$ owns an ordered olivine structure, orthorhombic space group *Pnma*. Its systal constants of a, b and c are 1.033, 0.601 and 0.4693μm respectively. Fig.2 and 3 show the crystal structure of $LiFePO_4$, an ideal model and actual structure. The framework of $LiFePO_4$ consists of FeO_6-octahedra and PO_4-tetrahedra. FeO_6-octahedra and PO_4-tetrahedra contact each other by sharing oxygen vertices in *b-c* plane. The FeO_6-octahedra then links another PO_4-tetrahedra by sharing a edge. All the PO_4-tetrahedra don't touch each other. Lithium atoms are situated in the interstitial voids of the framework, forming infinite chains along the *c*-axis in an alternate *a-c* plane. Li atoms occupy M1 site and Fe M2 site. The Fe atoms occupy zigzag chains of corner-shared octahedra running parallel to the *c*-axis in the other *a-c* planes. O arranges in terms of hexagonal close packed structure with a slight distortion.

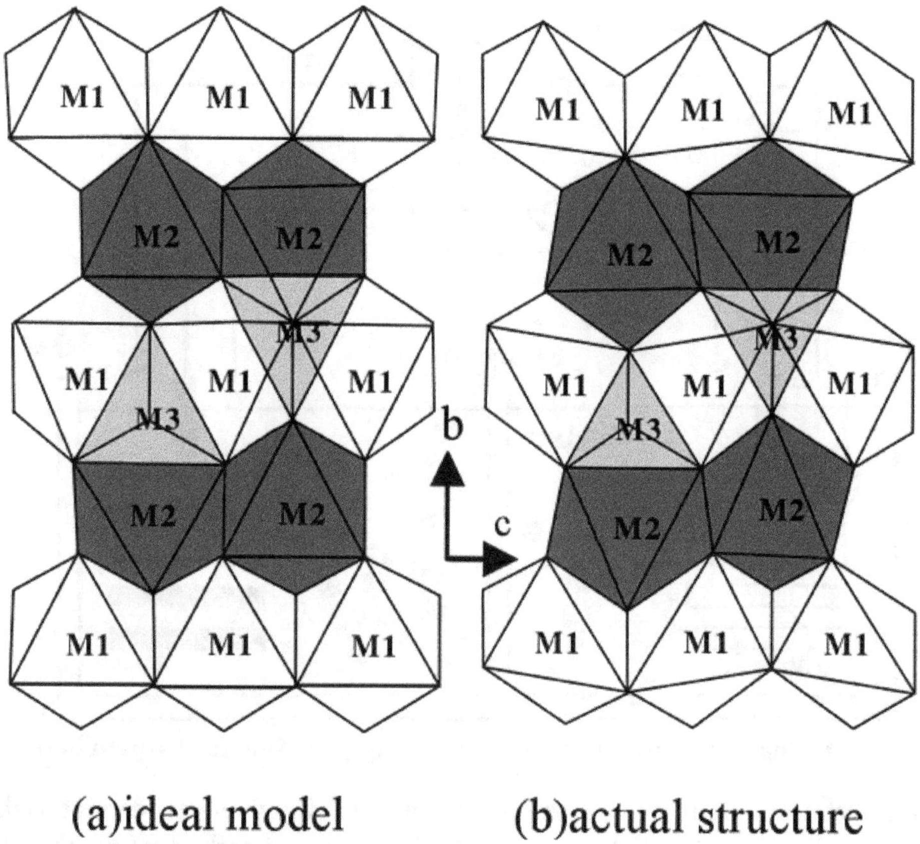

Figure 2. Crystal structure of $LiFePO_4$ (a) ideal model and (b) actual structure.

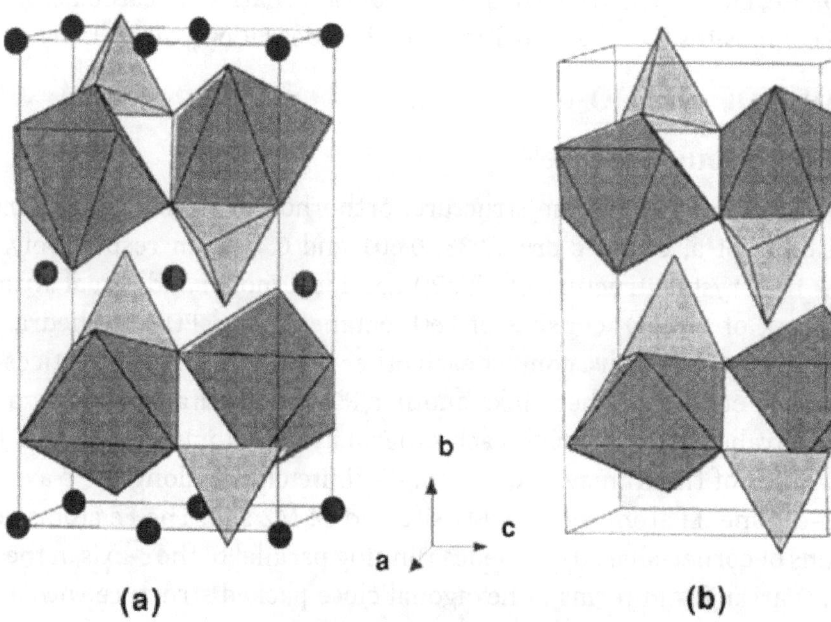

Figure 3. Crystal structure of (a) $LiFePO_4$ and (b) $FePO_4$

The strong P-O covalent bond forms the 3D delocalizating chemical bond, herein $LiFePO_4$ is thermodynamically and dynamically stable even at temperature above 200ºC.

A.K.Padhi pointed out that $LiFePO_4$ and $FePO_4$ almost possessed the same structure (Fig.3), both of them were of orthorhombic system (A. K. Padhi et al, 1997 a, 1997b). The small distinction between the two compounds results in only slight volume change, so then it won't cause crystal structure damage during charge and discharge process. Not like other cathode materials, the unique olivine structure of $LiFePO_4$ can assure an excellent stability, therefore its lifetime is much longer.

Fig.4 illustrates the theoretical insertion/extraction process of $LiFePO_4$ which is different from others. During lithium ions insert into a cathode particle (discharge), the surface region of particles becomes lithiated. And a phase interface emerges between two distinct phase regions (a lithiated phase and a delithiated phase region). The interface shrinks with charge process until the particle becomes one phase region. As discharge proceeds, the surface region becomes delithiated. The phase interface between lithiated phase and delithiated phase propagates inwards until disappear so that a complete charge and discharge procedure is finished.

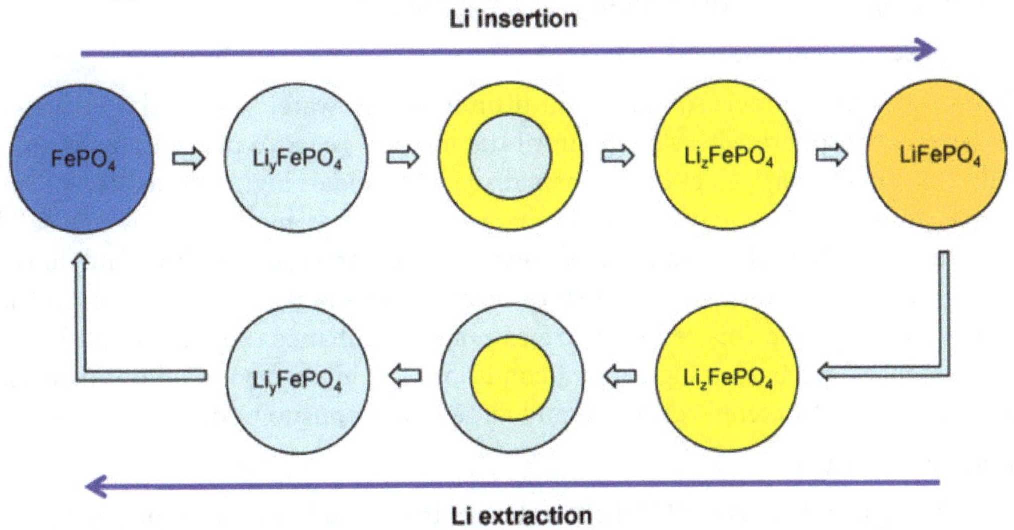

Figure 4. Complete Li insertion and de-insertion cycle of a spherical $LiFePO_4$ particle.

That distinct mechanism can explain the phenomenon of super flat voltage plateaus which are related to the surface Li concentration of particles. Since the incoming Li leads only to inner shift of the phase interface, the Li concentration of surface region is constant. That unique flat voltage plateaus are caused by unusual insertion/extraction process of this material.

2.2. Progress in preparing material

$LiFePO_4$ occurs in nature as the mineral triphylite. However, its impurity could worsen the electrochemistry properties so that triphyllite can't be utilized efficiently. Synthesis of $LiFePO_4$ is another way to obtain better materials. Preparation methods decide the microstructure of compounds which will carry a big weight in the performance of $LiFePO_4$.

2.2.1. Solid and liquid state reaction synthesis technology

1. High temperature solid state method

It processes as follows: mix the precursor-salts and lithium salt with a definite ratio, sinter the mixture first at a much lower temperature (300~400ºC) to remove volatiles second at a higher temperature (500~800ºC) to crystallize $LiFePO_4$, and then cool, grind and sieve to prepare the product. Solid state reaction method has been a developed technology and been used much frequently since it's simple to synthesis and easy to make mass production. The need for high temperature unfortunately upgrades cost, what's more, the product size always can't be small. So other measures are added to circumvent those problems, such as slightly lowdowning temperature or adding carbon granules.

2. Co precipitation method

Blend the soluble precursor-salts and lithium salt in water, adjust the PH value under N_2 atmosphere with stirring until the mixture react to precipitate $LiFePO_4$. Filter, wash and dry $LiFePO_4$ so then dispose it under high temperature. The temperature won't largely affect the microstructure as high temperature solid state method does.This technology needs a shorter reaction time and lower temperature compared with solid state reaction. The particle size can reach the nanometer level and this reduced size can help to enhance the charge-discharge performance especially at big current condition. The inadequacies inhibiting mass production are the complex process and large power consumption.

3. Sol-gel method

Complexing agent helps the Li^+, Fe^{2+} and $PO_4^{\ 3-}$ to form sol in water solution. Sinter the precipitation of sol-crystal at 400~900°C. The uniform distribution of ions in solution finally acts out a much uniform product that in nanometer size. Monitor the process is easy but industrialization of this technology is hard as the method is complex. Much waste water is produced at last also.

4. Hydrothermal method

The reactions of reactants that dissolve in water proceed in high pressure autoclave at a fixed temperature (<150°C). The product is then watered, filtered and dried to receive pure phase $LiFePO_4$, in small uniform size just like sol-gel does. This pure phase material can't behave excellent unless be doped with carbon. So then its capacity could reach to 137~160mAh g^{-1}. Rietveld analysis from S. Yang found that 7~8% Fe^{2+} occupy the Li^{+} sites, which is related to the weaker charge/discharge behavior (S. Yang et al,2002). LiOH as precipitator is added excessively to make sure the completeness of reaction which increases the cost. This system no more needs the protection from inert gas since O_2 is hardly dissolved in water. But operation is on a tiny scale constrained by the manufacture condition.

5. Carbothermal method

J. Barker adopt a new way to synthesize $LiFePO_4$ with the very inexpensive and readily Fe_2O_3 as available precursor source, instead of FeC_2O_4 et al (J. Baker et al, 2003). The precursors are intimately mixed and pelletized. The reaction mixture is then heated in a tube furnace (at ~750ºC) equipped with a flowing Ar atmosphere holding for about 8h. The transition metal reduction and lithium incorporation processes are each facilitated by the high temperature carbothermal reaction based on the transition from C to CO. The Fe^{3+} always can't be deoxidated absolutely and impurities can't be avoided virtually which will have a bad effect on battery performance.

6. High energy ball mill method

The reaction of Ball mill method is drived by the energy generated from steel balls in high speed revolution. A homogeneous solid solution with a small amount of contaminant, coming from the milling equipment, are obtained after 10~24h of high-energy milling of the initial mixture. A single calcination step goes on for no more than 1h. High energy ball mill method (also called Mechanochemical activation) is able to produce homogeneous powders with a particle size in the nanometer range, an increased specific surface area and typically, a high degree of activity. So the discharge capacity of $LiFePO_4$ can reach 150mAh g^{-1} at rate 0.5C as S.Franger have achieved (S.Franger et al, 2003).

7. Microwave synthesis method

The starting materials are mixed with ethanol in an agate mortar. The mixed powder is dried at ~60ºC (lower than that required for furnace heating) and pressed into pellets. Each pellet is covered with glass wool and placed in an alumina crucible with a lid to be heated in microwave oven. Electromagnetic energy is simultaneously absorbed by all parts of precursor so that uniform and rapid heating can be achieved within a short period of time. This distinctive character results in no temperature gradient within all parts. The initial discharge capacity was about 125mA g^{-1} at 60ºC by Masashi Higuchi et al which can increased to 161mAh g^{-1} if $LiFeOP_4$ is doped with carbon.(Min-Sang Song et al, 2007) This method can be used together with other synthesis approach so to largely reduce their process time.

8. Pulsed Laser Deposition (PLD)

Thin film electrode can relief the problem of poor conductivity. Pulsed Laser Deposition has been used to fabricate $LiFePO_4$ thin film electrodes. During the synthesis process of $LiFePO_4$, a laser with a fixed wavelength is used in the deposition. The films are deposited on platinum plates or basal planes of highly oriented pyrolytic graphite for minutes at room temperature. The films are then annealed at high temperature for hours under Ar or O_2 gas flow. The electric characters of thin film electrodes can be easily influenced by the articles size and fabrication process. Large particles on the surface of thin film electrode will lead to a bad capacity and poor cycling performance.

9. Electrostatic Spray Deposition (ESD)

A positive voltage of 8.0kV from a direct current power supply is applied to a stainless steel nozzle of a syringe to generate an aerosol. A syringe pump with a definite flowing rate is used to pump the precursor solution. The sample deposited on the substrate by ESD is calcinated at 500~750ºC for hours under pure Ar ambient. (A. Yamada et al, 2001) Electrostatic force shatters the solution into small drop of liquid with charge. So then the coulomb repulsion between drops can prevent them from collecting together. Dispersion degree is flexible by adjusting various voltage electrostatic fields. ESD method offers many advantages for thin-film deposition, such as low cost set-up, high deposition efficiency and easy control of composition of the deposited films.

10. Template-mediated approach

Templates such as silica and polycarbonate are used in the process. Charles R. Sides adopted polycarbon to manipulate the method as follows. An approximately $1cm^2$ piece of the polycarbonate filter is immersed in the electrode precursor solution of 1M $LiFePO_4$ prepared by sol-gel method in water for 24h. The impregnated template is then attached to a Pt current collector and dried in air at 80ºC for 10min. The template does not require tedious etching with HF or KOH to be removed since it is decomposed into carbon just for improved conductivity. The nanocomposite electrode prepared by Charles R. Sides (C. R. Sides et al, 2005) delivers almost 100% of its theoretical discharge capacity at the high discharge rate of 3C, and 36% of its theoretical capacity at the enormous discharge rate of 65ºC.

Lim et al reported the mesoporous $LiFeO_4$ using silica (KIT-6 and SBA-15) as template which is removed finally (Sunhye Lim et al, 2008). Both the nanowire and hollow $LiFePO_4$ cathodes demonstrate excellent rate capability, showing 137mAh g^{-1} at 10ºC rate. However such ex situ templating methods are both expensive and inefficient.

Biological systems offer capabilities for environmentally benign materials synthesis. An M13 virus–based biological toolkit has been developed for the design of nano-architectured structures and materials. Yun Jung Lee fabricated high-power nanowire batteries materials using M13 bacteriophage (phage or virus) organized on a polymer surface (Yun Jung Lee et al, 2009). The gene VIII protein (pVIII), a major capsid protein of the virus, is modified to serve as a template for amorphous iron phosphate (a-$FePO_4$) growth. The first discharge capacity of electrode generated by this method at 10ºC reaches 130mAh g^{-1}. And also this nanowire $LiFePO_4$ shows the stable capacity retention. Cycling at 1ºC, up to 50 cycles, virtually no capacity fade is observed.

11. Other novel methods

Ultrasonic spray pyrolysis (USP)

An aqueous solution containing sugar and raw material is ultrasonically nebulized and a flow of argon carries the droplets through a heated tubular reactor where solvent evaporation and precursor decomposition occur. The pyrolysis product is collected in a series of water bubblers at the reactor outlet where the salt byproduct dissolves leaving $LiFePO_4$. The manufacturing process is one-step process and can prepare the well crystallized and homogeneous small particles with the spherical shape, pure phase. $LiFePO_4$ powders, reported by Mu-Rong Yang, can get the initial discharge capacity of 150mAh g^{-1} at ºC/10 and 50ºC. (M. R. Yang et al, 2006)

Freeze-drying

The procedure includes four steps: preparing the precursor solution, freezing the solution, drying the congelation, and calcining the drying product. The products of freeze drying are pure amorphous nanopowder without hard agglomerations. Nanocrystals are subsequently obtained by calcination.

Coupling technique

Quantities of methods have been developed but each generated with different characteristics. The combination of two and several techniques can take the merits and discard the defects.

Microwave hydrothermal method

Compared with Hydrothermal method, the process is almost the same but only the the type of heating is changed. The mixture in water is treated in a vessel using a microwave digestion system (Maria Cristina D'Arrigo et al, 1998). Microwave-hydrothermal method can be used to crystallize nanophase materials with very high surface areas in a matter of a few minutes.

Microwave-Co. precipitation method

K. S. Park reported a capacity of 151mAh g^{-1} of $LiFePO_4$ at small current density through this coupling technique (K. S. Park et al, 2003). The use of microwave greatly cut down the reaction time for Co. precipitation method so that can save the power.

Ball-milling followed by solid-state reactions

Byoungwoo Kang successfully synthesized high-rate material though ball-milling and solid-state reaction. At 50ºC, corresponding to a time of 72s to fully discharge the capacity, the material achieves about 80% of its theoretical capacity (Byoungwoo Kang et al, 2009). Precursor is synthesized by ball-milling, then heating the mixture at 350ºC for 10h. The sample is cooled, ground and pelletized manually then heated at 650ºC for 10h under argon. The excellent rate property is thought to be caused from the strategy to facilitate transport across the surface by creating a poorly crystallized layer with high Li^+ mobility.

Various approaches have been developed to improve electrochemical performance. In addition, the performance can be enhanced though the optimization of the synthesis processes, adjusting weight and ratio of the raw materials even with common routes. Yanyi Liu fabricated nanocomposite film cathodes by sol–gel processing with excessive polymer additive (l-ascorbic acid $C_6H_8O_6$) while carbon serving as both defects and conductive nanocoating on the surface of $LiFePO_4$ particles. High electrochemical performance with initial discharge capacity of 312mAh g^{-1} is observed (Yanyi Liua et al, 2011). Capacities higher than theoretical limit observed in other nanostructured electrode materials have also been reported in literatures by other authors (Colm O'Dwyer et al, 2009; Chunhai Jiang et al, 2007; Dawei Liu et al, 2009). On the other hand, different carbon sources for carbon coating around $LiFePO_4$ particles have been implemented at enhancing the intrinsic electronic conductivity of $LiFePO_4$. However, many obstacles have been encountered for these methods from a laboratory process to mass production because of the complicated synthesis techniques and the hard controlled synthesis situation. The lack of an excellent large-scale synthesis technique is the obstacle to the commercial application of $LiFePO_4$. Therefore, it is critical to develop economic and efficient synthesis routes for the practical application of $LiFePO_4$ materials.

2.3. Drawback analysis

Despite the above mentioned advantages, the observed electrochemical performances of $LiFePO_4$ are found to be less impressive at high rates as this material has intrinsically poor ionic and electronic conductivity. A report (Amin et al, 2007) on the electronic and ionic properties of $LiFePO_4$ single crystals pointed out that the Li^+ conductivity was nearly four orders of magnitude lower than the electronic conductivity along the *b*-and *c*-axes and many orders of magnitude lower along the *a*-axis, implying that mass transport of Li^+ was crucial for improving the kinetic issues. The low ionic conductivity can be attributed to the one-dimensional nature of Li diffusion in unique olivine $LiFePO_4$, as clearly shown by a recent first principles calculation (Chuying Ouyang et al, 2004a, 2004b). Normally speaking, there are three kinds of possible paths for lithium ions to transport, i.e. [010], [001], [101]. Fig.5 (b) directs that three move paths, denoted as A, B and C direction, respectively. Figuring out the diffusion pathways is always crucial in understanding the microscopic diffusion mechanism. Some articles investigate the diffusion via computing with sorts of methods. Saiful Islam calculated the activation energy of various passageways to estimate the most possible transport route using well-established atomistic modeling techniques (Saiful Islam et al, 2005). The lowest Li migration energy is found for the pathway along the [010] channel with a nonlinear, curved trajectory between adjacent Li sites. D Moran attained the diffusion constant of Li ions in different axis with density functional theory (DFT)-based calculations (D.Morgan et al, 2004) on $LiMPO_4$ (M Fe, Mn, Co, Ni), and have found low activation barriers for Li ion motion through one-dimensional channels. Different calculation methods lastly come into the same result as Tab.3 revealed.

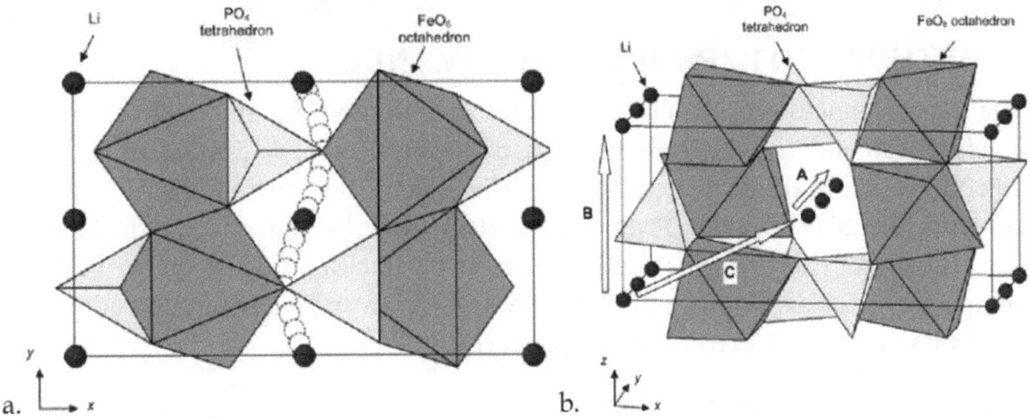

a. b.

Figure 5. Curved trajectories for Li ion migration between sites in the [010] direction. The diffusion path lies out of the x-y plane. (a,2D model. b,3D model, A, B, C denote [010], [001], [101] direction respectively)

This unique structure constrains the Li ions to move only in one wave-like chain since the polyhedra are slightly distorted (illustrated in Fig.5a). This "wavelike" trajectory for long- range migration results in lower migration energy than that of condition if the Li ion followed a direct, linear path. The different energy barriers for the three species are mainly due to the different atomic surroundings of the diffusion pathway, including the different ionic position and bonding length. Distortion of the *hcp* oxygen array has been related to the cation-cation coulomb repulsion across the shared edges.

Table 3. Description of Li+ migration path in $LiFePO_4$ crystal structure form some references

	D. Morgan			M. Saiful Islam		
	[010]	[001]	[101]	[010]	[001]	[101]
Space between Li-Li	3	4.7	5.7	3.01	4.67	5.69
Emig(eV)				0.55	2.89	3.36
Eact(eV)	0.27	"/2.5	1			
D(cm^2 s^{-1})	10^{-8}	10^{-45}	10^{-19}			

And within this crystal, all the FeO_6-octahedra can't form a share-edge network structure because polyanions PO_4 separate them. Therefore electrons also can't pass smoothly. In the layered compounds such as $LiCoO_2$, transition state of hybrid Co^{4+}/Co^{3+} anions make a contribution to the electron conductivity between layers, whereas that don't happen in olivine $LiFePO_4$ structure because $FePO_4$ was formed after extraction of Li^+.

The low temperature performance of $LiFePO_4$ is inferior. The capacity at 0.1C (156mAh g^{-1}) and 0.3C (148mAh g^{-1}) at 25ºC deteriorate largely, and only 91mAh g^{-1} and 65mAh g^{-1} are yielded at 0.1ºC and 0.3ºC at -20ºC respectively.

3. METHODS TO IMPROVE PERFORMANCE

The performances of $LiFePO_4$, however, are limited by its poor electronic conductivity and the sluggish kinetics of lithium ions to diffuse through the $LiFePO_4/FePO_4$ interface, which can result in a significant loss of capacity at high currents (Padhi et al., 1997 ; Prosini et al., 2002). And the high power density is the requirement of the power batteries for electric vehicles. Avenues of synthesizing composite materials, doping ions, nanocrystallization and others are introduced to clarify its improved electrochemical properties.

3.1. Composite materials

Owing to the poor inherent electronic conductivity, $LiFePO_4$ can be incorporated with conductive additives to form composite materials.

Carbon sources, such as acetylene black, glucose, sucrose and other organics, are added into the raw material and then carbonized at high temperature to form $LiFePO_4$/ºC materials. The $LiFePO_4$ particals are tightly coated with carbon and the interface between crystallized $LiFePO_4$ and amorphous carbon is nicely observed in Fig.6. An excellent cyclical stability is also observed. The reasons of its outstanding electrochemical performances are that the addition of carbon does not only perform its electrical conductivity, but also enhances electrolyte penetration into the cathode (Aricò et al., 2005). And the uniform small grain size would be beneficial to high rate performance. While the density of carbon is 40% lower than that of $LiFePO_4$, adding too much of which would lead to low tap density and also influence volume energy density of the cathode. So a reasonable amount of it is preferred.

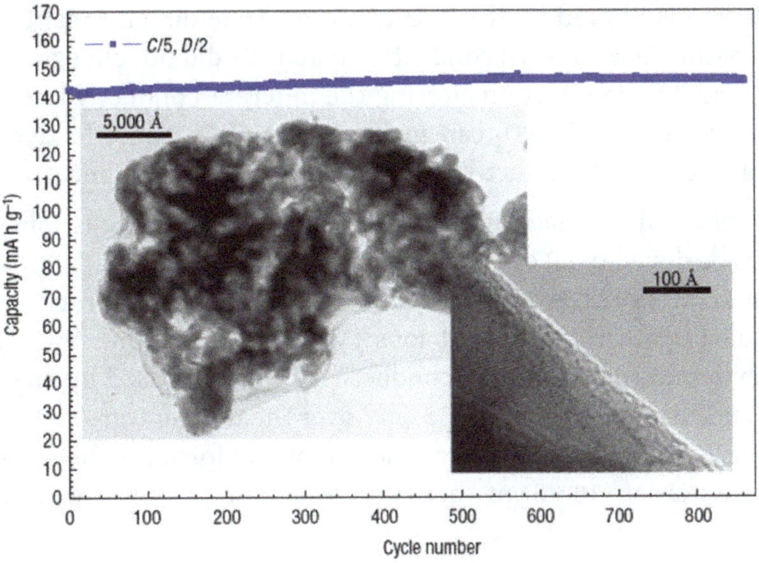

Figure 6. The performances of C-$LiFePO_4$ composite materials (Aricò et al., 2005)

Electric polymer organics (PAn, PPy, PTh, PPP and so on) work with inorganic cathode has emerged as one measure to address problem. Such as adding polyaniline (PAn) into the C-$LiFePO_4$, both the function of electronic conductive reagents and that of active materials are performed by adding it. The capacity of 87mAhg^{-1} can be performed by PAn at 0.1ºC, which can contribute to the specific capacity of the composites.

Some other materials like metals (Cu, Ag, Ni, etc.) can also be used to composite with semi-conducting $LiFePO_4$. TiO_2-$LiFePO_4$/ºC had higher electrochemical reactivity for lithium insertion and extraction than the un-doped $LiFePO_4$. The initial discharge specific capacity of the 30-min coating TiO_2-$LiFePO_4$/ºC material was about 161mAhg^{-1}, showing the potential of this material being used as a cathode material for Li-ion batteries. They decrease the charge transfer resistance and increase the surface electronic conductivity. Besides, the Fe dissolution might be simultaneously overcome by coating the $LiFePO_4$ particles with electrical conductive.

Compositing with additive can not only enhance the electronic conductivity and the penetration with electrolyte but also restrain the grain growth and the dissolution of Fe^{2+}/Fe^{3+} ions in the electrolyte. Above all, the electrochemical performances can be improved through forming the composite materials.

3.2. Doping

$LiFePO_4$ is a semiconductor with a band gap of 0.3eV, which is determined by its structure. The electrons transport is restricted by the strong Fe-O bonds and the Li^+ diffusion is limited by the Li-O bonds and one dimensional Li^+ migration pathways. Coating $LiFePO_4$ with conductive materials did not change the structure parameters and had no effect on altering the inherent conductivity of the lattice, while doping ions into $LiFePO_4$ can make it. It could be an effective method in increasing its electronic conductivity and Li^+ diffusion coefficient.

Many researchers have made numerous achievements. Various ions have been attempted to be doped in $LiFePO_4$. On the basis of different sites, it can be classified as doping at Li (M1) sites, Fe sites (M2) and O sites. Chung et al. reported chemical doping of $LiFePO_4$ with multivalent ions (Mg^{2+}, Ti^{4+}, Zr^{4+} and Nb^{5+}) into the Li 4a site. They found the electronic conductivity was increased by eight orders of magnitude and absolute values >10^{-3} S cm^{-1} over the temperature range from –20ºC to +150ºC (Fig.7). Doping it with supervalent ions can form p-type semiconductors with conductivities of ~10^{-2} S cm^{-1} arising from minority Fe^{3+} hole carriers (Chung et al., 2002).

The Li^+ ion diffusion could be optimized by doping F^- into the lattice of olivine structure. The capacity is increased after doping and the value varies with the doping amount. As is shown in Fig.8, the capacities are improved after doping especially with the amount of 2%F^-, achieving 156 $mAhg^{-1}$. The cycling performances are also enhanced. That could be attributedto the introduction of F- into the lattice of olivine structure, which result in the weakness of Li-O bonds (Sun et al., 2010). However, as is shown above, there is an optimum doping amount to make the materials exhibit the best electrochemical performances. When the ions are doped to a certain extent, it will increases the degree of disorder of ions and so lead to the enhancement of impedance (Fig.9). And the electrochemical performances will be ultimately affected.

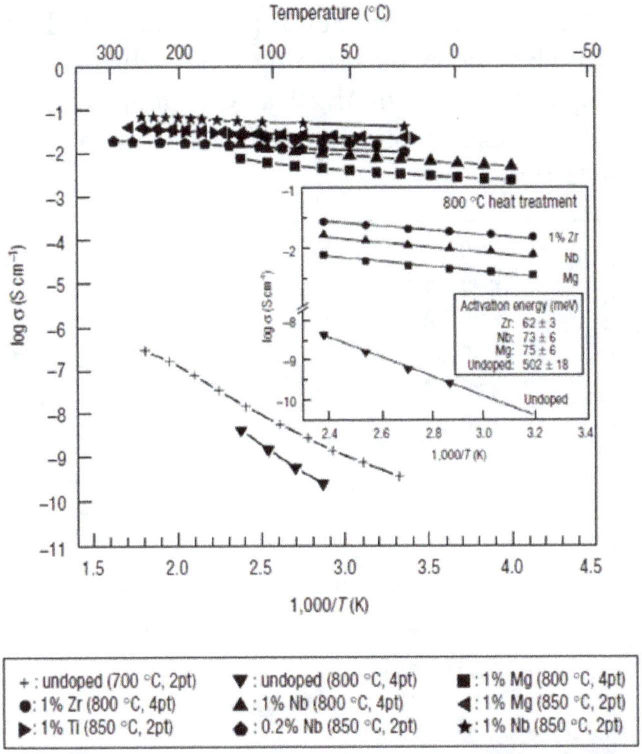

Figure 7. The electrical conductivity of Doped olivines of stoichiometry $Li_{1-x}M_xFePO_4$ M=Mg, Ti^{4+}, Zr^{4+} and Nb_{5+}) (Chung et al., 2002)

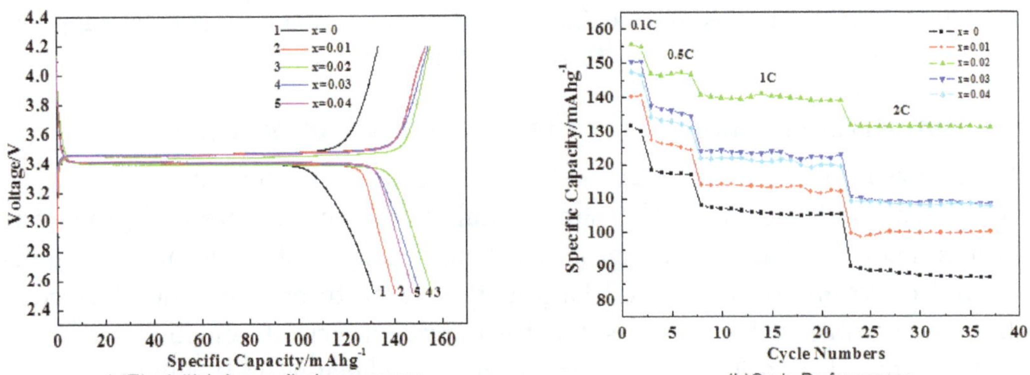

(a)The initial charge-discharge curves (b)Cycle Performances

Figure 8. The electrochemical performances of $LiFe(PO_4)_{1-x/3}F_x/C$(x=0, 0.01, 0.02, 0.03, 0.04)

Compare doping with one kind of ions, the co-doping with two or more would be much more beneficial to increase the electrochemical properties. It has been proved to be successful in $LiFe_{0.99}Mn_{0.01}(PO_4)_{2.99/3}F_{0.01}$/C. Mn^{2+} and F^- addition make the lattice parameter and the cell volume expanded which can facilitate the Li^+

diffusion between $LiFePO_4$ phase and $FePO_4$ phase(Yang et al., 2010). The Mn-Cl co-doped in $LiFePO_4$ also shows outstanding electrochemical properties, it can be achieved the capacity of 157.7mAhg^{-1} at 0.1C and nearly unchanged after 50cycles. For all these reasons, doping is an effect avenue to enhance the inherent conductivity of the lattice.

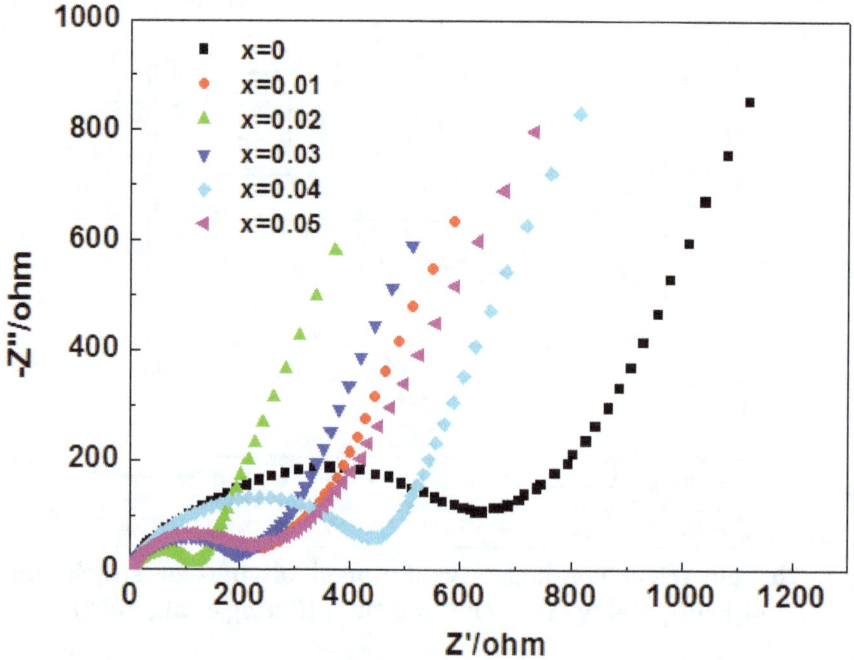

Figure 9. Electrochemical impedance spectra of $LiFe(PO_4)_{1-x/3}F_x/C$(x=0, 0.01, 0.02, 0.03, 0.04) cathodes at 25℃ (amplitude is 5mV in the frequency range of 10^5Hz~0.01Hz)

3.3. Nanocrystallization and preferential growth of particles

Nanoarrays have attracted significant attention for their applications in energy storage/conversion devices. The nanocrystallization and preferential growth of cathode materials have advantages, including (i) short path length for lithium-ion and electronic transport and large surface area to enhance the electrode/electrolyte contact. All of these result in the improved cycle life and higher charge/discharge rates (Aricò et al., 2005). For the nano-sized materials, the limiting factor for charge/discharge is the delivery of Li^+ and electrons to the surface rather than bulk diffusion (Kang & Ceder, 2009). So the inferior rate performance, caused by intrinsic low diffusion, can be perfected by synthesizing the coated nano-sized materisals, the ultrafast charging and discharging performances of which are remarkable to be applied on EVs (Fig.10).

The morphologies can be controlled by adopting specific synthetic routes and additive. Spherical particles, nanorods, flaky materials and nanowires are the common morphologies (Fig.11), the sizes of which are all nano level.

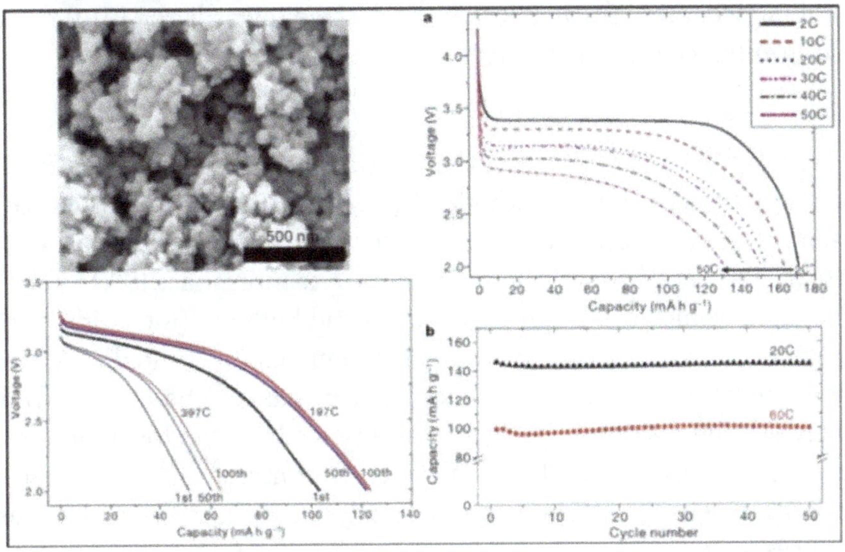

Figure 10. The high rate performances of nano-sized $LiFePO_4$. (Kang & Ceder, 2009)

Figure 11. The SEM micrograph of prepared $LiFePO_4$ with various morphologies: (a) Spherical particals(Kima et al., 2007), (b) nanorods(Huang et al., 2010), (c) flaky materials(Zhuang et al., 2005) and (d) nanowires(Wang et al., 2009)

The lithium ions can only extracted from $LiFePO_4$ and intercalated into $FePO_4$ in the [010] direction (Islam et al., 2005). Preferential growth of particles can shorten the (010) facet path and may increase the ratio of one-dimension tunnels in the bulk of the crystal. Hence, the diffusion across the surface towards the (010) facet can be increased to enhance rate capability.

3.4. Other means

To prepare the high power battery, the improvement of electrolyte and anode is also necessary, besides that of cathode. Especially at low temperature, the Li-ion cell containing liquid electrolyte can not cycle if the electrolyte is frozen. Ethylene carbonate (EC) is useful to form the solid electrolyte interphase (SEI) layers, but the high ratio of EC would result in high viscosity and high melting point. Adding low melting point electrolyte like Ethyl methyl carbonate (EMC) and diethyl carbonate (DEC) would increase the Li^+ ion diffusion performance. The $LiPF_6$ is wildly used as electrolyte lithium salt but its weak stability leads to the formation of HF that accelerates the Fe dissolution from cathode. By contrast, LiODFB can match the low-temperature electrolyte and forms steady SEI film, so it can enhance the performances of batteries.

4. APPLICATION

As of now, lithium-ion batteries have emerged as the primary power source due to their elevated electrochemical potential relative to Li/Li+, lightweight nature, design versatility, and exceptional energy density. Cost and safety remain significant factors constraining the spread of Li-ion battery applications. Li-ion batteries are used throughout various industries. Mobile phones, notebook computers, and cameras are examples of widely utilized technological products. Li-ion batteries are becoming utilized in electric vehicles due to developmental requirements.

4.1. HEV

Batteries make the consumer electronics convenient, even more after lithium ion batteries successfully enhance the power efficiency. This technology is now actively pursued for electric vehicle application. The lack of oil enhances the development of batteries, especially the one with high power and energy used in electric vehicle. High light is casted on Li-ion battery to look for hope.

Hybrid electric vehicle (HEV) is the most likely to be achieved as it combines the merits of electric vehicle (EV) and petrol-driven ones, i.e. HEV owns batteries and combustion engine simultaneously. According to the placement of combustion engine and electromotor, HEV is divided into series-type and parallel-type. S-type HEV is drove by batteries which are charged by combustion engine. P-type HEV uses electromotor to work during complicate and changeable working condition (launch, speed change, et al), and it shifts to combustion engine if condition is steady such as long-distant course in suburb. Both P and S-type avoid the loadswing and fast response of combustion engine whereas the fuel automobiles do which can lessen thermal efficiency. Related to mass application in HEV, the most appropriate power system should be splendid in terms of safety, cycle, calendar lifetime and cost. In addition, the availability and cost of the transition metals used in these compounds are unfavorable as the Wh/$ is a more important figure of merit than Wh/g in the case of large batteries to be used in an electric vehicle or a load-leveling system. Batteries are not so demanding in high energy and also capacity could not be high since engine can charge it consecutive. In HEV systems the operation windows would be defined much smaller (e.g. SOC=30–60%), according to power requirements, cold cranking and aging issues.

Low cost, long cycle life and non-toxic are the most obvious advantages of $LiFePO_4$. It's normal for $LiFePO_4$ to maintain almost sound structure after 1200 cycles at 1ºC. The power capability of olivine cells for very short-term pulse durations is nearly independent from SOC and SOC history. As a reference, the current price per unit of $LiFePO_4$ ranges from $1.90/Wh to $2.40/Wh. Although a little higher compared with $0.86/Wh for typical manganese-based Li-ion batteries, it is estimated that the price of $LiFePO_4$ will go down companying with the rapid development of technique. It is reported that the electrolyte decomposes completely below the limit of 5.0V with lithium cobalt and manganese oxides as cathodes due to the catalyses effects on the electrolyte/electrode interface. The overcharge test of $LiFePO_4$ doping with Al^{3+} appreciates a higher electrolyte decomposing voltage plateau that appeared between 5.20 and 5.45V (Hui Xie et al, 2006). It has been proved that $LiFePO_4$ can maintain the perfect olivine structure of the composite under overcharging conditions. Its thermal stability is superior as $LiFePO_4$ can endure condition under 400~500ºC (~200ºC for $LiCoO_2$ and $LiMn_2O_4$). $LiFePO_4$ as cathode material has become one of the most promising candidate for hybrid/electric vehicle propulsion.

4.2. Potential in future

LiFePO4 is versatile enough to function as a safety power source, allowing its application in several fields beyond transportation. The feasibility of designing a rubber-tyred container gantry crane without a diesel generating set is becoming viable due to the use of this innovative energy storage unit.The transfer of the rubber-tyred gantry crane can be effectively addressed with the implementation of lithium iron phosphate batteries for power supply. In light of the evolving trajectory of substation systems, characterized by heightened automation and service supply integration, the ferric phosphate lithium cell expedites the implementation of this trend. It can also improve the efficiency of green energy resources (solar, wind, etc.) by addressing the instability issues inherent in these systems, as electricity generated by solar and wind is not consistently reliable. LiFePO4 has garnered significant interest as a next-generation cathode material for lithium-ion batteries.

5. CONCISE SUMMARY

Understanding of LiFePO4 is increasing, and development is advancing rapidly. Lithium-ion batteries have emerged as the primary power source due to its elevated electrochemical potential relative to Li/Li+, lightweight nature, design versatility, and exceptional energy density. Numerous ways have been devised to achieve widespread practical use with advantageous features. Methods for manufacturing composite materials, doping ions, and nanocrystallization have been employed to enhance electrochemical characteristics. In addition to A123, Valence in the USA and Phostech in Canada, an increasing number of firms are focusing their efforts on the production of olivine cells, which are associated with LiFePO4 materials. The feasibility of large-scale production and widespread use is significantly enhanced by the durability, non-toxicity, high capacity, and energy density of LiFePO4. Iron-based olivine-type cathodes, primarily lithium iron phosphate (LiFePO4), are considered viable substitutes for cathodes composed of rare metal composites.

REFERENCES

1. Ritchie A. G. Recent development and future prospects for lithium rechargeable batteries. Journal of power Sources, 96 1June 2001 1 4 0378-7753
2. Padhi A. K. Nanjundawamy K. S. Goodenough J. B. Phospho-olivines as positive electrode materials for rechargeable lithium batteries. Journal of the Electrochemical Society, 144 4April 1997 1188 1194 0013-4651
3. Padhi A. K. Nanjundawamy K. S. Msaquelier C. Okada S. Goodenough J. B. Effect of structure on the Fe^{3+}/Fe^{2+} redox couple in iron phosphates. Journal of the Electrochemical Society, 144 5May 1997 1609 1613 0013-4651
4. Yamada A. Chung S. C. Hinokuma K. Optimized Li. Fe P. O. for Lithium. Battery Cathodes. Journal of The Electrochemical Society, 148 3March 2001 A224 A229 0013-4651
5. Aricò A. S. Bruce P. Scrosati B. Tarascon J. M. Schalkwijk W. V. Nanostructured materials for advanced energy conversion and storage devices. Nature Materials, 4 5May 2005 366 377 1476-4660
6. Byoungwoo Kang & GerbrandCeder.Battery materials for ultrafast charging and discharging. Nature, 458 12March 2009 190 193 0028-0836
7. Sides C. R. Croce F. Young V. Y. Martin C. R. Scrosati B. high-rate A. nanocomposite Li. Fe P. O4/carbon cathode. Electrochemical and Solid State Letters, 8 9July 2005 A484 A487 1099-0062

8. Chung S. Y. Bloking J. T. Chiang Y. M. Electronically conductive phospho-olivines aslithium storage electrodes. Nature Materials, 1October 2002 123 128 1476-4660

9. Chunhai Jiang, Itaru Honma, Tetsuichi Kudo & Haoshen Zhou. Nanocrystalline Rutile TiO_2 Electrode for High-Capacity and High-Rate Lithium Storage. Electrochemical and Solid-State Letters, 10 5March 2007 A127 A129 1099-0062

10. Chuying Ouyang, Siqi Shi, Zhaoxiang Wang, Xuejie Huang & Liquan Chen. First-principles study of Li ion diffusion in $LiFePO_4$. Physical Review B, 69 10March, 2004 1 5 0163-1829

11. Colm O'Dwyer. Vladimir Lavayen. David A. Tanner Simon. B. Newcomb Eglantina. Benavente Guillermo. González Clivia M. Sotomayor Torres. Reduced Surfactant Uptake in Three Dimensional Assemblies of VO_x Nanotubes Improves Reversible Li+ Intercalation and Charge Capacity. Advanced Functional Materials, 19 11June 2009 1736 1745 0161-6301X

12. Ouyang C. Y. Shi S. Q. Wang Z. X. Li H. Huang X. J. Chen L. Q. The effect of Cr doping on Li ion diffusion in $LiFePO_4$ from first principles investigations and Monte Carlo simulations. Journal of Physics-condensed matter, 16 13April 2004 2265 2272 0953-8984

13. Morgan D. Vander A. Ven Ceder G. Li conductivity. in Lix. M. P. O. . M= Mn. Fe Co. N. I. olivine materials. Electrochemical and Solid State Letters, 7 2 2004 A30 A32 1099-0062

14. Dawei Liu, Betzaida Battalla Garcia, Qifeng Zhang, Qing Guo, Yunhuai Zhang, Saghar Sepehri & Guozhong Cao.Mesoporous Hydrous Manganese Dioxide Nanowall Arrays with Large Lithium Ion Energy Storage Capacities. Advanced Functional Materials, 19 7April 2009 1015 1023 0161-6301X

15. Huang X. J. Yan S. J. Zhao H. Y. Zhang L. Guo R. Chang C. K. Kong X. Y. Han H. Electrochemical performance of $LiFePO_4$ nanorods obtained from hydrothermal process. Materials Characterization, 61 7July 2010 720 725 1044-5803

16. Hui Xie. Zhentao Zhou. Physical and electrochemical properties of mix-doped lithium iron phosphate as cathode material for lithium ion battery. Electrochimica Acta, 51 10February 2006 2063 2067 0013-4686

17. Islam M. S. Driscoll D. J. Fisher C. A. J. Slater P. R. Atomin Atomin-Scale Investigation of Defects, Dopants, and Lithium Transport in the LiFePO4 Olivine-type Battery Material. Chemistry of Materials, 17 20September 2005 5085 5092 0897-4756

18. Baker J. Saidi M. Y. Swoyer J. L. Lithium Iron. (II Phospho-olivines. Prepared by. a. Novel Carbothermal. Reduction Method. Electrochemical and Solid state Letter, 6 3January 2003 A53 A55 1099-0062

19. K.S.Park, J.T.Son, H.T.Chung, S.J.Kim, C.H.Lee & H.G.Kim. Synthesis of $LiFePO_4$ by co-precipitation and microwave heating. Electrochemistry Communications, 5 10October 2003 839 842 1388-2481

20. Kang B. Ceder G. Battery materials for ultrafast charging and discharging. Nature Materials, 458 12March 2009 190 193 1476-4660

21. Kima J. K. Cheruvally G. Choi J. W. Kima J. U. Ahna J. H. Chob G. B. Kimb K. W. Ahnb H. J. Effect of mechanical activation process parameters on the properties of $LiFePO_4$ cathode material. Journal of Power Sources, 166 1March 2007 211 218 0378-7753

22. Yang M. R. Teng Y. H. Wu S. H. Li Fe. P. $LiFePO_4$/carbon cathode materials prepared by ultrasonic spray pyrolysis. Journal of power Sources, 159 1September 2006 307 311 0378-7753

23. Maria Cristina D'Arrigo, Cristina Leonelli & Gian Carlo Pellacani.Microwave-Hydrothermal Synthesis of Nanophase Ferrites. Journal of the American Ceramic Society, 81 11November 1998 3041 3043 0002-7820

24. Min-Sang Song, Yong-Mook Kang, Jin-Ho Kim, Hyun-Seok Kim, Dong-Yung Kim & Hyuk-Sang Kwon.Simple and fast synthesis of $LiFePO_4$-C composite for lithium rechargeable batteries by ball-milling and microwave heating. Journal of power Sources, 166 1March 2007 260 265 0378-7753

25. Padhi A. K. Nanjundaswamy K. S. Goodenough G. B. Phospho-olivines as Positive-Electrode Materials for Rechargeable Lithium Batteries. J Electrochem Soc, 144 4April 1997 1188 1194 0013-4651

26. Prosini P. P. Lisi M. Zane D. Pasquali M. Determination of the chemical diffusion coefficient of lithium in $LiFePO_4$. Solid State Ionics, 148 1-2May 2002 45 51 0167-2738

27. Ruhul Amin, Palani Balaya & Joachim Maier.Anisotropy of Electronic and Ionic Transport in LiFePO4 Single Crystals. Electrochemical and Solid State Letters, 10 1 2007 A13 A16 1099-0062

28. Yang S. Song Y. Zavalij P. Y. Whittingham M. Reactivity stability. electrochemical behavior. of lithium. iron phosphates. Electrochemistry communications, 4 3Mar 2002 239 244 1388-2481

29. Franger S. Le Cras F. Bourbon C. Rouault H. Comparison between different $LiFePO_4$ synthesis routes and their influence on its physico-chemical properties. Journal of power Sources, 119-121 1June 2003 252 257 0378-7753

30. Shoufeng Yang. Yanning Song. Peter Y. Zavalij Stanley M. Whittingham Reactivity. stability electrochemical behavior. of lithium. iron phosphates. Electrochemistry Communications, 4 3March 2002 239 244 1388-2481

31. Sun C. S. Zhang Y. Zhang X. J. Zhou Z. Structural and electrochemical properties of Cl-doped $LiFePO_4$/C. Journal of Power Sources, 195December 2009 3680 3683 0378-7753

32. Sunhye Lim, ChongS. Yoon & Jaephil Cho.Synthesis of Nanowire and Hollow $LiFePO_4$ Cathodes for High-Performance Lithium Batteries. Chemistry of Materials, 20 14June 2008 4560 4564 0897-4756

33. Wang G. X. Shen X. P. Yao J. One-dimensional nanostructures as electrode materials for lithium-ion batteries with improved electrochemical performance. Journal of Power Sources, 189 1April 2009 543 546 0378-7753

34. Yang L. Jiao L. F. Miao Y. L. Yuan H. T. Synthesis and characterization of $LiFe_{0.99}Mn_{0.01}(PO_4)_{2.99/3}F_{0.01}$/C as a cathode material for lithium-ion battery. J Solid State Electrochem, 14 6July 2009 1001 1005 1432-8488

35. Yanyi Liua, Dawei Liua, Qifeng Zhanga, Danmei Yua, Jun Liuc & Guozhong Cao.Lithium iron phosphate/carbon nanocomposite film cathodes for high energy lithium ion batteries. Electrochimica Acta, 56 5February 2011 2559 2565 0013-4686

36. Yuan Gao. Dahn J. R. Synthesis and characterization of $Li_{1+x}Mn_{2-x}O_4$ for lithium ion battery application. Journal of the Electrochemical Society, 143 1 1996 100 144 0013-4651

37. Yun Jung. Lee Hyunjung. Yi-Jae Woo. Kim Kisuk. Kang Dong. Soo Yun. Michael S. Strano, GerbrandCeder & AngelaM.Belcher. Fabricating Genetically Engineered High-Power Lithium-Ion Batteries Using Multiple Virus Genes. Science, 324 22May 2009 1051 1055 0036-8075

38. Zhuang D. G. Zhao X. B. Cao G. S. Mi C. H. Tu J. Tu J. P. Morphology and reaction mechanism of $LiFePO_4$ prepared by hydrothermal synthesis. The Chinese Journal of Nonferrous Metals, 15 12 2005 2034 2039 1004-0609Borong Wu, Yonghuan Ren and Ning Li

12 Chapter A COMPREHENSIVE ELECTRIC VEHICLE CURRICULUM

1. INTRODUCTION: AN OVERVIEW

Electric Vehicles (EV) have been present on the market for the past 110 years. In the initial phase of vehicle development, there existed just two competitors: the internal combustion engine (ICE) and electric vehicles (EV). The electric vehicle was the predominant choice over internal combustion engines until 1930; thereafter, the landscape shifted due to the advancement of gasoline technology, the mass production of the Ford Model T, the superior performance of internal combustion engines, and their reduced cost. The aforementioned factors and an inadequate energy infrastructure resulted in a deficiency of interest and advancement in EV technology (Chan & Chau, 2001).This forgotten research area for near 40 years came back in the early 70´s with more strength since the appearance and continue development of advanced semiconductor devices, new storage technologies, sophisticated materials, advanced modeling and simulation techniques, real time implementation of complex control algorithms, maturity of power electronics and motor drives area. Since it is second big pushed to EV, a lot of improvements have been achieved by the constant effort of physics, chemical, mathematics, mechanical, computer, electrical and electronics specialists committed to develop a highly energy efficient device of transportation (Chan & Chau, 1997).

Nowadays, the term EV includes plug-in hybrids, extended range EV and all-EV, (Department of Energy of the United States of America, 2011). One big step forward to the mass introduction of all-EV has been the introduction of hybrid electric vehicle (HEV) in several automobile companies. The mass introduction of HEV started in 1997 by Toyota with the Hybrid-Prius, a parallel configuration integrated with a Toyota Hybrid Systems (THS). The THS-C was implemented later to the Estima Hybrid, (a THS combined with a continuous variable transmission (CVT)). Following this trend, a Toyota Hybrid Systems for Mild hybrid system (THS-M) was implemented in the Crown. In 2004, the THS II was installed in a new Prius, which had the main characteristic to increase the power supply voltage. This electric drive train added a direct current to direct current (DC/DC) converter, between the low voltage battery pack (276-288V) and the traction motor (500V or more), to use a smaller battery pack and more powerful motors compared with its previous version. In addition the THS name was modified to Hybrid Synergy Drive (HSD) to allow its use in other vehicles′ brands (Pyrzak, 2009). It is necessary to say that Toyota is not the only vehicles′ manufacturer to develop hybrid technology other brands include Ford, GM, Honda, Nissan, etc.

Today, the $12 billion investment to develop vehicle technologies given by the Department of Energy (DOE) from the United States of America (USA) has opened a third stage in the development of EV. It is foreseen that the classical high vehicle costs, performance predicaments, and safety issues claimed in EV sector; will be overcome in the near future motivated by the American Recovery and Reinvestment Act and DOE's Advanced Technology Vehicle Manufacturing (ATVM) Loan Program. Those programs will support the development, manufacturing, and deployment of the batteries, components, vehicles, and chargers necessary to put on America's roads millions of electric vehicles in 2015. Accordingly with USA's Vice President Joe Bide in 2015 the cost of batteries for the typical all-EV will drop almost 70% from $33,000 to $10,000, and the cost of typical PHEV batteries will fall in the same rate from $13,000 to $4,000 (Department of Energy, United States of America, 2011).

Currently, there is little doubt that electric vehicles (EVs) play a crucial role in society, and their growth is anticipated, particularly in the social, economic, and industrial sectors, driven by environmental concerns. In addition to the significance of electric vehicles, there are several global bachelor's, undergraduate, and postgraduate programs that strive to integrate all aspects related to electric vehicle design into a unified curriculum (See Section 1.4). Conversely, the advancement of electric vehicles has been regarded as a discrete application of prior knowledge in electric machines, power electronics, energy systems, chemical engineering, or mechanical structures. Currently, the integration and nuances of various facets of this inherently multidisciplinary application are often overlooked, resulting in missed opportunities for developing high-efficiency electric vehicles that are both effective and cost-efficient, due to a deficiency in experience and expertise.

1.1. Typical EV electrical architecture and energy storage unit

Current electric, hybrid and plug-in electric vehicle (EV, HEV, PHEV) power trains comprise at least of one on-board energy generation unit, energy storage, traction drive and peak power unit (Wirasingha & Emadi, 2011). The correct power management of those different sources increase the energy efficiency and reduces the overall fuel consumption (hence cost and emissions) (Kessels et al., 2008). In general the advantages of EV are higher energy efficiency and regenerative braking (Lukic & Emadi, 2004) compared with conventional ICE. Since electric motor efficiency is higher than the heat engine, overall significant efficiency fuel consumption can be achieved by assigning electric motor or engine for the propulsion depending on driving cycle. In addition, some EVs are able to generate electricity and recharge battery without any external supply (Emadi & Ehsani, 2001).

At the present moment, different HEV has been reported for instance vehicle to the grid (V2G), V2G plus vehicle-to-load, V2G plus vehicle-to-home, V2G plus vehicle-to-premise, V2G plus vehicle-to-grid-net metered, V2G plus advanced vehicle-to-grid (Tuttle & Baldick, 2011). The main characteristic of those proposals are the use of a particular power electric drive train for each specific applications.

In contrast all-EV traction train configuration proposed in literature are simpler than HEV and they can use for example battery (B), fuel cell (FC), photovoltaic (PV) as their main energy generation/energy storage unit. Additionally several arrays of B, FC and PV linked with supercapacitors (SC) in all-EV has been reported (Emadi, 2005), (Pay & Baghzouz, 2003), (Schofield, 2005), (Solero et al., 2005), (Intellicon, 2005). Figure 1 shows the most common configurations.

Today in the all-EV there are two main energy generation units, B and FC; both of them with the following characteristics,

1. They produce current just when it is supplied by its fuel/energy storage unit.
2. They achieve a high energy efficiency between 40-60%, which its load dependent.
3. B-EV and FC-EV produces zero or almost zero pollution and noise.
4. Li-ion battery and Proton Exchange Membrane (PEM) fuel cell are best candidate for vehicular applications due to its high power density, small volume and low temperature.

In contrast to the B-EV, the FC-EV particularities such as load dependency, incapacity to accept regenerative energy, intolerance to the input ripple current, start-up time, and slow load response, make unviable the single use of FC in traction applications. Therefore different FC-SC configurations have been proposed, i.e. characteristics of configuration i*)* are,

1. The use of only one power electronic converter (PEC).
2. The use of a SC as a peak power buffer during EV acceleration.
3. The SC accepts the regenerative power for the EV breaking period.
4. There is an inherent decoupling between the peak and average EV power. As a result the power converter just deals with the average power. This behavior is translated in a small size and weight of the PEC.
5. The PEC needs to operate in a wide input voltage operation region caused by the FC load dependency.
6. It is necessary to implement a Power Management Strategy for the appropriate operation of the overall system.

It has been reported in literature different power converter that can be used as a step-up/down converter for configuration i). For example Boost, Buck/Boost, Boost interleaved, Half Bridge, Full Bridge, Full Bridge Zero Voltage Switching (ZVS) and/or Zero Current Switching (ZCS) or Push-Pull, (Profumo et al., 2004). Their main differences are the conversion ratio, power ratio, current ripple, uni/bidirectional capacity, efficiency and isolation (Blaabjerb et al., 2004) (See Section 1.3).

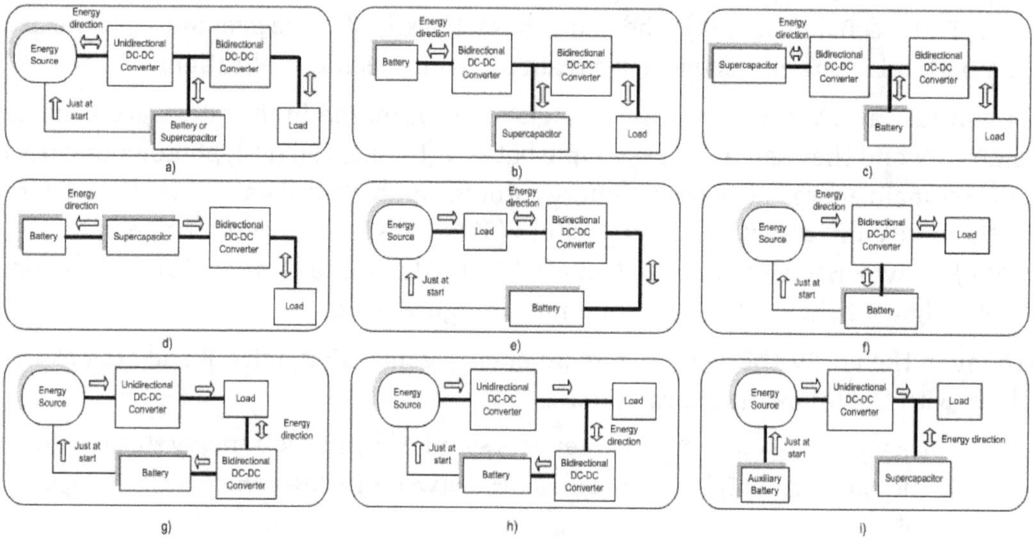

Figure 1. Different all-EV configurations reported in literature.

1.2. Mechanical drivetrain EV

The basic mechanical architecture of EV, HEV and PHEV found in the market consists at least of one ICE and one electric motor where the torque produced by the engine is transmitted to the wheels by using a lossy and heavy mechanical shaft directly coupled to the rear or front wheels. Figure 2 a) shows a typical four wheel all-PEV with mechanical differential.

In this configuration, it is used a mechanical differential to produce different speed to each wheel during cornering, the closer wheel to the curve will run slower compared with the outer wheel. However such relationship is usually fixed and it does not depend of the steering angle and a rollover phenomena can be produced (a similar action is produced in the three wheel configuration). The trend for advanced vehicle architecture is to remove the traditional mechanical drive shaft and differential, and replacing it with an Electric Differential (ED) implemented by electric motors directly coupled to the wheels (using one fixed gear). Another trend is completely removing the gear and allocating the motor inside the wheel; this configuration is known as in-wheel motors, the in-wheel motors can be brushless or permanent magnet (Tabbache et al., 2011).

Additional features of ED are a) no mechanical link between the wheels, b) it is applied lees power to the inner wheel in a turn, c) there is synchronization between the wheels during straight paths and d) it uses a virtual masterfor relative speed synchronization (Perez-Pinal, 2009). Figure 2 b) shows a typical four wheel all-PEV with ED.

The main characteristic of ED is the use of one PEC for each motor and the increment of vehicle´s safety during cornering and risky maneuvers compared with its mechanical counterpart. Those advantages are achieved by two reasons: a direct torque control in the wheel and on-the-fly change in the differential ratio.

1.3. Modern EV design

At the beginning EV were directly adapted from ICE, such replacement was achieved by replacing the combustion engine and the fuel tank by an electric motor and a battery pack. In this kind of conversion usually were remained the overall components (Ehsani et al., 2004; Miller, 2004). However, low performance was a characterization of those EV.

The vehicles´ mechanical operation (ICE or EV) are based in fundamental mechanical laws, the inital design variables are two, static and dynamic. The initial static characteristics are a desired acceleration, stop, driving and turning angle. The dynamic characteristics include the aerodynamic resistance, the rolling resistance, and the traction force (Emadi, 2005a).

Nowadays, to design a modern EV are involved chemistry, mechanical, electronics, computer engineers and business' guys (Ehsani et al., 2004), in other words an EV has evolved from a pure mechatronic system to a more chemechatronic system (the word chemistry plus mechatronic). The term chemechatronic was firstly employed in 1991 by the company Tosoh to describe its research efforts in the area of biotechnology and pharmaceutics (Tosoh, 1991). In addition (이시우, 2003) used the same term to describe a system on a chip that includes in a single device chemical, mechanic, electronic, control system and computer science technology, it can be noticed that in essence an EV is chemechatronic system. Along this chapter the chemechatronic term refers to the approach that integrates areas of chemistry, control theory, computer science, electrical and electronics within a product development with the main aim to enrich and/or optimize its functionality.

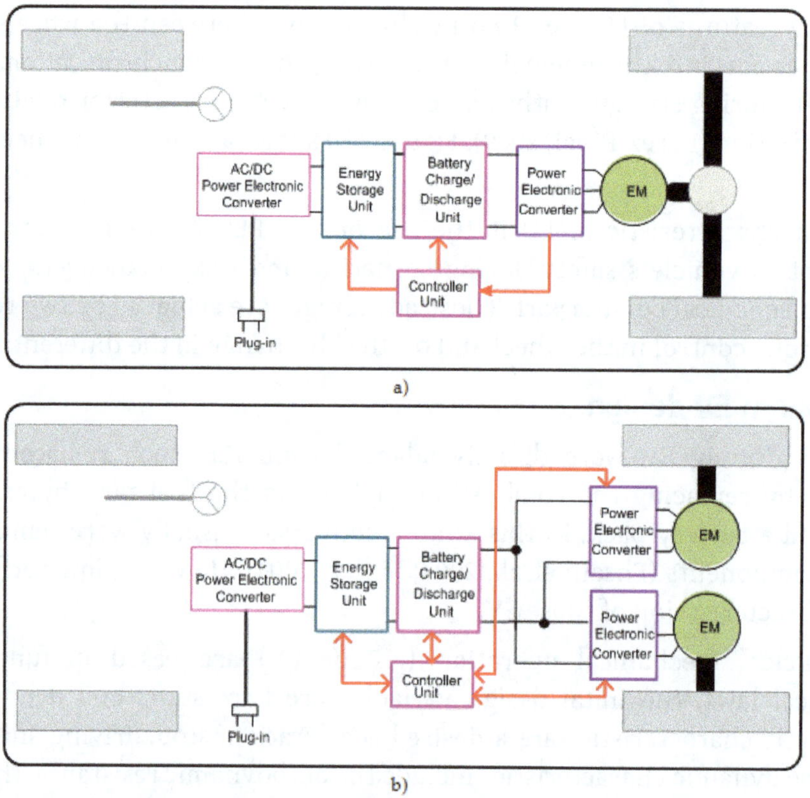

Figure 2. Typical four wheel all plug-in electric vehicle a) with mechanical differential, b) with electric differential.

Accordingly with (Perez-Pinal, 2006) a lot of research has been done in order to develop accurate guidelines to design EV, some text book and classical papers can be found in literature (Ehsani et al., 2004; Miller, 2004; Emadi, 2005b; Ehsani et al., 1997; Husail & Islam, 1999). The main three characteristics required for modern EVs are,

1. Low weight.
2. High energy efficiency.
3. High torque response.

In addition, modern EV performance is evaluated in terms of,

1. Acceleration performance
2. Acceleration time.
3. Acceleration distance.
4. Maximum cruise speed.
5. Gradeability.
6. All the last characteristics inside a driving cycle.

The first step to design an EV is to determine the relationship between the mechanical torque and the power electronic stage including the electric motor (Perez Pinal, et al., 2006). There exist two different techniques to initially design the power stage of an EV. The first technique determines the maximum mechanical power needed by the EV based on a driving cycle. The second technique finds the average mechanical power needed in terms of an initial speed, acceleration time and the maximum speed, for both techniques once the mechanical power is determined.

The second step sizes the maximum electric power needed for the power stage; in this step it must be considered the kind of electric motor and power losses. The kind of motor is generally chosen in terms of the base speed, maximum mechanical speed, power losses, and control topology.

The third step determines the main source and DC- bus voltage. In this stage there are many possibilities in terms of energy source and energy storage unit. The main motivation to choose one or another are based on the environment of the final product, sell point, and performance (Ehsani et al., 2004), this step is related with the selection of the PEC to step up the energy source unit. Here, it can be found several architectures related with the PEC, some criteria to select one or another are related with the power range, isolation requirement, efficiency and cost. However, the most important criterion to select one PEC configuration is to supply the deficiencies of the power source unit. For instance, a PEC for a FC power source unit should fulfill the following characteristics,

1. An efficient increment of the low output voltage from the FC to the motor drive.
2. A low input current ripple.
3. A unidirectional power direction between the power source unit and the motor drive.

As it can be implied from the list of requirements, there are several PEC architectures that satisfy those needs, the most usual are the following (Profumo et al., 2004), (Blaabjerb et al., 2004).

1. Boost converter,
2. Buck/boost converter.
3. Interleaved boost converter.
4. Half bridge and full bridge converter.
5. Full bridge converter with zero voltage-zero current switching (ZVS-ZCS).
6. Push-pull converter.

Table 1 summaries the overall characteristics of the PEC, it can be observed that several PECs can be used for the DC/DC power stage.

The general characteristic of the isolation architectures is that an input current reduction can be achieved at the expenses of increasing the inductors' values, or increasing the switching frequency. However an increment of the switching frequency produces an increment of the semiconductors switching losses. Isolation architectures are suitable for applications with high conversion ratio or where isolation is mandatory i.e. Japan and USA. In order to select the appropriated topology for any EV, it is necessary to perform a comparison of the device losses, power density, and efficiency. Recently there is a trend to use paralleled or interleaved topologies; some advantages of those topologies are an inherent power sharing between the number of cells, an inherent robustness, and an increment of the switching frequency (Chan & Pong, 1997).

Table 1. Overall characteristics of different DC/DC converters.

Converter	Conversion ratio	Current ripple	Power direction	Efficiency	Power range	Isolation
Boost	Up to 5 times	High	Unidirectional	Medium	< 3kW	No
Buck/boost	Up to 2 times	High	Unidirectional	Medium	< 3kW	No
Boost interleaved	Up to 5 times	Low	Unidirectional	High	< 10kW	No
Half bridge	Variable with Transformer	High	Bi-directional	Medium	< 10kW	Possible
Full bridge	Variable with Transformer	High	Bi-directional	Medium	< 10kW	Possible
Full bridge ZVS-ZCS	Variable with Transformer	High	Bi-directional	High	< 10kW	Possible
Push-pull	Variable with Transformer	High	Unidirectional	High	< 10kW	Yes

After it has been determined the size and characteristics of the power source and storage unit, the following step is to select the motor drive. The final drive depends on the selected motor, which can be direct current (DC) or alternating current (AC). For example, the available topologies considering a three - phase induction motor are,

1. Hard-switching voltage source inverter (VSI).
2. Hard-switching current source inverter (CSI).
3. Resonant phase leg inverter (RPLI).
4. Active clamp resonant dc link inverter (ACRDI).
5. Auxiliary resonant commutated pole inverter (ARCPI).
6. Push pull.

Additionally, it can be integrated the step-up converter and inverter in a single stage, i.e. the Z converter (Blaabjerb et al., 2004). Once again, the most important criterion to select one or another is the energy efficiency, power density and cost.

1.4. Current curricula efforts

There are different programs in the area of EV and HEV implemented up to now, Table 2 shows a list of current programs available in the market, (Center for Automotive Research, 2003; CSU Ventures, 2009; Ferdowsi, 2010; Hammerstrom & Butts, 2011; Heinz & Schwendeman, 2011; Michigan Technological University, 2011; Purdue University, 2010; Rizkalla et al., 1998; Simon, 2011; The National Alternative Fuels Training Consortium, 2009; University of Detroit Mercy, 2009).

Additionally to these programs other universities and companies offer courses in the EV and HEV such as the Department of Automotive Engineering Cranfield University, the company Georgia Power, The Illinois Institute of Technology (IIT), The University of Manchester (UMIST), among others.

Table 2. Current HEV, EV programs.

Year	Program Title / University	Level	Area
1998	A new EE curriculum in electric vehicle applications, Purdue School of Engineering and Technology at Indianapolis	Undergraduate, Graduate	EV
2003	Center for Automotive Research, The Ohio State University	Certificate Program, Graduate	EV, HEV
2007	Designing a Multi-Disciplinary Hybrid Vehicle Systems Course Curriculum Suitable for Multiple Departments, Minnesota State University, Mankato	Graduate	EV, HEV
2009	The National Alternative Fuels Training Consortium, West Virginia University	Colleges, Undergraduate	EV, HEV
2009	Certificate engineering program in Advanced Electric Vehicles (AEV), University of Detroit Mercy	Undergraduate, Graduate	EV, HEV
2009	Advanced Electric Drive Vehicle Education Program: CSU Ventures, Colorado State University (CSU), Georgia Tech (GT), Ricardo, MRI, KShare, Arapahoe Community College, Douglas County Schools	Colleges, Undergraduate	EV, HEV
2009	J Sargeant Reynolds Community College	Certificate, Undergraduate	EV, HEV
2010	Advanced Electric Drive Vehicles –A Comprehensive Education, Training, and Outreach Program, Missouri University of Science and Technology, University of Central Missouri, Linn State Technical College, St. Louis Science Center	College, Undergraduate, Graduate	EV, HEV

Year	Program Title / University	Level	Area
2010	Electric Vehicles part 1 and 2, Portland State University	Undergraduate, Graduate	EV, HEV
2010	Indiana Advanced Electric Vehicle Training and Education Consortium, (I-AEVtec), Purdue University, NotreDame University, IUPUI, Ivy Tech, Purdue-Calumet, Indiana University -Northwest	Technician, Undergraduate, Graduate	EV, HEV
2010	Development and Implementation of Degree Programs in Electric Drive Vehicle Technology, Macomb Community College, Wayne State University, NextEnergy	Certificate, Undergraduate, Graduate	EV, HEV

From Table 2, it can be observed that only three programs have a link between college and graduate studies. One similarity in those programs is a permanent effort between regional Colleges, Universities and vehicles' companies. For example, the program from The University of Detroit, Mercy's College of Engineering and Science in conjunction with Engineering Society of Detroit is founded by Ford. This program is focused on electric and hybrid drivetrain technology, and it is expected to open seven new courses related to the automotive and defense ground vehicles industries.

Another similarity between those programs is to prepare and recruiting technician and automotive engineers starting in the high school level by conducting seminars and summer camps. In addition, it is expected to develop education material and video demonstration about EV and HEV to inform the general public by using internet as their main platform. After analyzing those programs and its references were identified eight different areas related with EV, Figure 3.

It must be mentioned that overall areas from the technician to the PhD level proposed in this chapter are related with Figure 3 (see Section 2). In general the area of technician is related with the maintenance and repair of the end user product, in this stage the understanding of each particular area and a general appraise of each stage is not fundamental. This level is related to know how work the overall EV´s devices and it is not emphasized to answer why they behave in a certain or different way. Those questions are further explained in the undergraduate and graduate levels, where a fully understanding and generation of novels ideas to the state of the art is expected in the final levels.

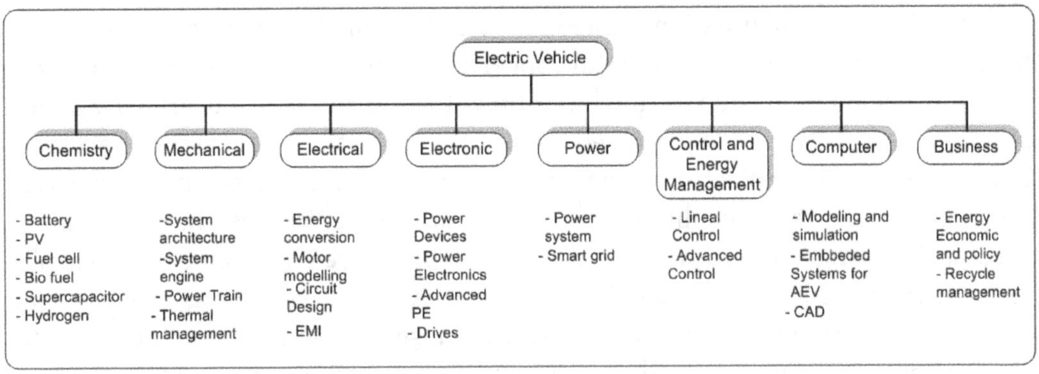

Figure 3. Typical areas covered by Electric Vehicles.

1.5. Organization of the chapter

In order to come out with an integrated curriculum, different active learning techniques and curriculum strategies were compared and integrated in this proposal. The chapter begins (Section 2) with the overall description of the curricula in the following levels: Technician, Bachelor in Technology, Bachelor in Science, Master in Engineering/Science and Doctor in Philosophy (Ph. D.). Moreover, the objective of each level, its requirements, expected results, and overall recommendations are also given. This section provides the mandatory and final elective courses in each level. In Section 3, it is presented the proposed teaching model based on inquiry-based learning and active learning techniques widely developed in McMaster University. The inquiry process is about exploring, discovering, and ultimately, reaching a higher level of understanding. Here, it is addressed the recommended methodology to lecture this topics and a general flowchart is provided. Finally some concluding remarks, future directions, and particularities are given in Section 5.

2. CURRICULA DESCRIPTION

It is widely know that the design of a curriculum is not an easy task. The curriculum itself is the fundamental part of any institution, from basic to graduate level, in the design of a curriculum can be given the desired requirements and characteristics for admission and graduation. In addition, it can be addressed the general requirements and difficulty of each course, textbook, interrelation to other courses, lab session, credits, duration, syllabus, etc.

The design of a curriculum in engineering has been performed before in other areas. For example in the area of electronic engineering was proposed a power electronics (PE) curriculum after a meeting sponsored by the National Science Foundation (NSF), (Batarseh et al., 1996). As a result of that meeting, new directions and activities to increase the recruiting of students was pointed out i.e., to use EV as a catch, the intensive use of multimedia, state of the art lab facilities, open houses for research labs and environmental concerns. Those activities were summarized and they were a basic step in the development and growth of this area. However, several changes have been produced around the globe the last years in the area of engineering i.e. globalization, financial reorganization, advances in information technology and resource limitation. Those are some factors that motivate a substantial change in the design of a curriculum in the areas of engineering (Faculty of Engineering, 2009). Additionally to those facts, the area of EV is broader than PE, and it is in essence a multidisciplinary area, see Section 1. Therefore in order to come out with an integrated curriculum, in this section is proposed a modular curriculum oriented from the basic understanding of EV to the development and researching of more advanced applications. This proposal has been inspired by tools introduced in the Development of a Curriculum (DACUM, 2011), and it was complemented to the new and expected needs in the area of EV.

Accordingly with (DACUM, 2011), the main characteristic of DACUM are a natural relationship from its early stages between a desired competence or module, measurement on performance, and the curriculum designed to fulfil that competence; that basic idea has been preserved in this work. However, that idea has been completed with the following methodology (Schmal & Ruiz-Tagle, 2008): an identification of a module, module sequence, structuring of module, revision of each module, revision of curriculum and construction of syllabus for each module. As it can be noticed from this process the curriculum is an active entity, which needs to be adjusted and updated in a regular time-basis. Additionally, it has been emphasized the competency-based in all the stages of this curriculum and the permanent link between industry and academia. Figure 4 shows the three key areas interrelated in this proposal: experience, infrastructure and collaboration.

Experience from academics is one fundamental requirement in the practice of any curriculum. This experience and expertise must be reflected in the number of papers, books, patents, projects, etc., summarized for the overall academia involved in the curriculum implementation. However the isolated knowledge of the engineering area is just one requirement, for a good practice of this curriculum; it is recommended to implement a mandatory training in learning and lecturing in higher education. The main idea of that mandatory course is to increase the understanding of student learning, to improve the academic teaching expertise, and develop information for educational improvement at the level of courses at overall programme (McMaster University, 2010).

Another important area is infrastructure which is related with collaboration. There is no doubt that economic constraints have produced a new way to accomplish the learning activity. Today it is not longer attractive to have one laboratory per module or per academic faculty this way of organization is impractical and expensive. In this work, it is proposed the use of share resources at four different levels, industry, government, departments and universities. Through this scheme a more efficient way to achieve the learning scheme can be accomplished, see section 3.

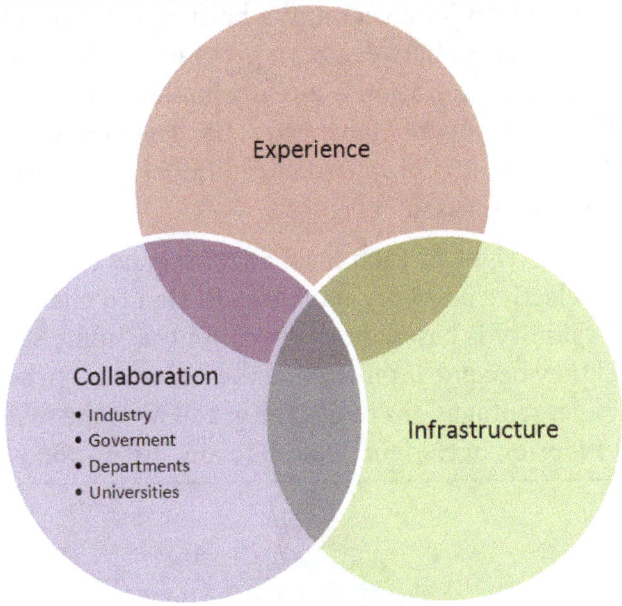

Figure 4. Areas to match.

Based on the premises discussed previously, Figure 5 shows the modular EV curriculum. Here, it is proposed at the beginning a three year studies finishing with a technician degree. This technical level is mainly focused to the maintenance and service of EV; areas covered in this level are fundaments of mechanics, battery management and disposal, circuits, fundamentals of electronics and others.

The second stage comprises two possible degrees the first part is a two year Bachelor in Technology, which can be updated to a traditional Bachelor in Science with an additional two years studies and mandatory one module section. The main characteristic of this level is the emphasis in hands-on experience in the first two years and the optional module complete the knowledge in math and engineering required for continuing with the Bachelor in Science. The difference between the Bachelor in Technology and Science is that the second option is more design oriented rather than maintenance or diagnostic. Both programs can be delivered in the form of lectures, tutorials, seminars and laboratories. Nowadays, a similar program is being adopted by Mohawk College and McMaster University, Canada; those programs offer university level courses, work in industry-focused lab and mandatory co-op work experience (McMaster-Mohawk, 2010). The main difference with the current system in McMaster University-Mohawk College and this proposal, it is the natural link between technician, bachelor level and graduate level proposed here, which is not currently offered.

A similar two year program is proposed in the graduate studies with two options, Master in Engineering and Master in Science. Here, it is proposed a 180 credits program for the first option (one year and a half) and 180 credits for the second one (two years),, the different between both programs is the teaching or research oriented emphasis. This organization is already implemented with good results in universities like The University of Manchester, UK. The final stage proposed in the graduate level is PhD, here it is proposed a traditional three year course oriented to research in the areas discussed in Figure 3.

It can be noticed in the right of each level a transversal module. Those modules are proposed to be elective and they must be satisfied to change from one grade to another. This flexibility is based on the premise that some students start from the know-how and they become interested in the know-why. In addition, it has not been provided any percentages or credits per grade with the main aim to provide flexibility for the adoption of this curriculum to any institution.

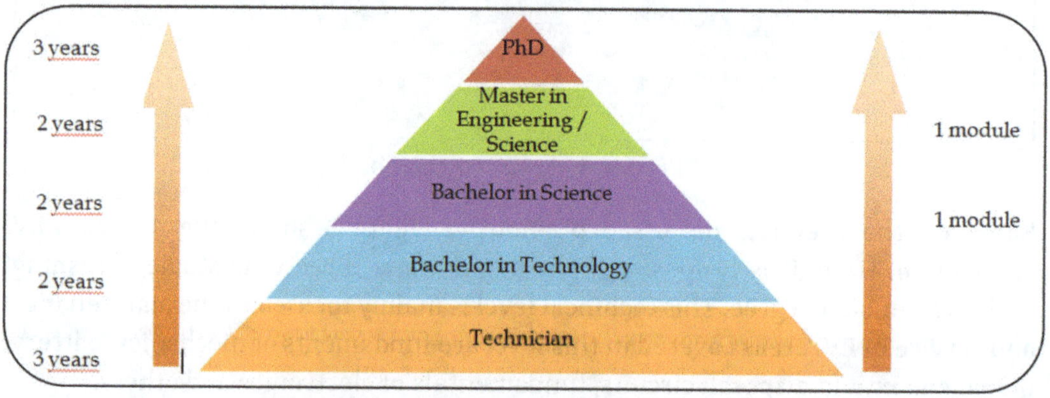

Figure 5. Proposed modular model.

2.1. Technician curricula

The main objective is to bring the students the knowledge of maintenance and repair of EV considering the different automakers philosophy and EV structure. In this level, the student will acquire training in basic dynamic, electric fundamentals, computing, safety, equipment, tools, and software related with the diagnostic of EV. The student will be able to deal with user and maintenance manuals, to detect failures in the areas of mechanics, electric and electronics. In addition, the student must fulfill preventive and corrective maintenance for the different EV automakers.

This level is organized in two terms per year and five courses per term. A common core is proposed for the first four terms based on chemistry, physics, computing and mathematics. Table 3 shows the common core following by a list of optional third year courses.

In order to obtain industry experience before completing the technician level; it is proposed a mandatory four month internship or co-op after completing the second term in year two. This practical experience will help the student to probe their skills before completing the third year and it will help them to further select their final years′ areas of interest. In addition, it is proposed to review the technical program every three years for possible updates. As mentioned earlier, it is proposed in the final terms elective courses following the main areas shown in Figure 3.

Table 3. Technician level organization.

Year 1		Year 2		Year 3	
Term 1	Term 2	Term 1	Term 2	Term 1	Term 2
Math 1	Math 2	Electric Circuits 1	Electric Circuits 2	Optional	Optional
Computer Science 1	Computer Science 2	Electronics 1	Electronics 2	Optional	Optional
Physics 1	Physics 2	Mechanics 1	Mechanics 2	Optional	Optional
Chemistry 1	Chemistry 2	Introduction to EV	Automotive software	Optional	Optional
Reading and writing workshop 1	Reading and writing workshop 2	Health and Safety	Management	Optional	Optional

Area Chemistry

1. Introduction to Energy Storage Unit.
2. Maintenance and repair of Energy Storage Unit.
3. Administration and Recycle of EV materials.

Area Mechanics

1. Introduction to ICE.
2. Introduction to Diesel motor.
3. Maintenance and repair of Suspension.
4. Maintenance and repair of Braking System.
5. Maintenance and repair of Automatic Transmission and CVT.
6. Maintenance and repair of ICE.
7. Maintenance and repair of Diesel motor.

Area Electrical

1. Introduction to Electric Machines.
2. Maintenance and repair of Electronic and Control Unit.

3. Maintenance and repair of Electric System.
4. Maintenance and repair of Electric Machines.
5. Maintenance and repair of Charging Station.

It can be noticed from Table 3 that the first and second year gives to the student the basic tools that they will use in more advanced courses. In addition the working co-op experience will provide to the students a real-world experience for a better choice of specialization. In addition, it would provide to the academic a state of the-art feedback from their student resulting in a better understanding of the market needs.

2.2. Bachelor in technology / science curricula

The main objective of the Bachelor in Technology (B. Tech.) is to provide the knowledge of analysis, operation and planning in the maintenance and repair of EV considering the different automakers philosophy and EV structure. In this level, the student will acquire advanced training in mechanics, electric systems and software related with EV. The student will be able to deal with different automaker′s maintenance manuals to detect errors and implementing upgrades in the areas of mechanics, electric and electronics. Additionally, after completing the Bachelor in Technology, the students have the option to take in the summer a mandatory module required to pursuit a Bachelor in Science (B. Sc.) degree.

It is necessary to say that the Bachelor in Science is a design oriented program rather than maintenance in the areas shown in Figure 3. In particular, emphasis is given in: power source, materials, manufacturing, electric and electronic systems, charging infrastructure, control systems, embedded systems, management and quality control. Table 4 shows the core for both programs following by a list of optional second year's courses.

In order to obtain industry experience before completing the Bachelor in Technology and Bachelor in Science; it is proposed a mandatory four month internship or co-op after completing the second term in year two, respectively. In a similar way that in the Technician level, this practical experience will help the student to master their skills before completing the second year and it will help them to further select their final years′ courses. In addition, it is proposed to review both programs every two years for possible updates.

As mentioned earlier, it is proposed in the second year several elective courses for the Bachelor in Technology and Sciences following the main areas shown in Figure 3.

Table 4. Bachelor level organization.

Year 1		Year 2	
Term 1	Term 2	Term 1	Term 2
Math 1	Math 2	Elective	Elective
Mechanics 1	Mechanics 2	Elective	Elective
Chemistry 1	Chemistry 2	Elective	Elective
Electronics 1	Electronics 2	Elective	Elective
Electric Circuits 1	Electric Circuits 2	Elective	Elective
Year 3		**Year 4**	
Term 1	Term 2	Term 1	Term 2
Math 3	Math 4	Elective	Elective
Mechanics 3	Mechanics 4	Elective	Elective
Chemistry 3	Chemistry 4	Elective	Elective
Electronics 3	Electronics 4	Elective	Elective
Electric Circuits 3	Electric Circuits 4	Elective	Elective

Elective Year 2. Bachelor in Technology

Area Chemistry

1. Energy Storage Unit.
2. Advance Material.

Area Mechanics

1. ICE and Diesel Motor.
2. Heat Transfer.
3. Thermodynamics.
4. Steering and Suspension.
5. Introduction to Mechatronics.

Area Electrical

1. Energy Conversion.
2. Electric Drive in EV.
3. Electromechanics.

Area Electronic

1. Electronic Control Unit.
2. Power Electronics.

Area Power

1. Power System Distribution.
2. Renewable Energy.

Area Control and Management

1. Automatic Control of Dynamic System.

Area Computer

1. Vision Systems.
2. DSP Programming.

Area Business

1. Administration and Recycle of EV Materials.
2. Business Logistic and Supply Chain.
3. Quality Control of EV.
4. Project Management.

Elective Year 2. Bachelor in Science

Area Chemistry

1. Production and Storage Hydrogen.
2. Production and Storage Biofuel.
3. Fuel Cell and Supercapacitor Technology.

Area Mechanics

1. Modeling and Design of Steering and Suspension.
2. Modeling and Design of Advanced Braking System.
3. Modeling and Design of CVT and Transmission.
4. Computer-aided Design, (CAD).

Area Electrical

1. Advanced Theory of Electric Machines.
2. Electromagnetic Interference in EV.

Area Electronics

1. Embedded Systems.
2. Design of Hardware in the Loop Automotive Systems.
3. Modeling of PE.
4. Control of PE.

Area Power

1. Design of Charging Station.
2. Power Protection.
3. Smartgrid.

Area Control and Management

1. Advanced Control.
2. Digital Control.

Area Computer

1. Design of Navigation System.
2. Finite Element Analysis.
3. Dynamic Programming.

Area Business

1. Energy and Sustainability Management.
2. Human System Integration in EV.

Table 4 indicates that the first year provides students with foundational information applicable in subsequent advanced courses. The necessary course transitioning from Bachelor of Technology to Bachelor of Science is proposed in relation to Mathematics for Engineering. Upon finishing the Bachelor's degree, students may pursue careers in fields such as electric vehicle (EV) design and component manufacturing, quality assurance, and the creation of electronic, electrical, and software systems connected to EVs.

2.3. Master in engineering / science curricula

In this document a Master degree is understood like a postgraduate study to specialize in some area related with EV, it is proposed a Master in Engineering (M. Eng.) and Master in Science (M. Sc.) postgraduate studies. The following are the common structure for both degrees: two year length, full or part-time, lectures, assignments, exams, laboratory and one year common core. The difference between both degrees is on the second year where the students have to select among a professional oriented program M. Eng. and a research intensive program M. Sc.

The objective of the M. Eng. to provide the students with in-depth skills in a particular area of EV. Once completing this program, the student will be able to propose new designs, to lead projects and to manage people under its supervision in the area of EV. In order to graduate from this program, it is necessary to submit a teaching- based project report.

In contrast, the objective of the M. Sc. is to provide the students with research skills in a particular area of EV. Once completing this program, the student will be able to propose and develop innovative solutions for new designs and carry on projects in the area of EV. In order to graduate from this program, it is necessary to submit a research thesis,two research papers in a major conference of the area, or one paper in an ISI transaction.

Table 5 shows the proposed structure program. Once again in order to select a project can be used the areas shown in Figure 3.

Table 5. Master Degree level organization.

Year 1		Year 2 MEng	
Term 1	**Term 2**	**Term 1**	**Term 2**
Storage System 1	Storage System 2	Seminar	Seminar
Control System 1	Control System 2	Management 1	Management 2
Computer Design 1	Computer Design 2	Business 1	Business 2
Advanced Power Electronics	Automotive motor drives	Project MEng	Project MEng
Mechatronics Systems	Mechatronics Systems 2	Year 2 MSc	
		Seminar	Seminar
		Project MSc	Project MSc

2.4. PhD curricula

The degree of Ph. D. is proposed to be a minimum of three year research oriented program, with the main aim to provide original results in one or more areas related with EV, Figure 3. Here, it is proposed to follow the traditional scheme and presenting after the first year a comprehensive report to the supervisory committee outlining the proposed line of research, timetable, expected minimum deliveries, etc. Once completing this program, the student will be able to propose and develop novel solutions for new designs and carry on independent projects in the area of EV. In order to graduate from this program, it is necessary to submit a research thesis, and least one paper in an ISI transaction.

3. SOME IMPLEMENTATION GUIDELINES

There is no doubt that the era of Information and Technology (I&T) has arrived in the classroom, in fact our students are more active and visual that they used to be just five years ago. Today, we face in the lecture or classroom the Y generation; so far Facebook, Twitter, Blogs, wikis, instant messaging are just some of the several tools currently used by our students to share information. The use of a computer or smartphone with several ads-on for everyday activity is familiar to our students and the students expect from the faculty to be familiar with those tools and they also expect an inclusion of those technologies in the classroom (McMaster University, 2010b). Therefore, for a better practice of this curriculum is recommended to include those new tools in the design of the overall courses. This will provide a natural way to engage the student´s interest in the subject. For example, it can be included a twitter account for the course administrated by the faculty, where the students can check any last minute announcement.

In addition, another change in the classroom is the increment of students per academic faculty, in the first world universities is a common practice the use of large auditoriums for lecturing. That fact has reduced to a minimum the classical relationship between the student and instructor and the learning activity has become almost anonymous. Those constrains have opened a new paradigm in the area of research and development in academia and industry, today is not longer valid the exclusive use of blackboard and chalks for the academic intercourse. Based in that scenario, it is recommended to implement new teaching techniques in the proposed curriculum, the students learn by doing, making, writing, designing, creating and solving (McMaster University, 2010b). Therefore, it is proposed for a successful implementation of this curriculum the adoption of active learning techniques, which contributes to the student motivation and curiosity to learn new material. Active learning techniques have been widely applied in McMaster University by the Centre for Leadership in Learning. Some examples of active learning strategies are a) to capitalize on student´s interest, b) to collect students´ feedback regarding what makes their classes more or less motivating, c) to increase motivation and curiosity.

Figure 6 shows a proposed flowchart based on active learning techniques, which can be implemented to any level by giving emphasis to the engineering or science degree. It is necessary to say that the academic faculty can develop their own flowchart based on their teaching style and needs.

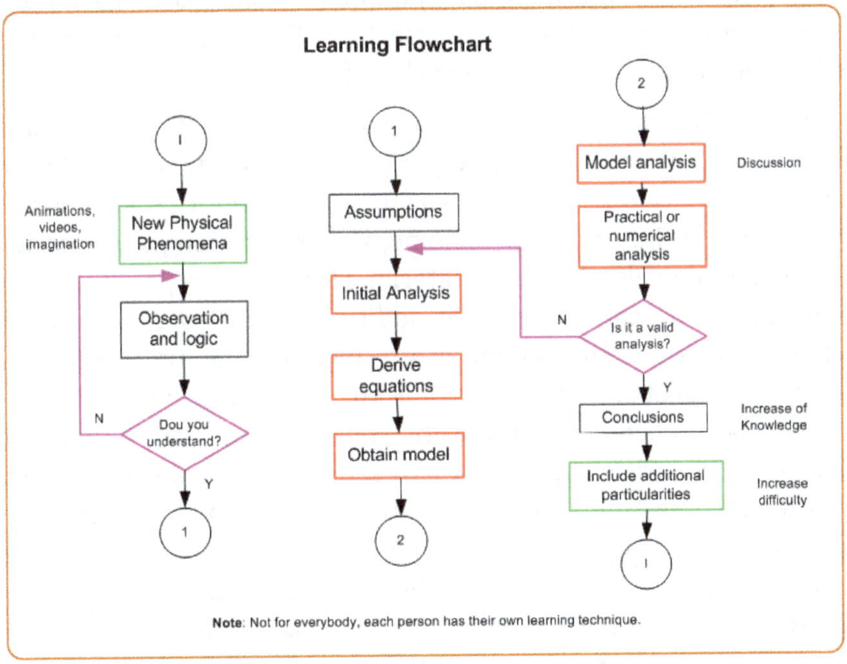

Figure 6. General learning flowchart.

3.1. Course webpage

In addition to the active learning techniques included in the lecture or classroom; it is necessary to prepare a well-organized course and friendly webpage. Those actions will increase the interest in the students providing them with all the required information in one single place; and it will help the academic faculty to reduce his time delivering new material related to the course, Figure 7 shows a proposed web page per faculty and teaching course (Perez-Pinal, 2011). It is necessary to say that there is in the market software oriented for delivering courses such as Blackboard, Avenue, Moodle, etc. That software is known like Course Management System (CMS), also known as a Learning Management System (LMS) or a Virtual Learning Environment (VLE), those are applications that instructors can use to create effective online learning sites (Blackboard, 2011). Objectives of those platforms are the same that the course website, which are to connect more effectively to the students with their instructor to keep the student, informed, involved and collaborating in the course.

Figure 7 shows a proposed course webpage, which is divided in three main sections, left menu, center part to display information and right menu to provide the course in-depth details.

In the left section, it is given a menu to select the information regarding the instructor, i.e. background, expertise, awards and citation, news, contact etc. This menu will provide all the information to the student about his instructor, providing confidence about the instructor′s expertise. In addition, at the center section it is displayed all the information selected in the left menu.

In particular, the teaching course section has a submenu titled "Further details," this submenu option will display a password protected menu displayed on the right, Figure 7. This new menu provides all the information regarding the particular course, for instance course home, syllabus, readings, labs, assignments, exams, tools, and download course material. Here it is proposed to publish the announcement in the course home in addition to the course description and course characteristics. In this section is also included the information regarding the textbook. The syllabus sections provides the information of the term, teaching assistant, lab staff, schedule, prerequisite, course description, course objectives, assessment criteria, written work and late submissions, academic integrity, and notes. The reading section gives information on the course's lecture sessions; here are posted the lectures′ slide, complementary notes, animations, and simulations presented in the lectures. The labs section provides information on the laboratory sessions schedule, laboratory manuals and laboratory policy, and safety considerations. The assignment section provides information regarding the assignments topic and schedule, tutorial calendar and slides. In addition, here it is proposed to include some practice problems with solutions. The exam section contains the current term's exams, i.e. midterm, final and test samples. The section tools contain the tutorials, multimedia and simulation resources for the course. Finally, the option "course materials to download" contain the same content as the online version in a single file.

Figure 7. Web page model one.

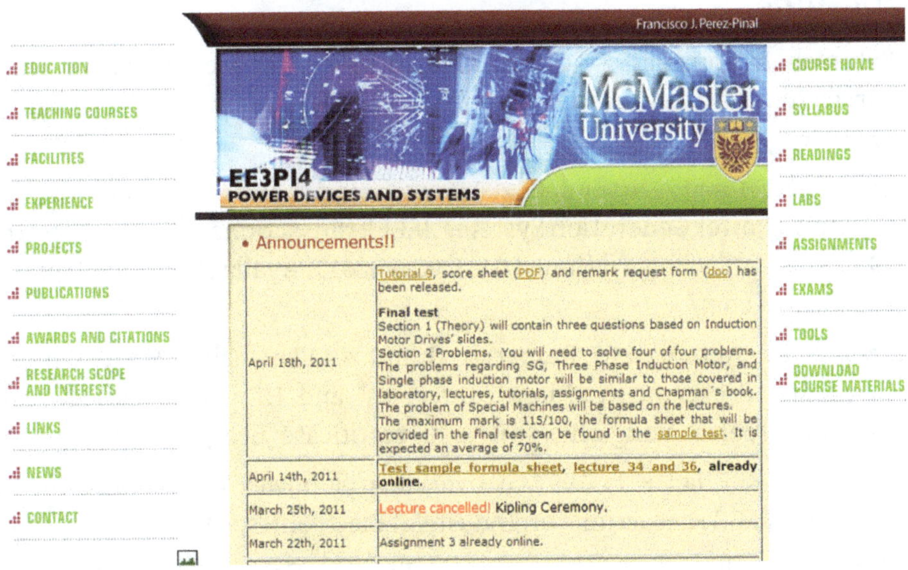

Figure 8. Web page model, two.

The proposed homepage design can be enhanced by incorporating a Twitter account, a question and answer part, and a blog to facilitate immediate feedback from students. Additionally, a component of video lectures may be incorporated to offer off-campus services.

4. CONCISE SUMMARY

This work provides an overview of electric car technologies. A typical electric vehicle (EV) electrical architecture and energy storage unit, the mechanical drivetrain, design principles for EVs, and an overview of current curricular initiatives have been presented. The conclusion drawn is that the electric vehicle is evolving into a mechatronic system, and this tendency is anticipated to persist in the field. Furthermore, an integrated curriculum has been proposed that focuses the primary domains of electric vehicles, extending from technical education to graduate degrees. The primary objectives at each level, together with their requirements and several areas of specialty, were provided here. Eight areas have been identified, along with other possible subareas of specialization. Furthermore, certain overarching standards for the proper execution of the suggested curriculum were introduced, grounded in active learning methodologies. An example of a homepage design was also shown, consolidating all material pertaining to the course in a one location. It is essential to note that numerous questions remain unresolved in the domain of electric vehicles and their curricular development. This evolving field of research and development must seamlessly integrate cutting-edge tools and techniques in software, animation, and learning methodologies to ensure the transportation needs of current and future generations are met.

REFERENCES

1. Batarseh I. Gonzalez A. Qu Z. Khan A. 1996Proposed power electronics curriculum, Proceedings Conference Record Southcon/96,, 251 262June 25-27, 1996
2. Blaabjerb F. Chen Z. Kjaer S. B. 2004Power electronics as efficient interface in Dispersed Power Generation Systems, IEEE Transactions on Power Electronics, 19 1187 1194Blackboard (June 2011). 07.06.2011, Available from: http://www.blackboard.com/
3. Center for Automotive Research (June 2011The Ohio State University, Certificate Program, Graduate course, EV and HEV, 07.06.2011, Available from http://elearn.eng.ohio-state.edu/certificate_progs_05.pdf
4. Chan C.C. & Chau K.T. 1997An overview of power electronics in electric vehicles. IEEE Transactions on Industrial Electronics, 44 1 3 13
5. Chan C. C. Chau K. T. 2001 Modern Electric Vehicle Technology, Oxford Science Publications, 0-19850-416-0UK

6. Chan C. H. Pong M. H. 1997Input current analysis of interleaved boost converters operating in discontinuous-inductor- current mode. IEEE Proceedings Power Electronics Conference, 392 398

7. CSU Ventures 2009Advanced Electric Drive Vehicle Education Program, Colorado State University (CSU), Georgia Tech (GT), Ricardo, MRI, KShare, Araphoe Community College, Douglas County Schools. 07.06.2011, Available from http://www1.eere.energy.gov/vehiclesandfuels/pdfs/merit_review_2011/technology_integration/arravt031_ti_ebron_2011_p.pdf

8. DACUM 2011Developing a Curriculum. 07.06.2011, Available from http://www.dacum.org/

9. Department of Energy, United States of America 2011One Million Vehicles by 2015, February 2011 Status report. 07.06.2011, Available from www.energy.gov/media/1_Million_Electric_Vehicle_Report_Final.pdf

10. Ehsani M. Rahman K. Toliyat H. 1997Propulsion system design of electric and hybrid vehicles. IEEE Transactions on Industrial Electronics, 44 1February 1997.

11. Ehsani M. Yimin G. Gay S. E. Emadi A. 2004 Modern Electric, Hybrid Electric, and Fuel Cell Vehicles: Fundamentals, theory and design, CRC Press, 2004.

12. Emadi A. Ehsani M. 2001Chapter 21 More electric vehicles, Handbook of Power Electronics, CRC Press, Nov. 2001.

13. Emadi A. 2005aTutorial Notes. Modern automotive systems: power electronics and motor drive opportunities and challenges. IEEE, Proceedings of International Electric Machines and Drives Conference, Laredo Texas, USA May 15 18

14. Emadi A. 2005b Handbook of Automotive Power Electronics and Motor Drives. Edited by Ali Emadi, CRC Press, 2005.

15. Faculty of Engineering 2009Engineering a sustainable society. MacMaster University, Strategic Plan 2009 2014Available from http://www.eng.mcmaster.ca/strategicplan/StrategicPlan.pdf

16. Ferdowsi M. 2010Advanced Electric Drive Vehicles-A Comprehensive Education, Training, and Outreach Program, Missouri University of Science and Technology,University of Central Missouri, Linn State Technical College, St. Louis Science Center, 07.06.2011, Available from Linn State Technical College, St. Louis Science Center, 07.06.2011, Available from http://www1.eere.energy.gov/vehiclesandfuels/pdfs/merit_review_2011/technology_integration/arravt034_ti_ferdowsi_2011_p.pdf

17. Hammerstrom D. Butts M. 2011Electric Vehicles part 1 and 2, Portland State University. 07.06.2011, Available from http://web.cecs.pdx.edu/~strom/ev1/ev1.html

18. Heinz R. Schwendeman-K L. 2011Advanced electric drive vehicles," J. Sargeant Reynolds Community College. 07.06.2011, Available from http://www1.eere.energy.gov/vehiclesandfuels/pdfs/merit_review_2011/technology_integration/arravt039_ti_schwendeman_2011_p.pdf

19. Husail I. Islam Mo. 1999Design, modeling and simulation of an electric vehicle system. SAE, International Congress and Exposition, Detroit Michigan, March 1 4

20. Intellicon, 2005Development of a smart DC/DC converter for light traction applications. White paper, 07.06.2011, Available from http://www.intellicon.info/

21. T. B. A. Koot M. W. T. Van den Bosch. P. P. J. Kok D. B. 2008Online energy management for hybrid electric vehicles. IEEE Transactions on Vehicular Technology, 57 6 3428 3440Nov. 2008.

22. Lukic S. M. Emadi A. 2004Effects of drivetrain hybridization on fuel economy and dynamic performance of parallel hybrid electric vehicles. IEEE Transactions on Vehicular Technology, 53 2 385 389

23. McMaster University, 2010Teaching at McMaster, Centre for Leadership in Learning, 2010 2011New Faculty Handbook, White paper, 07.06.2011. Available from: http://cll.mcmaster.ca/resources/pdf/NewFacultyHandbook_2010.pdf

24. McMaster University, 2010bLecturing Short Course Notes, October 21November 11, 2010, Centre for Leadership in Learning. Whitepaper.

25. McMaster-Mohawk 2010Bachelor of Technology. White paper, 07.06.2011. Available from http://mybtechdegree.ca/

26. Michigan Technological University 2011Hybrid Electric Vehicle Engineering, Available from http://www.doe.mtu.edu/hybrid_vehicle_engineering/

27. Miller J. 2004 Propulsion systems for hybrid vehicles, IEE, 2004 Pay S., & Baghzouz Y. (2003). Effectiveness of Battery- Supercapacitor Combination Electric Vehicle. IEEE Proceedings Power Tech Conference, 2003 Bologna, 3 23 26June 2003.

28. Perez-Pinal F. J. Nunez C. Alvarez R. Gallegos M. 2006Step by step design procedure of an Independent-Wheeled Small EV applying EVLS. IEEE Proceedings of 32nd Annual Conference on Industrial Electronics, IECON 2006, 1176 1181Nov. 2006.

29. Perez-Pinal F. J. Cervantes I. Emadi A. Stability of an electric differential for traction applications. IEEE Transactions on Vehicular Technology, 58 7 3224 3233

30. Profumo F. Tenconi A. 2004Fuel cells for electric power generation: Peculiarities and Dedicated solutions for power electronic conditioning systems. IEE Proceedings of European Power Electronics- Power Electronics and Motion Control, Riga, Latvia, 2004.

31. Purdue University, 2010Indiana Advanced Electric Vehicle Training and Education Consortium, (I-AEVtec. Purdue University, NotreDame University, IUPUI, Ivy Tech, Purdue-Calumet, Indiana University-Northwest. White paper, 07.06.2011 Available from http://www1.eere.energy.gov/vehiclesandfuels/pdfs/merit_review_2010/technology_integration/tiarravt032_caruthers_2010_o.pdf.

32. Pyrzak K. 2009Development of hybrid electric technology, Procedings of Electric Vehicles, Renewable Energies, Monaco, March 26 29

33. Rizkalla M. E. Yokomoto C. F. Sinha A. S. C. El -Sharkawy M. Lyshevski S. Simson J. 1998A new EE curriculum in electric vehicle applications. Proceedings Circuits and Systems, 1998, 186 189Aug 1998.

34. Schmal R. Ruiz-Tagle A. 2008A methodology for competency oriented curricula design," Ingeniare Revista chilena de ingeniería, 16 1 147 158

35. Schofield N. 2005 Course Notes Electric Vehicle Systems. The University of Manchester UK, 2005.

36. Simon K. Y. 2011Development and Implementation of Degree Programs in Electric Drive Vehicle Technology. Macomb Community College, Wayne state University, NextEnergy, White paper, 07.06.2011, Available from http://www1.eere.energy.gov/vehiclesandfuels/pdfs/merit_review_2011/technology_integration/arravt035_ti_ng_2011_p.pdf

37. Solero L. Lidozzi A. Pomilio J. 2005Design of multiple-input power converter for Hybrid Vehicles, IEEE Transactions on Power Electronics, 20 5September 2005, 1007 1015

38. Tabbache B. Kheloui A. Benbouzid M. E. H. 2011An adaptive electric differential for Electric vehicles motion stabilization, IEEE Transactions on Vehicular Technology, 60 1 104 110Jan. 2011.

39. The National Alternative Fuels Training Consortium 2009West Virginia University, Undergraduate and Colleges. White paper, 07.06.2011, Available in http://www.naftc.wvu.edu/

40. Tosoh 1991White paper, Available on http://www.icis.com/Articles/1991/02/07/11676/japans-careful-steps-into-global-niche-markets.html

41. Tuttle D. P. Baldick Ross. 2011The evolution of plug-in electric vehicle-grid interactions. White paper, 07.06.2011, Available from http://users.ece.utexas.edu/~baldick/papers/PEV-Grid.pdf,

42. University of Detroit Mercy 2009Certificate engineering program in advanced electric vehicles (AEV). White paper, 07.06.2011, Available from http://www.udmercy.edu/governance/proposals/adoptedpolicies/pdf/Elec_Vehicles_Grad_Certificate.pdf

43. Winstead V. 2007Designing a multi-disciplinary hybrid vehicle systems course curriculum suitable for multiple departments, Proceedings of American Society of Engineering Education,

44. Multiple Departments, Proceedings of ASEE 2007American Society of Engineering Education, Honolulu, Hawaii, June 24-27, 2007

45. Wirasingha S. G. Emadi A. 2011Classification and review of control strategies for Plug-In Hybrid electric vehicles. IEEE Transactions on Vehicular Technology, 60 1 111 122

46. 午.卯.庚.戊.亥.酉.子.戊. Vehicles. IEEE Transactions on Vehicular Technology, Vol. 1 111 122Jan. 2011

47. 이시우, 2003System on chip. White paper, 07.06.2011, Available on: www.postech.ac.kr/soc/presentation/sw-rhee.pdf

INDEX

A

Algorithms 165

Aluminum 63

Armature 103

Asymmetric 225

Automakers 64, 69

Automobile 62

Automotive 58

B

Barrel 34

Battery 4, 8, 10, 16, 19-20, 29, 32, 50, 52, 59-60, 62-63, 67, 71, 75, 77, 79, 82-84, 92, 95-100, 103, 106-111, 113, 116, 135, 145, 149, 188, 190-191, 199-200, 224, 242, 264, 273

Bridge 225, 274, 277

Business 38, 202, 204

C

Capacitors 152-154, 156

Carbon 1, 30, 81, 156, 256, 258

Carbonate 264

Cathodes 266

Charging 19, 28, 37, 47, 51, 99, 108, 173, 175, 188-189, 191-192, 196, 200, 202, 208-209, 212

Chemistry 59, 69, 107, 110, 275

Chopper 163

Circuit 188, 227-228, 231, 233

Climate 7, 30

Combustion 1, 12, 96, 114-115, 264, 271

Complex 14

Components 32, 111, 206

Congestion 39

Converters 15-16

Cooling 30

Credits 284

Current 103, 157, 159, 165, 188, 217, 226-228, 231, 233, 237, 239, 241-242, 244, 278-279

Customers 71

Cycles 154

D

Density 106, 109-110, 124, 259

Diesel 33, 112

Diffusion 257, 260, 262

Dimension 204

Dioxide 81, 86

Domain 237

Doping 260

Drying 255

E

Electric 2, 4-5, 10, 12, 14-15, 20-21, 24, 26-31, 33-36, 38, 40-42, 45-58, 60-61, 65-68, 70-79, 82-84, 92, 95-98, 105, 108, 112-114, 116, 119, 123, 134-137, 139-140, 148, 151-152, 166-167, 171-173, 175, 188-189, 193, 196-200, 214, 217, 224, 264, 271-272, 274, 277, 286, 294-298

Electrical 154, 161, 225

Electrodes 154

Electronic 231, 256, 259

Elements 205

Emissions 23, 26, 28, 30-31, 64-65, 69, 81, 84, 92, 119-120, 172-173, 179-181, 193, 196

E-mobility 209-212, 214

Energy 1-4, 6-8, 10, 12, 20-21, 24, 30, 32, 34, 38, 41-42, 45, 51, 76-77, 83-84, 86, 88, 90, 93, 95, 97, 106, 108-111, 113, 115, 117, 123, 135, 149, 152-154, 160-162, 166-167, 171-173, 176-180, 186-188, 193, 196-197, 199-204, 208, 247-248, 253, 257, 265-266, 271-273, 277, 295

Engine 1, 12, 33, 58-59, 111, 264-265

Evolution 187

F

Feedback 236

Feedstock 70

Fossil 88, 90, 193, 196

Frequency 159, 165, 278

G

Garages 47, 189

Gasoline 32-33, 45-49, 51, 53-58, 61-65, 70, 73, 88, 90-92

Gearbox 111

Generator 10, 99, 114-115

Geothermal 178, 182

Greenhouse 84, 89, 91-92

H

Highways 73

Hybrids 49, 71, 75, 152, 166

Hydraulic 139

Hydrogen 5, 20, 70, 87-88, 120-123, 135-136, 139, 141, 149

I

Industry 36

Instructor 292

J

Journal 215, 266, 268-269

L

Lattice 260

Lithium 107, 110, 116, 248-249, 264-269

M

Magnetic 218

Marine 133

Market 5, 24, 27-28, 34, 36-37, 46-47, 49, 60, 63, 75, 128, 173, 186

Measuring 157

Melaina 57

Methanol 87, 125
Migration 257
Mining 86
Mobile 208
Modeling 288
Models 24, 202, 204
Module 282
Motors 14, 104, 217, 274

N

Nickel 109

O

Operation 45, 50
Operator 212

P

Patent 117
Petroleum 91
Pipeline 87
Plants 30, 179, 181-182
Plug-in 41, 71, 74-76, 83, 95, 197-198
Points 189
Policy 41-42
Pollutants 29-30
Population 23
Propulsion 12

R

Regulator 234, 236
Reverse 227
Roaming 210-212

S

Shuttles 82
Signal 228, 233
Simulation 146
Sodium 67
Solution 254-255
Stations 19, 58, 152
Structure 223, 249-251, 258, 260
Subsidy 83
Symposium 78
Systems 122-123, 137, 141, 169, 172-173, 217, 265, 286, 288, 297

T

Technology v20, 24, 34, 38, 60, 74, 77, 82-83, 93, 110, 124-125, 133, 136, 181-185, 189, 193, 252, 286, 295-297
Template 254
Torque 14, 97-98, 106, 162, 165, 217-218, 220, 234-237, 241, 244

V

Variable 183
Vehicle v, 9-10, 12, 20-21, 27, 31-33, 35, 38, 47-50, 53-54, 56-63, 68, 70, 72-78, 82, 84, 88, 90-91, 93, 95-97, 128, 135, 137-138, 145, 148, 161, 188-189, 198-200, 218-220, 242, 264, 271-272, 294, 296-297
Voltage 15, 100, 103, 141, 145, 154, 157, 159-160, 163, 165, 226-231, 242, 248

W

Warming 30, 151
Webpage 292

www.ingramcontent.com/pod-product-compliance
Lightning Source LLC
LaVergne TN
LVHW062223200926
848484LV00006B/1461

* 9 7 8 1 7 8 7 1 5 4 1 0 0 *